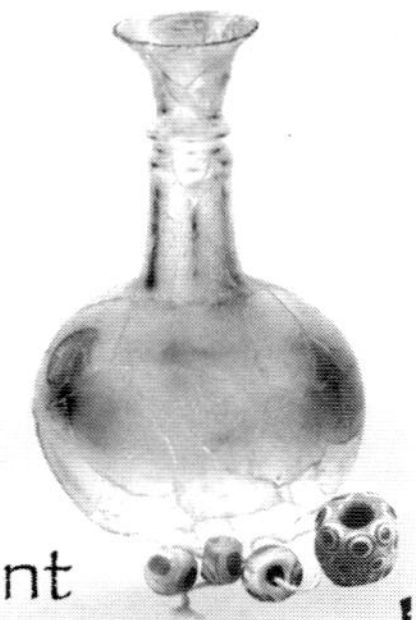

Ancient Glass Research along the Silk Road

Ancient Glass Research along the Silk Road

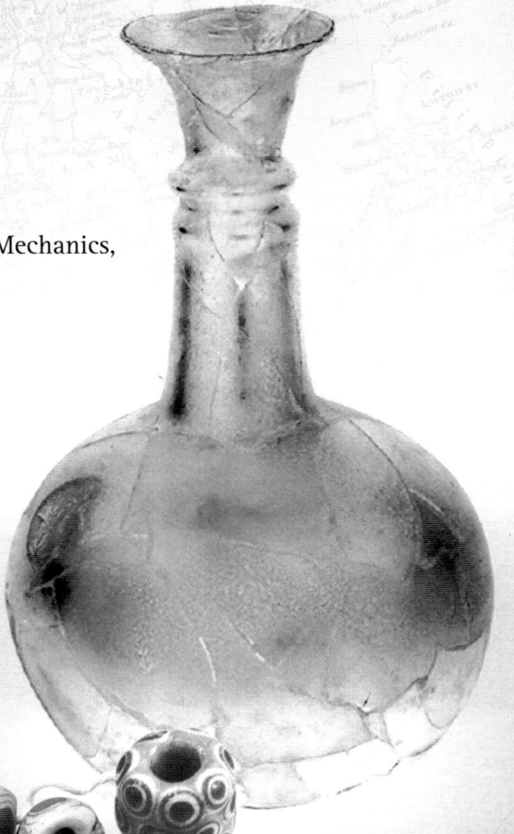

Editor-in-Chief

Gan Fuxi
Shanghai Institute of Optics and Fine Mechanics,
Chinese Academy of Sciences, China
Fudan University, China

Co-editors

Robert Brill
The Corning Museum of Glass, USA

Tian Shouyun
Shanghai Institute of Optics and
Fine Mechanics,
Chinese Academy of Sciences, China

World Scientific

NEW JERSEY • LONDON • SINGAPORE • BEIJING • SHANGHAI • HONG KONG • TAIPEI • CHENNAI

Published by

World Scientific Publishing Co. Pte. Ltd.
5 Toh Tuck Link, Singapore 596224
USA office: 27 Warren Street, Suite 401-402, Hackensack, NJ 07601
UK office: 57 Shelton Street, Covent Garden, London WC2H 9HE

British Library Cataloguing-in-Publication Data
A catalogue record for this book is available from the British Library.

ANCIENT GLASS RESEARCH ALONG THE SILK ROAD

ISBN-13 978-981-283-356-3
ISBN-10 981-283-356-0

Typeset by Stallion Press
Email: enquiries@stallionpress.com

Printed in Singapore by Mainland Press Pte Ltd

Preface

Glass, as one of the important artificial materials and a major vehicle for East–West cultural and technical exchange, has played a great role in the course of human civilization. Its origin and evolution attract the attention of archeologists and glass scientists worldwide.

The research on ancient Chinese glasses in China started in the middle of the last century. During the past 50 years, glass artifacts have been discovered frequently in excavated ancient tombs and ruins dating from the Qin and Han Dynasties to the Tang, Song, Yuan, Ming and Qing Dynasties, providing us with very important evidence and material for further study of ancient Chinese glasses. Chinese art historians and archeologists have systematically summarized the unearthed ancient Chinese glass artifacts and studied their excavation, historical background, shaping and emblazonry art, glass character, etc., and Chinese glass scientists have also become involved in the scientific research on unearthed ancient glass samples, not only through chemical composition analyses but also through technological studies, glass weathering and conservation, etc. Since the 1980s, several symposia have been held in China on the origin, technological provenances, and development of ancient Chinese glass, and many scientists and experts in glass archeology, both from home and abroad, have attended the symposia, which have made contributions to the ancient Chinese glass research in a worldwide context.

Many more ancient glasses were unearthed in the Yellow River and Yangtze River valleys, and these glasses have been studied in greater detail. Many ancient glasses were unearthed in the south and southwest of China, as well as in the north and northwest. They are closely correlated with the ancient glass exchange between China and foreign countries along the Northwest, the Southwest and the Maritime Silk Road. These glasses had previously not been studied much, and in recent years we have emphasized research on them.

The Symposium on Ancient Glasses in Southern China was held in Nanning, Guangxi, on 16–19 December 2002. After the symposium more than 70 ancient glass artifacts and samples, provided by the museums and institutes of cultural relics and archeology in the south and southwest of China were measured and analyzed by the nondestructive analytical method in Shanghai. The proceedings of this symposium, entitled *Study on Ancient Glasses in Southern China*, was published by *Shanghai Scientific and Technical Publishers* in 2003. For the same purpose, the Symposium on Ancient Glasses in Northern China was held in Urumchi, Xinjiang, from 29 August to 6 September 2004. This symposium was supported by the Basic Research Division of the Chinese Academy of Sciences and the Cultural Heritage Bureau of Xinjiang Uygur Autonomous Region, and organized by the Special Glass Technical Committee, the Chinese Ceramic Society, the Xinjiang Turfanology Research Society and other related institutions. Art historians, archeologists and experts in the natural sciences from the following institutions attended the symposium and gave their presentations and research reports on the ancient glass artifacts excavated in the north and northwest of China: the Xinjiang Institute of Cultural Relics and Archeology, the Museum of Xinjiang Uygur Autonomous Region, the Cultural Heritage Bureau of Turfan, the Guyuan District Museum of Ningxia, the Ningxia Institute of Cultural Relics and Archeology, the Qinghai Institute of Culture Relics and Archeology, the Liaoning Institute of Cultural Relics and Archeology, the Inner Mongolia Museum, the Shanghai Institute of Optics and Fine Mechanics (CAS), the Shanghai Institute of Ceramics (CAS), Fudan University and Beijing University of Science and Technology. The discussion of the spread of ancient glass and its distribution along the Northern (Desert) Silk Road and cultural exchange between the East and the West proceeded actively. The symposium has further promoted the collaborative research on the ancient glass in the north of China.

"The Silk Road" is a name given to a group of cultural, political and technological exchange routes linking up the East and the West. It played a significant role in ancient times. Zhangqian's travels to the Western Regions were a magnificent undertaking that influenced the East–West exchange at that time, but long before his travels westward there had been primitive trade roads in the Euro-Asian region. Conservatively, it can be estimated that this occurred in the 10th century BC, between the Shang and the Zhou Dynasty in China. After Zhangqian's travels, new transportation routes were explored between China and the outside world. China was the center of the Silk Road in Asia, but not the terminal. The Silk Road was extended from China to the Korean Peninsula, Japan and Southeast Asia. A few years ago, UNESCO identified four main routes of the Silk Road: (1) the Northern (Steppe) Route, (2) the Northwestern (Oasis) Route, (3) the Southern Maritime Route and (4) the Southwestern (Buddhist) Route.

Under the auspices of the Chinese Ceramic Society and the Technical Committee of Archaeology of Glass, the International Commission of Glass (TC-17, ICG), the Shanghai International Workshop on Archeology of Glass was held on 12 April 2005, in conjunction with the 2005 Shanghai International Symposium on Glass. The topic of the workshop was "Ancient Glass Along the Silk Road." The purpose of this symposium was to bring together the archeologists, art historians and natural scientists interested in glasses found along the Silk Road, to learn from each other, to exchange ideas, and to plan for collaboration in the future. The participants in the workshop came from the Corning Museum of Glass (USA), the Pusan Museum (Korea), the National Academy of Arts of Uzbekistan, the Institute of Archeology, the Chinese Academy of Social Sciences, the Xinjiang Institute of Cultural Relics and Archeology, the Shanxi Institute of Archeology, the Shanghai Institute of Ceramics (CAS), the Shanghai Institute of Optics and

Fine Mechanics (CAS), Shanghai University, Fudan University and Beijing University of Science and Technology. Scientific papers and reports were also submitted by the China National Institute of Cultural Property, the Hepu Museum of Guangxi, the Guizhou Provincial Museum, the Inner Mongolia Museum, the Sichuan University Museum, etc. At this fruitful workshop, scientific materials and research results concerning the excavation background, historical profile, shaping art, outside character and chemical composition of ancient glass samples along the Northern (Oasis) Silk Road and Southern Maritime Silk Road were reported. A book containing the proceedings of both of the meetings mentioned above was published in Chinese by the Fudan University Press in June 2007. It reflects the newest research results on ancient glasses in Asia along the Silk Road.

I am very grateful to World Scientific for publishing the English edition of the above-mentioned book. To have more readers understanding the ancient glass research, an English version is necessary. So I invited Dr R. H. Brill of the Corning Museum of Glass to serve as a coeditor of this book to help me. I thank him for his active response and valuable support, which enhanced my confidence in accomplishing this work.

Based largely on the Chinese edition, I have made an effort to add some new advances to the contents and to provide as much information as possible in this book. In addition, six papers presented at the 2004 International Congress of Glass (held in Kyoto), which have not been published before, are included in this book. All these make the English edition more substantial and up to date.

Acknowledgment is made to the authors of this book for their contribution of papers and color photographs of unearthed glass artifacts. More than 80 color photos of ancient glass artifacts are shown in this book for the reader's reference and appreciation.

Thanks are due to my colleagues at the Shanghai Institute of Optics and Fine Mechanics for their assistance and cooperation, especially to Prof. Tian Shouyun, who served as a coeditor, checking and editing all the manuscripts, and to Prof. Gu Donghong, and also Mrs Zhao Hongxia, who took part in the work of organization, communication and computer processing. Without their patient efforts it would have been impossible to publish this book. Finally, I wish to express the memory of my wife, Prof. Deng Peizhen, a materials scientist, who accompanied me for nearly 50 years and gave me full support in every respect. The editing and publication of this book were also supported under the Research Grant of the National Natural Science Foundation of China, and the Intellectual Innovation Project of the Chinese Academy of Sciences.

Gan Fuxi
Shanghai, December 2007

Contents

List of Contributors

Abdugani A. Abdurazakov
National Institute of Arts and Design Named After K. Bekhzod
St. Academic Rajabiy
77, 700031 Tashkent
Uzbekistan

An Jiayao
The Institute of Archeology
Chinese Academy of Social Sciences
Beijing 100710
China

Robert H. Brill
The Corning Museum of Glass
Corning
New York 14830
USA

Cheng Huansheng
Institute of Modern Physics
Fudan University
Shanghai 200433
China

Feng Xiaoni
Department of the Museum
Fudan University
Shanghai 200433
China

Fu Xiufeng
Shanghai Institute of Optics and Fine Mechanics
Chinese Academy of Sciences
Shanghai 201800
China

Gan Fuxi
Shanghai Institute of Optics and Fine Mechanics
Chinese Academy of Sciences
Shanghai 201800;
Fudan University
Shanghai 200433
China

Gao Menghe
Department of the Museum
Fudan University
Shanghai 200433
China

Gu Donghong
Shanghai Institute of Optics and Fine Mechanics
Chinese Academy of Sciences
Shanghai 201800
China

Akiko Hokura
Department of Applied Chemistry
Tokyo University of Science
Shinjuku
Tokyo 162-8601
Japan

Hu Yongqing
Henan Research Institute of Cultural Relics and Archaeology
Zhengzhou 450000
China

Huang Xueyin
The Capital Museum
Beijing 100045
China

Jiang Jie
Famen Temple Museum
Shaanxi 722201
China

Takayasu Koezuka
Nara National Research Institute for Cultural Properties
Nara 630-8577
Japan

Insook Lee
Busan Museum
Korea
210 UN Street
Nam-gu
Busan 608-812
Korea

Li Qinghui
Shanghai Institute of Optics and Fine Mechanics
Chinese Academy of Sciences
Shanghai 201800
China

Lu Chi
Shanghai Institute of Visual Art
Fudan University
Shanghai 200433
China

Lu Lipeng
Archeology Team
Xinjiang Uygur Autonomous Region Museum
Urumchi 830000
China

Ma Bo
School of Information Science and Engineering
Fudan University
Shanghai 200433
China

Ma Qinglin
China National Institute of Cultural Property
Beijing 100029
China

Izumi Nakai
Department of Applied Chemistry
Tokyo University of Science
Shinjuku
Tokyo 162-8601
Japan

Qian Wei
Institute of Historical Metallurgy and Materials
University of Science and Technology
Beijing 100083
China

Takashi Sawada
Department of Applied Chemistry
Tokyo University of Science
Shinjuku
Tokyo 162-8601
Japan

Shen Shifang
Chongqing Museum
Chongqing 400015
China

Yoko Shindo
Section of Islamic Archeology and Culture
The Middle Eastern Culture Center in Japan
Suginami
Tokyo 167-0042
Japan

Hiroshi Shirahata
Muroran Institute of Technology
Muroran 050
Japan

Song Yan
China National Institute of Cultural Property
Beijing 100029
China

Takashi Taniichi
Okayama Orient Museum
9-31 Tenjin-cho
Okayama 700-0814
Japan

Wang Bo
Archeology Team
Xinjiang Uygur Autonomous Region Museum
Urumchi 830000
China

Wang Weizhao
Hepu County Museum
Guangxi Zhuang Autonomous Region
Hepu 536100
China

Xiong Zhaoming
Archaeological Team
Guangxi Zhuang Autonomous Region
Nanning 530022
China

Xu Yongchun
Shanghai Institute of Optics and Fine Mechanics
Chinese Academy of Sciences
Shanghai 201800
China

Kazuo Yamasaki
Professor Emeritus
Nagoya University
Nagoya 464-860
Japan

Zhang Ping
Xinjiang Institute of Cultural Relics and Archeology
Urumchi 830001
China

Zhang Zhiguo
China National Institute of Cultural Property
Beijing 100029
China

Chapter 1 第一章

Origin and Evolution of Ancient Chinese Glass

Gan Fuxi

Shanghai Institute of Optics and Fine Mechanics,
Chinese Academy of Sciences,
Shanghai 201800, China

1. Outline of the Study of Ancient Chinese Glass

In ancient Chinese writings, there are some descriptions of glass. The earliest Chinese terms for glass are *"miaolin langgan," "liulin," "liuli," "boli,"* etc., appearing in historical books such as *Mutianzi Zhuan* (*Biography of King Mu*), *Shangshu — Yugong* (*Book of Ministers — Yugong*) and *Shanhaijin — Zhongshan* (*Book of Mountains and Seas — Zhongshan Mountain*). However, these words were used as general terms for natural gemstones and artificial glasses. After the Han Dynasty, the terms *"liuli"* and *"biliuli"* were often used in some historical literature, such as *Yantie Lun* (*Discourses on Salt and Iron*), *Xijing Zaji* (*Notes of the Western Capital*), *Hanshu* (*History of the Han Dynasty*), *Houhanshu* (*History of the Later Han Dynasty*) and *Suishu* (*History of the Sui Dynasty*). Following the introduction of Western glassware to China during the Han Dynasty, the glasses from the West were called *"boli,"* while the glasses made in China were called *"liuli."* Other terms, such as *"yaoyu," "xiaozi"* and *"liaoqi,"* were also used. Later, after the Song Dynasty, the terms *"liuli"* and *"liuliwa"* were specially used to indicate the bricks and tiles made by multicolor glazed pottery at low temperature; then the terms *"boli"* (glass) and *"liuli"* (glaze) were gradually resolved. During the reign of Kangxi of the Qing Dynasty, the Manufacturing Bureau of the Court, Internal Affairs Ministry, named the site for making glazed tiles *"liuli chang"* (glaze works) and the site for making glass *"boli chang"* (glass works) respectively, thus making the terms distinguishable. Confusion of the terms leads to misunderstanding of the essence of glass materials.

The term "glassy state" now in technical dictionaries both in China and abroad is defined as the cooled melt becoming solid-state while maintaining its molten structure at room temperature. It belongs to the noncrystalline state. Conversely, the minerals, jades and gemstones, which largely existed in the natural world, belong to the crystalline state, including polycrystalline and single crystals. The glassy materials, in addition to a few natural glasses, such as obsidian and tektite, are all artificially synthesized materials. While the synthetic crystals that appeared in the 20th century are a small part of the crystalline materials, most of them are natural ones.

Before the glassmaking technique came into being, the primitive people started off with faience and frit. Faience is made of sintered silica sand coated with glaze, and frit is a mixture of silica sand and glass. Both of them are not fully amorphous, SiO_2 being their main component (90% by weight). The earliest faience and frit as well as glass were the man-made products imitating jade. Most of them were made into beads and they were always strung together with quartz crystal beads and jade beads and tubes. A necklace composed of faience eye beads and jade discovered in Egypt (1500 BC) and a necklace composed of rhombic faience beads and jade tubes unearthed from a tomb at Zhengshan, Suzhou (mid-to-late Spring and Autumn period, 600–500 BC) are examples (photos 1.1 and 1.2).

Due to the confusion of the glass terms and essence mentioned above, we have to employ scientific examination to identify artificial glass, faience and frit, as well as natural jade and gemstones, in order to study scientifically the origin and evolution of ancient Chinese glass.[1]

Photo 1.1. Necklace composed of faience eye beads and jade (1500 BC; Nation Museum of Egypt).

Photo 1.2. Rhombic glass beads unearthed in Suzhou (Middle-to-late Spring and Autumn).

Introduction of the ancient Chinese glass and discussion of its origin started in the 1930s in modern history. But most of the works were based on analysis and introduction of historical writings. During the past 50 years, sectors dealing with Chinese cultural relics and archeology have analyzed and discussed the shapes, patterns and essences of ancient glasses covering different times and different regions in China. A prevalent cognition was that the ancient Chinese glass artifacts and techniques for making them were introduced from the West along the Silk Road, via Xiyu (the Western Regions) starting from the Han Dynasty. It is true that some Chinese historical literature, such as *Weishu* (*History of the Wei Dynasty*), *Xiyuzhuan* (*Memoir of the Western Regions*), *Taipingyulan* (*Taiping Imperial Commentary*), *Beishi — Darouzhi Zhuan* (*History of the North — Memoir of Great Yen Chin*) and *Jiutangshu* (*Previous Book on the History of the Tang Dynasty*), provides some records about the inflow of glassware and glass-making techniques from the West. Also, a large number of glass artifacts showing typical ancient Roman, Persian or Islamic culture were unearthed in China. Therefore, a general cognition reached by Chinese and foreign scholars for a long time was that the origin of the ancient Chinese glassmaking technique was from outside of China, starting from Zhangqian's travels to the Western Regions. The "exotic hypothesis" is widely accepted. Meanwhile, a number of scholars are in disagreement with this viewpoint.

Ancient Chinese writings like *Huainanzi — Laminxong* (by Liu An of the Western Han Dynasty) and *Lunheng — Shuaixinpian* (by Wang Chong of the Eastern Han Dynasty) have a record like this: "Melting five-color stones, making wares by casting." In the early 1960s, Shen Chongwen put forward a viewpoint based on his investigation of ancient Chinese glass relics. He said in his paper entitled "Discussion on the History of Glass Technology" that "glass-making technology in China was evolved from making small bead ornaments into making small piece engraved objects; this process was completed no later than 2200 years ago, which was the Warring States period."[2] In the 1970s, Gan Fuxi *et al.*, on the bases of searching the literature and preliminary technical measurements, reached a "self-invention" hypothesis on the origin of ancient Chinese glasses. This touched off a dispute in scholarship.[3] Yang Beda supported the viewpoint of "self-invention" according to his analysis of the unearthed objects and documents.[4] Later on, the discussion of this problem attracted attention and reports outside of China.[5]

Chinese historical relics had frequently been run off since the Opium War. The West started to carry out scientific archeology on ancient Chinese glasses in the 1930s. They conducted chemical analysis and investigations of the samples of collections. Among them, the most attractive work is considered to be that done by Seligman *et al.*[6] They measured the chemical compositions of ancient glasses (collections) unearthed in Henan province, dating from the pre-Han Dynasty to the Tang Dynasty, and found them mostly belonging to the lead barium silicate glass system containing PbO and BaO. This system is quite different to the compositions of the majority of Western ancient glasses (West Asia, Egypt and Rome) — the soda lime silicate glass containing Na_2O and CaO. But they still insisted on adopting the viewpoint of "glasses of the Far East originated from the West," only according to the patterns, colors and designs of the ancient glass beads.[7] From the late 19th century to the early 20th century, some Western explorers, such as Sven Hedin and Aurel Stein, excavated and took away a lot of cultural relics, including ancient glasses from the Xinjiang area of China (i.e. the Xiyu Regions in ancient China), most of which belonged to the Han Dynasty afterward. The analysis of glass chemical compositions was conducted in succession after the 1950s. Results show that most of them are soda lime silicate glasses. Therefore, the viewpoint on the origin of Chinese glasses mainly focused on the "exotic hypothesis."

Scientific research on ancient glasses in China started in the mid-20th century and has developed quickly since the 1980s, promoted by the following three aspects:

(a) Reports show that more ancient glasses have been discovered from more than 500 excavation sites during the past 50 years. Archeology and cultural relics sectors sorted out the unearthed ancient Chinese glasses systematically, and studied them in every respect, including cultural exchange, historical background, comparison of the relics, etc.
(b) Researchers of glass science and technology in China have joined in the scientific study of ancient Chinese glasses. They

analyzed the chemical compositions of more than 300 samples and studied weathering, preservation and making technology of ancient glasses.

(c) Overseas scientific glass archeologists, mainly from the Corning Museum of Glass, USA, made analyses and investigations of more than 100 samples of the collected ancient Chinese glasses.

Therefore, the experts and scholars both from China and abroad in the above-mentioned three fields could get together to discuss the origin and evolution of ancient Chinese glasses. The important events include the Archeology of Glass Sessions of the 1984 International Symposium on Glass in Beijing; the Symposium on Archeology of Glass of the 17th International Congress on Glass in Beijing, 1995; and the Topic Meeting on Ancient Glass Along the Silk Road of the 2005 International Symposium on Glass in Shanghai. The proceedings, both in Chinese and in English, were published after the meetings, and resulted in active effects.[8–10]

As an enthusiast and an amateur of ancient Chinese glass research, the author, at 70 years of age, is still scheduled as a part-timer to meet Chinese scholars and experts in the ancient glass field, to discuss and analyze further the systematic development of ancient Chinese glasses. He sponsored and organized the Symposium on Ancient Glasses Unearthed from Southern China (2002; Nanning, Guangxi) and the Symposium on Ancient Glasses Unearthed from Northern China (2004; Urumqi, Xinjiang); and edited and published the proceedings.[11,12] He invited researchers of glass science and technology and researchers of archeology and cultural relics working together to write a book entitled *Development of Ancient Chinese Glass*.[13] All these efforts have advanced the understanding of and insight into ancient Chinese glasses.

The ancient glass artifacts found in China can be divided into the following three aspects; (a) the glass artifacts made by self-invented glass-making technology and using local raw materials; (b) the glass artifacts made by foreign glass-making technology and using local raw materials; (c) the glass artifacts imported from abroad. It should be pointed out that the provenances of ancient glasses are different between Inner China and Xiyu (the Western Regions). "Inner China" specifically indicates the Yellow River, Yangtze River and Pearl River valleys, while "Xiyu" indicates northwestern China, mainly the Xinjiang area. This article focuses on the origin and development of ancient glass in Inner China. The ancient glass in the Western Regions and its relationship with the ancient Silk Road will be discussed in the next article.

2. Development of Ancient Chinese Glass and Evolution of Its Chemical Compositions

In *Development of Ancient Chinese Glass*, we have introduced and analyzed the ancient glasses of different periods and different areas in China, including their shapes, patterns, histories, essences, chemical compositions and structures. Also, over 500 excavation sites where the ancient Chinese glasses were discovered and the chemical compositions of more than 500 glass samples are collected and edited in the Appendix of this book. From the shapes and patterns of the unearthed ancient Chinese glasses, one can find the typical Chinese characteristics of ancient glass objects, such as a *bi* (ritual disk), an ear pendant, a bottle for Buddhist body ash and a *han* (a bead put in the mouth of the dead). The scriptures and patterns on them could provide some information about their making period, and from their history and background and the C^{14} isotopic analysis on excavated sites and cofunerary objects, their making dates can be traced. The chemical composition of ancient glasses is an important indication for identifying where they come from. Although the history of glass making in ancient Egypt and West Asia is much earlier than that in China, the chemical composition of their glasses was not diversified, mainly belonging to the soda lime silicate glass system (Na_2O–CaO–SiO_2). Additional components and content of K_2O, MgO, Al_2O_3, etc., could be used to determine where (plateau or coast) this type of glass was produced. The main chemical

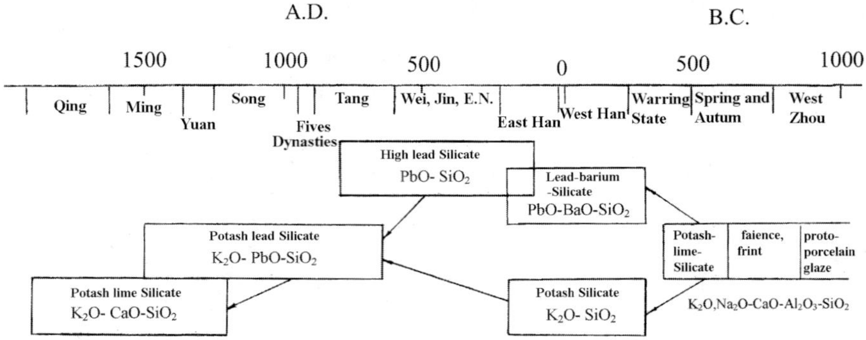

Fig. 1.1. Development of the chemical compositions of ancient Chinese glasses.

compositions of ancient Chinese glass in its history of development are quite different to that of the West. Figure 1.1 shows an evolution sketch of the chemical compositions of ancient Chinese glasses. We can see from the figure that the development of ancient Chinese glass can be divided into five stages according to the evolution of the glass composition:

(1) From the Spring and Autumn period to the early Warring States period (800–400 BC), the K_2O–CaO–SiO_2 system, where $K_2O/Na_2O > 1$;
(2) From the Warring States period to the Eastern Han Dynasty (400 BC–200 AD), the BaO–PbO–SiO_2 and K_2O–SiO_2 systems;
(3) From the Eastern Han Dynasty to the Tang Dynasty (200–700 AD), the PbO–SiO_2 system;
(4) From the Tang Dynasty to the Yuan Dynasty (600–1200 AD), the K_2O–PbO–SiO_2 system;
(5) From the Yuan Dynasty to the Qing Dynasty (1200–1900 AD), the K_2O–CaO–SiO_2 system.

The shapes, patterns and essences, as well as excavation background and history of the glass findings in each historical period, can be seen in detail from Ref. 13. The following parts will deal with the excavation background and chemical composition of the glass objects in each period.

2.1. *Early Chinese faience and frit (from the Western Zhou to the Spring and Autumn period, 1100–800 BC)*

Faience and frit were the products made before people could make glass. Owing to the fact that the available furnace temperature was not high enough, the materials could not be melted into glass completely. Chinese faience and frit were mainly unearthed in Shanxi and Henan provinces (the Yellow River valley), dating from the Western Zhou to the Spring and Autumn period. Archeological researchers often called them *"Liao-qi,"* confusing them with glass artifacts. Artifacts of this kind have been unearthed in large quantities. For example, more than 1000 pieces have been unearthed from the tomb of Yube and his wife, dating back to the mid-Western Zhou Dynasty (10th century BC); this shows that they could be produced locally at that time. Table 1.1 lists the collected data on faience beads of the Zhou Dynasty. Recently a small quantity of frit has been discovered along the Yangtze River valley, dating back to the early Spring and Autumn and Warring States periods; it reveals that the furnace temperature was increased. Table 1.2 shows the analytical results of those frit beads and tubes. Also, inlaid beads (eye beads) could be produced, such as the stringed faience beads unearthed from the Zenghouyi tomb in Suixian of Hubei province (see photo 1.3). The SiO_2 content (weight) in the faience is higher than 90%, and a little amount of alkali oxides (R_2O), such as Na_2O and K_2O, is present. Chinese faience and frit are characterized by a high content of K_2O, which is higher than the Na_2O content ($K_2O/Na_2O > 1$ in weight).[14] R. Brill found that for the glass phase in ancient Chinese faience containing high K_2O (up to 15% w.t.), the potash source might be leached plant ash or saltpeter (KNO_3).[15] Figure 1.2 shows the ratio relationship of K_2O and Na_2O contents in Chinese faience and in Egypt faience. Plant ash could be used as a flux agent for faience making. It was also used in making protoporcelain glaze in China, so its ratio of $K_2O/Na_2O > 1$ too. Natural natron (Na_2CO_3) was often used as a flux agent in early faience and glass making in West Asia

Table 1.1. Chemical composition of *liuli* beads (faience) of the Zhou Dynasty.

				Chemical composition (wt%)								
Unearthing place	**Artifact**	**Time**	**Method**	**SiO_2**	**Al_2O_3**	**Fe_2O_3**	**CaO**	**MgO**	**K_2O**	**Na_2O**	**CuO**	**MnO**
Henan	Bead	WZ	EP	> 90	0.3	0.2	0.4	0.3	3.4	1.2	1.6	0.3
Shanxi	Bead	WZ	EP	94.0	0.7	0.4	0.4	0.2	0.3	0.3	0.8	
Henan	Tube	WZ	EP	most		0.33	0.35	0.15	1.30	0.64	1.2	PbO 0.23
Shanxi	Bead	WZ	EP	92.4	4.4	0.8	1.7	0.24	0.5	0.0	0.1	
Henan	Bead	SA	EP	94.11			0.18	0.06	1.19	0.44		
				> 90			0.3	0.11	3.2	0.86		
Henan	Bead	Early SA	XRF	88.7	3.7	0.9	5.0	0.6	1.0	0.3	< 0.1	—
Ancient Egypt	Faience	(1500 BC)		96.73	0.2	0.14	0.70	0.1	0.1	0.63		

WZ — Western Zhou period (1000–800 BC); SA — Spring and Autumn period (800–500 BC); WS — Warring States period (500–200 BC); EP — electron probe; XRF — X-ray fluorescence.

Table 1.2. Chemical composition of some faience and frit beads and tubes unearthed in China (wt%).

No.	Sample	Time	Na_2O	Al_2O_3	SiO_2	K_2O	PbO	BaO	CaO	Cl	Fe_2O_3	CuO	Method
WJ-09A	Beads unearthed at E-Ji-Na-Qi, Inner Mongolia	Western Zhou		3.52	84.33	1.16			0.99	0.97	0.46	6.20	PIXE
WJ-09B				2.64	89.55	1.04		TiO_2 0.06	0.84	0.96	0.29	2.95	
WJ-09C				3.35	89.07	1.31			0.48	1.56	0.18	3.20	
GSU-1	Light-green tube unearthed at Lixian, Gansu province	Warring States		3.27	51.84	0.17	28.89	8.46	1.30	P_2O_5 2.43	0.68	2.81	PIXE
QH-1	Toothlike faience bead unearthed at Datong county, Qinghai province	Eastern Han Dynasty	2.07	4.81	89.11	1.15				MgO 0.68	1.67		XRF

PIXE — proton-induced X-ray emission.

Photo 1.3. String of frit beads unearthed from the Zeng Houyi tomb, Suixian, Hubei (early Warring States).

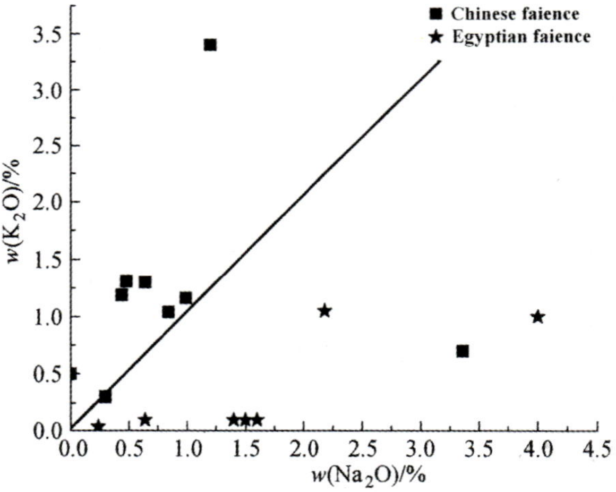

Fig. 1.2. Comparison of the K_2O/Na_2O ratio of ancient Chinese faience with that of ancient Egypt faience.

and Egypt, so its Na_2O content is higher than its K_2O content. Wad-El-Natrum is a place famous for producing natron. Chinese faience and frit making with plant ash is closely related to Chinese proto-porcelain glazes.

2.2. Early Chinese alkali lime silicate glass (pre-Qin Dynasty, 500–400 BC)

The earliest Chinese glass, to be an essential glassy state, is alkali lime silicate glass. Most of the discovered objects are monochromatic glass beads and ornaments dating back to the late Spring and Autumn and early Warring States periods (500–400 BC). There are not many of these unearthed objects, and they are mainly distributed in the Yellow River and Yangtze River valleys. Examples are the sword pommel ornament glasses of King Wu and King Yue, and the monochromatic glass beads excavated from the Chu tomb in Hubei province. These findings have not been studied in detail, being confined by experimental conditions during the excavation. Their chemical compositions measured by different methods gave different results. Table 1.3 shows the chemical compositions of those glass samples. However, all of them belong to the alkali lime silicate glass system (R_2O–CaO–SiO_2), including two types differing from the molecular ratio K_2O/Na_2O, and the content of CaO is about 3–8% (by weight). The chemical composition of the glass beads unearthed from Gushihou Pill and the Chu tomb at Xujialin in Xichuan (photo 1.4), Henan province, shows the ratio $K_2O/Na_2O < 1$, which should be the soda lime silicate glass. These glass beads should come from the West (to be discussed in another paper). The chemical compositions of the blue glass beads unearthed from the Chu tomb at Jiudian, Jiangling of Hubei province, and the sword pommel ornament glass of Goujian's sword (king of Yue state) unearthed from the Wangshan No. 1 tomb, $K_2O/Na_2O > 1$, have not been found in Western glasses for the same period.

Yue King Goujian's sword is one of the top national cultural relics. On its body, the inscription "Yue King Goujian's self-made and self-used sword" is engraved and black rhombic patterns are covered. It is very sharp and exquisite. In the pommel of this sword, blue glass ornaments are inlaid on one side (two pieces were still on it during the excavation — see photo 1.5), and turquoise inlaid on the other side.[16] This sword is very prestigious and famous in Chinese history, revealing that inlaid glasses were precious at that time.

Table 1.3. Chemical composition of early ancient glasses.

Name of group	Name of sample	Date	Unearthing site	Measuring method	Chemical composition of glass (mass percentage)									
					SiO_2	Al_2O_3	Fe_2O_3	CaO	MgO	BaO	PbO	K_2O	Na_2O	CuO
1.3-1. Earliest ancient glasses in China	Eye bead	500 BC	Tomb of Gushihou, Henan				0.65	9.42	0.35			0.52	10.9	
	Glass inlaid on pommel of King Wu's (Fuchai) sword	495–473 BC	Tomb of early Warring States, Huixian, Henan		xxx			xx						
	Glass inlaid on pommel of King Yue's (Goujian) sword	496–464 BC	Tomb of Wangshan, Jianglin, Hubei	PIXE	xxx		x	xx				xx		x
	Eye bead	~400 BC	Tomb of Zenghou Yi, Suixian, Hubei	CA	56.1	1.4	1.0	4.1	2.2	0.1	2.8	2.6	7.0	0.4
				XRF	xxx			xx				xx		

(*Continued*)

Table 1.3. *(Continued)*

Name of group	Name of sample	Date	Unearthing site	Measuring method	Chemical composition of glass (mass percentage)									
					SiO_2	Al_2O_3	Fe_2O_3	CaO	MgO	BaO	PbO	K_2O	Na_2O	CuO
1.3-2. Ancient glasses unearthed in Spring and Autumn to Warring States period	White *bi* (ritual disk)	Warring States	Chu tomb, Changsha		36.6	0.5	0.15	2.1	0.2	10.1	44.7	0.1	3.7	0.1
	Semitransparent gray bead	Late Spring and Autumn to early Warring States	Bozhou, Anhui		47.2	9.5	0.9	1.6	0.3	12.1	22.5	1.7	2.9	0.8
	Hexahedral green glass bead	Late Warring States	Jiangchuan, Yunnan		81.4	2.7		1.8				14.3		
1.3-3. Ancient silicate glasses unearthed in Southeast Asia	Glass bead	300–600 BC	Hastinapur, India		80	<1	2.6	3.9	<1		–	10.7	<1	–
	Glass bead	68 BC	Dingcun, Vietnam		mostly	1.3	0.8	2.8	0.5			18	0.2	1.3

Photo 1.4. Glass eye bead of the early Warring States period, unearthed at Xujialing in Xichuan, Henan.

Photo 1.5. Yue King Goujian's sword with inlaid blue glasses, unearthed from the No. 1 tomb at Wangshan, Jiangling, Hubei (496–464 BC).

The glasses inlaid on the sword pommel were measured together with the measurement of the sword body by the proton-induced X-ray emission (PIXE) technique in the 1980s.[17] Its PIXE spectrum was found recently, as shown in Fig. 1.3(a). It belongs to the potash lime silicate glass system. Table 1.3-1 shows its chemical composition.

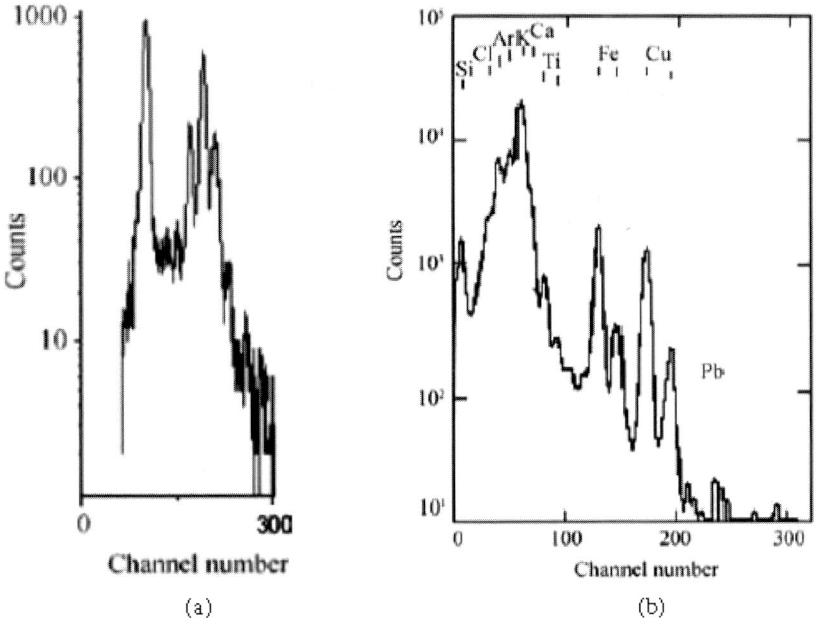

Fig. 1.3. PIXE spectral diagrams of ancient glass: (a) glass inlaid on the sword pommel of the king of Yue state (Goujian), Wangshan, Jiangling, Hubei; (b) glass bead fragment found in Chu state, Wangshan, Jiangling, Hubei.

By using the PIXE technique and the energy-dispersive X-ray fluorescence (EDXRF) method, the glass beads unearthed from the Chu tomb at the same place (Wangshan, Jiangling, Hubei province), but dating back a little later (450–400 BC), were analyzed[18]; Table 1.4 gives the measured chemical compositions. They belong to the alkali lime silicate glass system with high K_2O content. Figure 1.3(b) shows the PIXE spectrum of this glass sample. The spectrum in Fig. 1.3(a) is quite similar to that in Fig. 1.3(b), which confirms that the glass inlaid in Yue King Goujian's sword belongs to the same glass system. We have never seen the same chemical composition system in ancient glasses of Egypt and Babylon. Comparing the glass unearthed at Jiudian, Jiangling of Hubei province with the protoporcelain glaze having low CaO content from Jiangxi province (Table 1.5 lists the chemical composition of Chinese protoporcelain

Table 1.4. Condition and chemical composition of most early alkali-lime-silicate glasses found in Henan and Hubei.

A. Condition of glass samples

Number of sample	Name of tomb	Unearthing site	Description of sample	Date
HB-1	Xujialing Chu tomb No. 1	Xichuan, Henan	White eye in blue bead body (frit)	Early Warring States (500–400 BC)
HB-3	Jiudian Tomb M533	Jiangling, Hubei	Broken blue bead piece	Warring States (400–300 BC)
HB-6	Zenghou Yi tomb	Suixian, Hubei	Small blue glass piece	Late Spring and Autumn to Warring States (500–400 BC)

B. Chemical composition

Number of sample	Chemical composition of glass samples														Measurement method
	SiO_2	Na_2O	CaO	MgO	K_2O	Al_2O_3	PbO	BaO	CuO	Fe_2O_3	TiO_2	MnO	P_2O_5	SO_3	
HB-1	76.9	6.03	8.18	0.48	0.69	3.26		0.11	1.28	0.46			0.73	1.12	PIXE
HB-3	71.3	1.81	2.37	1.75	10.70	6.83	1.0	0.14	2.64	1.19	0.16	0.11	0.51	0.32	PIXE
	78.45		2.36	0.57	9.6	5.68	0.1		1.76	0.83		0.06			EDXRF
HB-6	79.67	7.97	6.03	0.53	0.6	3.07	0.02		0.2	1.3		0.01			EDXRF

Table 1.5. Chemical composition of ancient Chinese protoporcelain glaze.

Name	Name of sample	Date	Unearthing site	Chemical composition of glass (mass percentage)									
				SiO_2	Al_2O_3	Fe_2O_3	CaO	MgO	K_2O	Na_2O	MnO	TiO_2	P_2O_5
1.5-1. Protoporcelain glaze, Jiangxi	Protoporcelain glaze	Shang to Western Zhou	Fanchengdui, Qingjiang, Jiangxi	72.67	8.57	4.24	3.65	0.68	8.99	1.27	0	0.34	0
	Protoporcelain glaze	Late Shang	Jiaoshan, Yingtan, Jiangxi	61.69	17.97	5.00	4.49	1.72	7.43	0.47	0.05	0.96	0.22
1.5-2. Protoporcelain glaze	Protoporcelain glaze	Early Qin	Henglingshan, Boluo, Guangdong	63.94	15.41	1.73	11.23	3.18	2.28	0.53		0.48	

glaze[19]) their chemical compositions are in good agreement. This implies that the ancient glass-making technique in Inner China may have evolved from the protoporcelain-making technique followed by faience-making, and that their production areas were all along the Yangtze River valley. The glaze on protoporcelain was coated on pottery with glaze paste, and there was no necessity for a container. The most important technical improvement from making glaze to making glass was the use of a container — a refractory crucible needed for melting glass. Nevertheless, the bronze and lead metallurgy originating from the Shang Dynasty (16–12 centuries BC) had created necessary conditions.

2.3. *Early Chinese lead barium silicate glass and potash silicate glass (the Warring States period to the Han Dynasty, 400 BC–200 AD)*

In order to increase the transparency and lower the melting temperature of glasses, primitive Chinese people made efforts to improve the flux agent with various methods. Lead pills (lead salt, such as lead oxide) and saltpeter (KNO_3) as Chinese medicine materials had been well known since the Spring and Autumn period. They could also serve as a flux agent. So, lead barium silicate and potash silicate glasses were first developed along the Yangtze River valley during the Warring States period.

The technique of using lead in bronze metallurgy came very early in China. Lead can lower the melting temperature and increase flowability. The early bronze was a copper–tin–lead alloy. So the use of lead ores should be a long historical experience originating from bronze-making in the Shang and Zhou Dynasties. Lead ores like galena (PbS) and barium ores like barite ($BaSO_4$) are stored and produced in large quantities along the Yangtze River valley — Hunan, Anhui and Jiangxi provinces. It is understandable that the primitive people used them as a flux agent in melting glasses. The ancient Chinese lead barium silicate glass was discovered around the Yangtze River valley; it often fell into the same sites with the distribution of lead ores. A translucent glass bead was unearthed at

Bozhou, Anhui province, dating from the late Spring and Autumn period to the early Warring States period (6–5 centuries BC). This is the earliest lead barium silicate glass found till now. The chemical composition of early lead barium silicate glasses is shown in Table 1.3-2. Over 200 pieces of glass *bi* disks, beads, seals, sword tubes, etc., were unearthed at Zixin, Changsha of Hunan province, showing that they had become popular during that time. In the chemical composition of lead barium silicate glasses, the content of BaO is about 5–15%, PbO about 10–45%, Na_2O plus $K_2O < 15\%$, and others are mostly SiO_2 (35–65%). The lead barium silicate glasses dating back to the mid-to-late Warring States period were discovered in the south and southwest of China.[11] The chemical composition of PbO–BaO–SiO_2 glass is shown in Table 1.6.

Mold-casting technology was mainly used in ancient Chinese glass-making, which was copied from bronze metallurgy. During the Han Dynasty, large glass plates could be produced. For example, a plate glass of 9.5 cm × 4.5 cm × 0.3 cm dimension was found in the tomb of the King of Southern Yue state in Guangzhou in the early Han Dynasty; a glass *bi* disk with a diameter of 23.4 cm, a thickness of 1.8 cm and a weight of 1.9 kg was unearthed from the Maolin mausoleum of the Han Dynasty, in Shanxi province; and a big lead barium silicate glass plate of 32.5 cm × 14.8 cm × 3.5 cm dimension with a weight of 5.25 kg was unearthed at Jimo of Shandong province. The lead barium silicate glasses were widely spread and traded from the Warring States period to Han Dynasty, covering Guangdong and Guangxi in the south, Sichuan and Guizhou in the southwest, Qinghai, Gansu, Xinjiang in the northwest, and Liaoning and Inner Mongolia in the north (see Table 1.7). This kind of glass also serves as evidence of the ancient Chinese glasses spread to other countries.[13]

Primitive Chinese people increased the content of K_2O in making potash lime silicate glass, which was another approach to improving the glass flux agent. Saltpeter (KNO_3) was used as the flux agent instead of plant ash. The use of saltpeter has a long history in China. As a medicine material, it can be traced from the historical literature of the Western Han Dynasty. Saltpeter was also used in lead metallurgy, due to its low melting point (330°C).[20]

Table 1.6. Condition and chemical composition of glass samples from southern China.

Number of glass sample	Name of tomb	Unearthing site	Description of sample	Date
A. Condition of glass samples				
GD01	Songshan tomb	Zhaoqing, Guangdong	Eye bead with white eye and blue glass body	Warring States (400–300 BC)
GZH-2	Kele tomb M-91	Haozhang, Guizhou	Green glass bead	Late Warring States (00–250 BC)
M13-15	Haojiaping	Qingchuan, Sichuan	Eye bead with three-color eyes	Warring States (400–300 BC)
SX-14	Yujiaba, Sanxia reservoir	Kaixian, Chongqing	Deep blue eye bead with blue and white inlaid	Warring States (400–300 BC)
GZH-17	Zhongshan Liyuan M-42	Weining, Guizhou	Blue glass bead	Western Han–Warring States (250–200 BC)
44-18-35A	–	Lixian, Sichuan	Yellow glass bead	Western Han–Warring States (250–200 BC)
XJ-5A	Bao-Zi-Dong M-41	Wensu, Xinjiang	Blue and green bead fragment	Qin–Warring States (250–200 BC)

(*Continued*)

Table 1.6. *(Continued)*

B. Chemical composition														
Number of sample	Chemical composition of glass samples (mass percentage)													Measurement method
	SiO_2	CaO	K_2O	Al_2O_3	PbO	BaO	CuO	Fe_2O_3	TiO_2	MnO	Cr_2O_3	P_2O_5	SO_3	
GD01	49.95	2.87	1.81	9.01	18.58	15.82	0.06	1.43	0		0.03	0.45		PIXE
GZH-2	58.10	1.14	0.6	6.10	19.71	9.14	0.58	1.88	0	0.05	0	1.22	0.05	
M13-15	54.72	1.27	0.47	3.05	25.16	11.82	2.54	0.44	0		0.04			
SX-14	58.53	0.57	0.51	2.65	28.78	6.97	0.64	0.27				1.07		
GZH-17	77.38	0.48	10.83	6.8		0.18	0.20	1.41	0.27	1.09	0.03	0.25	1.09	
44-18-35A	73.87	7.04	9.75	3.44	0.36	0.03	0.01	1.07	0.17			0.19		
XJ-5A	75.13	1.74	18.40	2.47			1.09	0.74	0.09	0.04			0.04	

Table 1.7. Condition and chemical composition of glass samples from remote areas in China.

A. Condition of samples

Number of sample	Name of tomb	Unearthing site	Description of sample	Date
XJ-37A	Han tomb	Weiwu, Gansu	Green ear pendant	Han Dynasty (200–100 BC)
NM02-2	Zhalainuer tomb	Holabiar, Inner Mongolia	Green bead	
XJ-46	No. 14	Hami, Xinjiang	Blue squach-shaped bead fragment	
XJ-42	Ancient AKe-Si-Pi-Li Castle	Hetian, Xinjiang	Eye bead with black body and green inlaid	Western Han

B. Chemical composition

Number of sample	Chemical composition of glass samples (mass percentage)															Measurement method
	SiO_2	Al_2O_3	Fe_2O_3	CaO	MgO	PbO	BaO	K_2O	Na_2O	CuO	MnO	TiO_2	B_2O_5	Li_2O	ZnO	
XJ-37A	56.4	1.53	0.66	3.80		24.10	10.34	0.57		0.09	0.34					PIXE
NM02-2	49.91	3.22	0.37	2.91		35.04	0.08	0.49		0.26						
XJ-46	20.18	1.00	12.03	1.90	0.28	47.14	14.62	0.36	2.2	0.06	0.01					ICP
XJ-42b body black	43.52	3.65	9.96	1.50		31.28	1.43	0.76		5.28	0.03					PIXE
Green glass	49.35	3.19	0.82	2.52		34.24	7.87	0.34		0.64	0.01					

It has been found from chemical composition analysis of ancient Chinese glasses uncovered in Guangxi in the 1980s that the contents of Al_2O_3, CaO and Na_2O are all low (<3%), but the K_2O content is high (10%), belonging to typical potash silicate glass.[21] Most of these glasses were excavated from ancient tombs of the Han Dynasty, such as the ancient cemeteries of the Han Dynasty in Hepu of Guangxi province. So far, the earliest potash silicate glass found was unearthed from tombs of the Warring States period, such as the Chu tombs at Jiangling, Hubei and at Changsha, Hunan. The chemical composition of ancient potash silicate glasses is listed in Table 1.3-2 and Table 1.6. These glasses together with lead barium silicate glass beads were used as funeral objects. One may conclude from the above-mentioned that potash silicate glass and lead barium silicate glass were made nearly in the same period along the Yangtze River valley. During the early Western Han Dynasty, potash silicate glass also appeared in the southwest of China. The saltpeter is easily formed on the soil surface in warm places; in particular, hot weather after rainy seasons promotes its formation. Reasonably, southern China (Guangdong and Guangxi) became the major production sites for ancient potash silicate glasses, and from there this kind of glass of the Han Dynasty could be excavated.

Therefore, the earliest lead barium silicate and potash silicate glass products in the world can be said to have been made in Inner China. Table 1.3 gives the chemical compositions of typical ancient glass samples. The ancient glass objects all had special Chinese forms, like the *bi* (ritual disk), bead, ear pendant etc.; such examples are a white *bi* disk with nipple patterns (photo 1.6), a glass eye bead (photo 1.7) and a green ribbed bead (photo 1.8) dating back to the Warring States period unearthed from the Chu tomb at Changsha, Hunan, at Jianglin, Hubei and at Jiangchuan, Yunnan respectively. Another glass ear pendant of the Han Dynasty unearthed at the site of the present Nanchong Railway Station in Sichuan is shown in photo 1.9. Figure 1.4 shows the distribution of unearthing sites of ancient Chinese lead barium silicate and potash silicate glasses dating from the Warring States and the Han Dynasty.

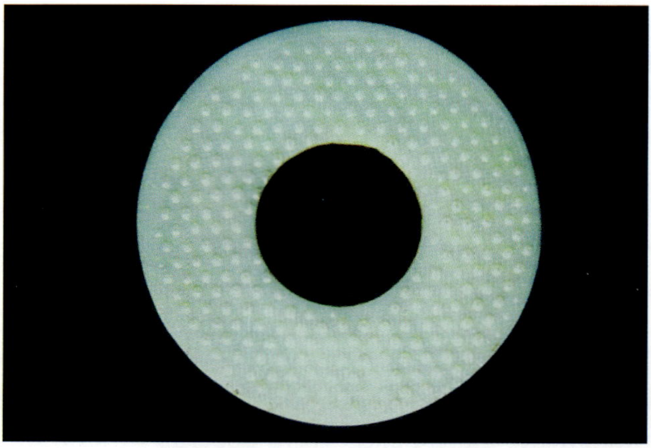

Photo 1.6. Glass ritual disk (*bi*) with a nipple pattern, unearthed at Changsha, Hunan (Warring States period).

Photo 1.7. Glass eye bead unearthed at Wangshan, Jiangling, Hubei (middle Warring States, around 329 BC).

2.4. *Early Chinese high lead silicate glass and potash lead silicate glass (Six Dynasties to Northern Song Dynasty, 200–1200 AD)*

Ancient Chinese glasses were mostly made into ritual utensils and ornaments to imitate jade ones. By using BaO, the resulted glass

Photo 1.8. Hexagonal glass bead unearthed from the tomb of the Warring States at Jiangchuan, Yunnan.

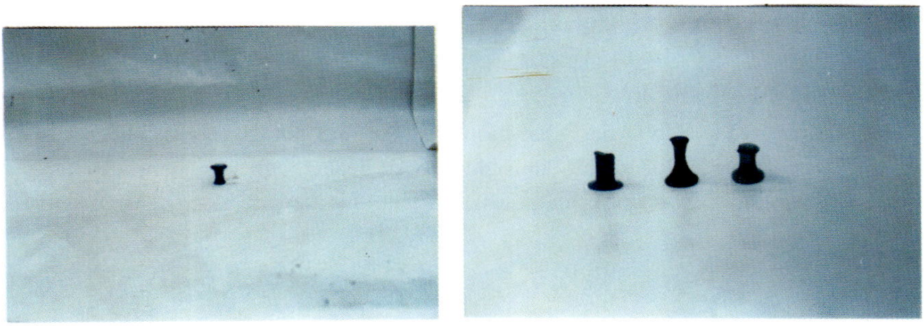

Photo 1.9. Ear pendant of the Han Dynasty, from the Nanchong railway station site, Sichuan.

became opaque and a turbid white; also, the melting temperature was decreased. *Bi* disks, beads, ear pendants, etc., were all made by the mold-casting method at that time. The experience accumulated from the long history of lead metallurgy was helpful in the preparation and application of yellow lead (PbO) and red lead (Pb_3O_4).[22] Therefore it is understandable that by increasing the PbO content instead of using BaO in the glass composition, transparent glasses could be obtained. The high lead silicate glass originated from the

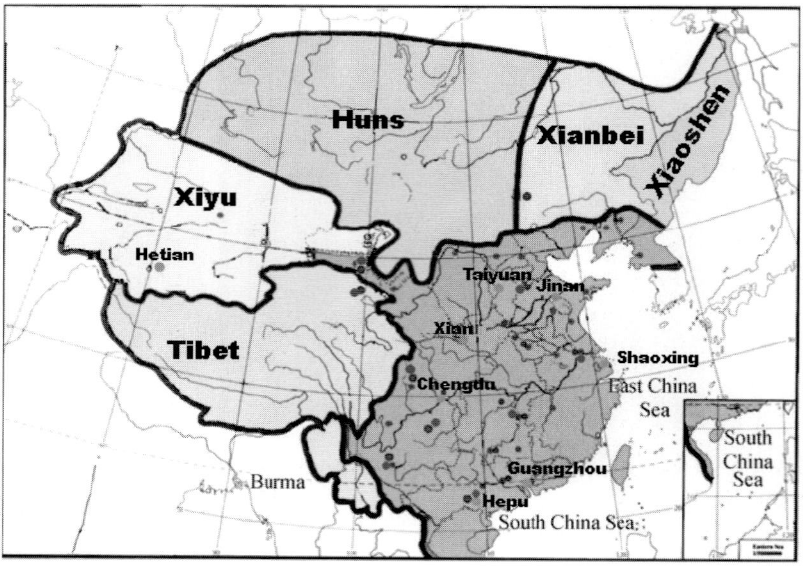

Fig. 1.4. Unearthing sites of ancient Chinese lead barium silicate glass and potash silicate glass from the Warring States and Han Dynasty.* • Lead Barium silicate glass • Potash silicate glass

* This old map of the Han Dynasty has been adapted from Tan Qixiang's, *Concise Historic Atlas of China*, (China Carto Graphic Publishing House, Beijing, 1991), and translated into the English by the author. Also, unearthing sites of ancient Chinese glass have been added to the map by the author.

Warring States period, evidenced by the beads unearthed at Luoyang, Henan, which have very a high content of PbO.[23] This kind of glass was popular till the Eastern Han Dynasty.

The inflow of glass-blowing techniques from the West to Inner China started in the Sui Dynasty (6th century AD), and was recorded in some historical writings, like *Beishi-Darouzhi Zhuan* (*History of the North — Memoir of Great Yen Chin*) and *Beishi-Hezhouzhuan* (*History of the North — Biography of Hezhou*). For blowing the glass into a vessel, a low changing rate of glass viscosity with temperature is expected. This was also the main reason for increasing the PbO content and not using BaO in ancient glass-making. High lead silicate glass caused severe corrosion of the refractory crucible used for melting glass, so K_2O was later used instead of partial PbO, gradually. Hence the potash lead silicate glass got into formation, while its low changing rate of glass viscosity remained. As mentioned above, primitive Chinese people had used KNO_3 very early, and accumulated experience on making potash silicate glass. So the development of potash lead silicate glass was an inevitable trend. This was also a process of ancient Chinese people understanding the relationship between chemical composition and physical property. We may say that the glass vessels made with high lead silicate glass and potash lead silicate glass by the blowing method were unique products of ancient China. Examples of the typical wares are a glass goblet unearthed from a tomb of the Sui and Tang Dynasties at Qingzhou, Guangxi (photo 1.10); a glass bottle with short neck unearthed from the Lishou tomb of the Tang Dynasty at Shanyuan, Shanxi; a bird-shaped glass ware unearthed from a pagoda base of the Northern Song Dynasty at Mixian, Henan (photo 1.11); a Buddhist body ash bottle, an egg-shaped ware, etc. All of them are characterized by Eastern culture.

Photo 1.10. Glass goblet unearthed from the tomb of the Sui Dynasty to the Tang Dynasty at Qinzhou, Guangxi.

Photo 1.11. Bird-shaped glassware from the pagoda base of the Northern Song at Mixian, Henan.

The chemical compositions of high lead silicate glass are Na_2O plus K_2O <5%, PbO 35–75%, SiO_2 35–75%, while those of potash lead silicate glass are Na_2O <1%, K_2O 7–15%, PbO 35–50%, SiO_2 30–60%, where the content of PbO varies a great deal. Table 1.8 shows the chemical composition of ancient glass samples with a high PbO content and with the potash lead silicate system. The high lead silicate glass discovered in China is not the earliest; the same type of glass was found in Nidmrund of the Mesopotamia area, dating back to the 6th century BC.[24] It is earlier than in China, but less was found afterward. Lead-containing glass was also found in ancient India, belonging to the same period as that in China.

Lead isotope analysis of the glasses containing lead is an effective method to determine where the ancient glasses were produced. R. H. Brill pointed out that, according to the results of lead isotope analysis carried out by the Corning Museum of Glass, USA on

Table 1.8. Chemical composition of ancient high lead silicate glasses.

			Chemical composition (wt%)								
Unearthing place	**Glass artifact**	**Date**	**SiO_2**	**Al_2O_3**	**Fe_2O_3**	**CaO**	**MgO**	**PbO**	**K_2O**	**Na_2O**	**CuO**
Baicheng, Xinjiang, China	Glass beads	~800 BC	64.31	1.36	1.10	0.01	2.67	9.01	2.42	—	0.01 Sb_2O_5 1.60
Luoyang, Henan, China	Beads	400–300 BC	18.20					74.01		3.29	
Guangxi, China	Glass cup	600 AD	34.92	1.57				62.1		1.43	
Pingba Machang, Guizhou, China	Glass beads	500–600 AD	49.38	2.61				35.52	7.48	3.69	0.63 Cl (0.66)
Pingba Machang, Guizhou, China	Glass beads	500–600 AD	47.91	2.41	0.1		1.1	38.68	7.36	1.78	Cl (0.59)
Lishou tomb, Shanxi, China	Green glass bottle	600–900 AD	36.16	2.42		1.09	2.84	46.65	0.95	10.01	
Chaoyang, Liaoning, China	Green glass beads	600–900 AD	26.32		0.16	0.13	0.1	50.31	10.09	0.29	0.13

(*Continued*)

Table 1.8. *(Continued)*

Unearthing place	Glass artifact	Date	Chemical composition (wt%)								
			SiO_2	Al_2O_3	Fe_2O_3	CaO	MgO	PbO	K_2O	Na_2O	CuO
Litai tomb, Hubei, China	Yellow glass bottle	600–900 AD	30.49	1.61	0.33	0.20	0.30	64.29	0.27	0.3	
Anhui, China	Green glass cup	960–1050 AD	27.88	0.32	0.20	0.22	0.04	66.86	0.53	0.13	2.96
				0.44	1.77	0.33	0.07	67.83	0.61	0.21	0.40
Hebei, China	Green glass bottle	960~1050 A.D	26.85		0.19	0.35	0.1	70.04	0.34	0.18	0.41
Gansu, China	Monk's bone ash bottle	960–1050 AD			0.19	0.13	0.1	50.31	10.09	0.29	0.13
Henan, China	Green glass goose	960–1050 AD			0.15	0.17	0.04	47.34	11.45	0.08	0.18
Henan, China	Red egg-shaped vessel	960–1050 AD	33.78	2.02	3.15	3.52	0.31	40.15	14.78	0.13	1.32
Hebei, China	Dark glass grapes	960–1050 AD	36.93	1.11	4.13	0.36	0.08	45.93	8.45	0.08	1.44

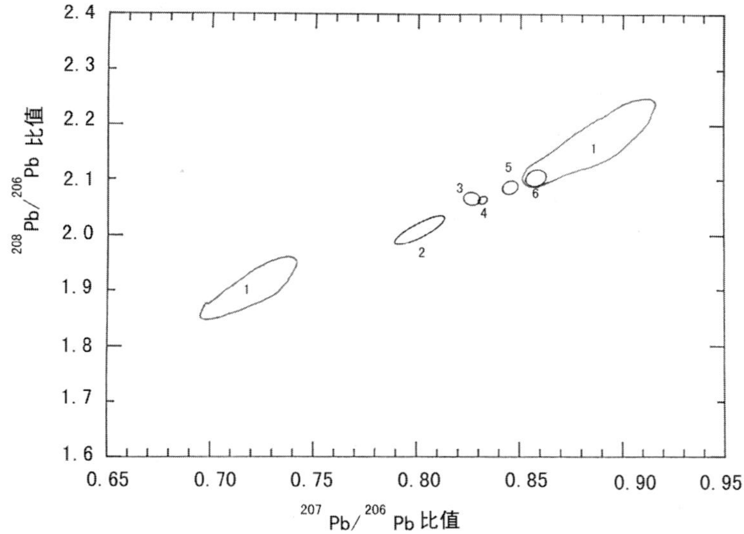

Fig. 1.5. Distribution of lead isotope ratios in lead-containing archeological objects from various places: (1) China, (2) Egypt, (3) Mesopotamia, (4) Greece, (5) England, (6) Spain.

various ancient glasses, the distributions of the isotope ratios $^{208}Pb/^{206}Pb$ and $^{207}Pb/^{206}Pb$ of ancient Chinese glasses differ from that of the ancient glasses and archeological objects containing lead elsewhere. The Chinese glasses fall into the high and low values in the distribution diagram shown in Fig. 1.5.

The ratios $^{208}Pb/^{206}Pb$ and $^{207}Pb/^{206}Pb$ of the most ancient Chinese glasses are 2.1–2.2 and 0.85–0.90 respectively. We plotted these ratios of Chinese lead ores in the same figure of Chinese glasses containing PbO (Fig. 1.6). One can find that the ratio of lead ores from southern China is lower than that from northern China, and the ratios of ancient Chinese glasses containing PbO all fall into the ratio scope of Chinese lead ores, also concentrated in central China. The earliest Chinese glasses containing PbO were unearthed mainly in central China, which was obviously correlated with the rich resource of lead ores in this area. Therefore a conclusion can be drawn that the ancient Chinese glasses containing PbO, including lead barium

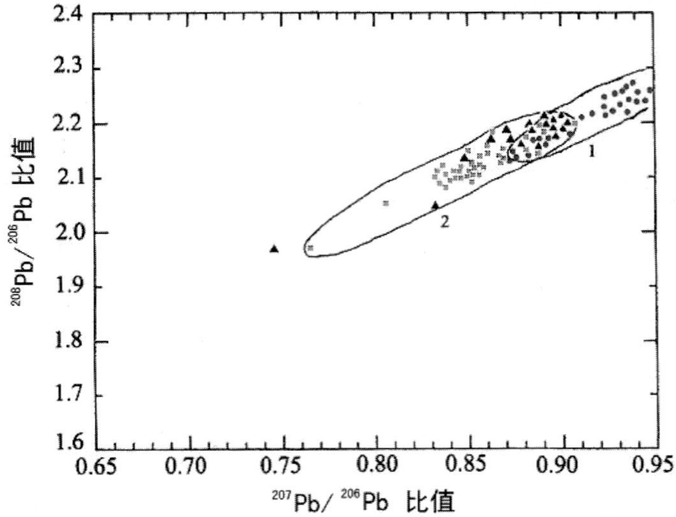

Fig. 1.6. Distribution of lead isotope ratios in lead ores and lead-containing ancient silicate glasses in China. ▪ Lead ores in southern and central China ▲ Ancient silicate glass containing lead, excavated in China • Lead ores in northern China.

silicate glass, high lead silicate glass and potash lead silicate glass, were all produced in Inner China, and then spread into peripheral regions such as East Asia, Southeast Asia and Central Asia.

2.5. *Early Chinese potash lime silicate glass (Yuan, Ming and Qing Dynasties, 1200–1900 AD)*

Soda lime silicate glass has been made and applied in Inner China since the Tang Dynasty, when the glass-blowing technique was introduced to China from the West. Na_2CO_3, $NaNO_3$ and $CaCO_3$ are more common minerals. By using these materials, the soda lime silicate glass was produced in Inner China, but was not very popular. The silicate glass containing calcium has a high chemical stability. Starting from the Song Dynasty, potash lead silicate glass evolved gradually by the use of CaO instead of PbO, and the

Table 1.9. Chemical composition of early Chinese potash lime silicate glasses.

			Chemical composition (wt%)								
Unearthing place	**Glass artifact**	**Date**	**SiO_2**	**Al_2O_3**	**Fe_2O_3**	**CaO**	**MgO**	**PbO**	**K_2O**	**Na_2O**	**CuO**
Boshan, Shandong, China	Green glass frit	1350–1400 AD	58.48	6.58	0.3	9.81	0.26	—	16.07 TiO_2 0.23	4.42 MnO 0.04	0.82 F4.99
Boshan, Shandong, China	Blue glass hairpin	1350–1400 AD	59.53	0.06	0.3	9.42	0.22	0.46	19.78	2.0	—
Boshan, Shandong, China	Glass pieces	1350–1400 AD	66.86	7.20	3.04	8.21	0.30	—	12.19 TiO_2 0.3	—	0.6
Sichuan, China	Glass hairpins	1350–1400 AD	67.42	2.28	0.57	11.36		0.1	17.31		0.03
Shantou, Guangdong, China	Blue glass hairpins	1400–1600 AD	66.25	5.60	0.30	10.54		0.04	16.45		0.68
Boshan, Shandong, China	Blue glass rod	1400–1600 AD	55.26	7.52	0.42	10.57	0.44	—	15.04	4.01	0.9
Beijing, China	Transparent bottle	1650–1700 AD	42.44	6.08	0.09	0.03	—	38.57	14.54	0.19	—
Beijing, China	Green cup	1750–1780 AD	74.80	1.63	0.15	0.19	0.04	0.25	20.89	0.18	0.49
Beijing, China	Transparent glass rod	1750–1780 AD	56.21	—	—	6.31	—	14.35	23.11	—	—
Beijing, China	Snuff bottle	1750–1780 AD	65.91	1.97	0.42	6.73	0.13	—	22.5	—	0.25

furnace temperature could reach a high degree (1400°C) at that time. So, from the Yuan Dynasty, the glasses made in Inner China mostly belonged to the potash lime silicate glass system. Its major manufacturing sites were the royal glassworks at the imperial palaces in Beijing, the *liuli* works at Zibo, Shandong and some other works later in Guangzhou. The chemical compositions of glasses made at the above-mentioned sites are SiO_2 60–70%, CaO 5–15%, K_2O 10–20%.[13] Table 1.9 lists the chemical composition of Chinese potash lime silicate glasses from the Yuan to the Qing Dynasty. The glassware made in the Yuan, Ming and Qing Dynasties is mostly small pieces of utensils, such as an ash tray (photo 1.12), hairpins (photo 1.13), a pot (photo 1.14) and a snuff bottle (photo 1.15).

It should be specially pointed out that the soda lime silicate glass system had been the common composition for worldwide glass-making from the 15th to the 19th century, and also that its main compositions are less varied; while, in Inner China, K_2O had always been used as the flux agent and the glass products were mainly made with potash lime silicate or potash lead silicate glass systems. This is a tradition of the application of K_2O and PbO as raw materials in Inner China.

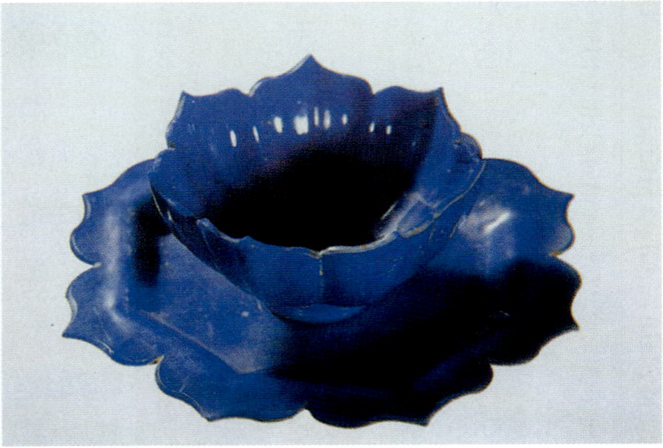

Photo 1.12. Glass lotus calyx and tray of the Yuan Dynasty, unearthed from the familial graveyard of Wang Shixian at Zhangxian, Gansu.

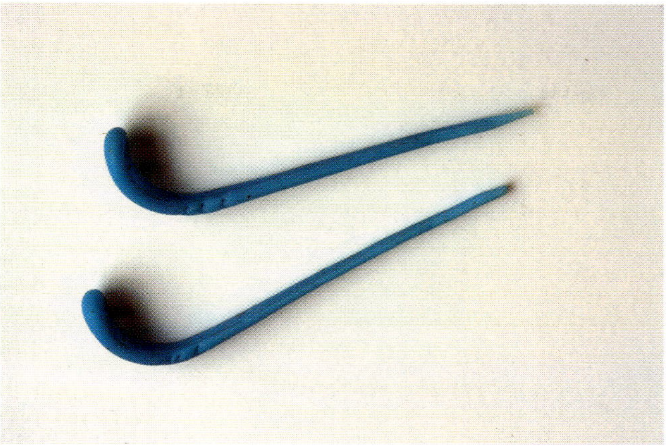

Photo 1.13. Glass hairpins from the tomb of the Ming Dynasty at Shantou, Guangdong.

Photo 1.14. Glass pot with chrysanthemum petal design, of the Yongzheng period of the Qing Dynasty.

3. Conclusion

As mentioned above, with a 3000-year history of self-made glasses in Inner China, the tradition of using K_2O and PbO as the main flux

Photo 1.15. Snuff bottle with flower and botanical design, of the Qing Dynasty.

agent has been inherited, and so the characteristics of ancient Chinese glasses in their chemical compositions can be displayed. This allows one to distinguish between the glass products made in China and those imported. It can be seen from the evolution of ancient Chinese glass composition that the ancient Chinese people consistently tried to improve glass-making technique and glass property. However, it should be noted that the specialty of the chemical composition of ancient Chinese glasses and the tradition of using materials confine the ancient Chinese glass products to the kinds of adornments and ritual utensils used until the Ming and Qing Dynasties. In addition, the Inner China people were accustomed to the use of porcelain — one of the earliest inventions of China. Both of them caused the ancient glass-making technique to develop slowly in China. That is really a pity.

References

1. F. X. Gan, Some viewpoints of ancient Chinese glass research, *J. Chin. Ceram. Soc.* (in Chinese) **32**, 181–185 (2004).
2. S.-G. Bi, Probing to the history of glass technology. In: Z. T. Li (ed.), *History of Glasses by Shen Chongwen* (in Chinese) (Wanjuai, 2005), pp. 1–5.

3. F. X. Gan, Z. F. Huang and B. Y. Xiao, The origin of ancient Chinese glass, *J. Chin. Ceram. Soc.* (in Chinese) **12**, 99–104 (1978).
4. B. D. Yang, Some problems of the research on ancient Chinese glass history, *Cultural Relics* (in Chinese) **5**, 26 (1979).
5. K. Yamasaki, History of Chinese glass: introduction of recent research, *Glass* (in Japanese) **8**, 2–5 (1980).
6. C. G. Seligman, P. C. Ritchie and H. C. Beck, Early Chinese glass from pre-Han to Tang times, *Nature* 138–721 (1936).
7. C. G. Seligman and H. C. Beck, Far Eastern glass: some western origins, *Bulletin of the Museum of Far Eastern Antiquities* **10**, 1–50 (1938).
8. F. X. Gan (ed.), *Research on Ancient Chinese Glass — Proceedings of the 1984 International Symposium on Glass* (Beijing, 1984). (China Architecture Press, Beijing, 1986), in Chinese.
9. R. H. Brill and J. H. Martin, *Scientific Research in Early Chinese Glass — Proceedings of the Archaeometry of Glass Session of the 1984 International Symposium on Glass* (The Corning Glass Museum, New York, 1991).
10. The Chinese Ceramic Society, *Proceedings of the 17th International Congress on Glass* (Beijing, 1995–6); Section: Archaeology of Glass.
11. F. X. Gan (ed.), *Research on Ancient Chinese Glasses of Southern China — Proceedings of the 2002 Symposium on Ancient Chinese Glasses of Southern China* (Nanning). (Shanghai Science and Technology Publishers, Shanghai, 2003) in Chinese.
12. F. X. Gan (ed.), *Study on Ancient Glasses Along the Silk Road — Proceedings of the 2004 Urumchi Symposium on Ancient Glasses in Northern China,* and *2005 Shanghai International Workshop of Archaeology of Glass* (Fudan University Press, Shanghai, 2007), in Chinese.
13. F. X. Gan *et al.*, *Development of Chinese Ancient Glass* (Shanghai Science and Technology Publishers, Shanghai, 2005), in Chinese.
14. X. F. Fu and F. X. Gan, Chinese faience and frit, *J. Chin. Ceram. Soc.* (in Chinese) **34**, **4**, 35–39 (2006).
15. R. H., Brill, Chemical composition of a faience bead from China, *J. Glass Studies* **31**, 11–15 (1989).
16. Z. Y. Chen, Date and buried dead of the Wangshan No. 1 tomb, in *Proceedings of the 1st Annual Meeting of Chinese Archaeology Society* (Cultural Relics Press, Beijing, 1978), in Chinese.

17. J. X. Chen, H. K. Li and C. G. Ren *et al.*, PIXE research with an external beam, *Nucl. Instrum. Methods* **168**, 437–440 (1980).
18. Q. H. Li, B. Zhang, H. S. Cheng and F. X. Gan, Application of proton-induced X-ray emission technique in chemical composition analysis of ancient Chinese glasses, *J. Chin. Ceram. Soc.* (in Chinese) **31**, 39–43 (2003).
19. H. J. Luo, J. Z. Li and L. M. Gao, Study on chemical composition and micro-structure of proto-porcelain, *J. Chin. Ceram. Soc.* (in Chinese) **24**, **1**, 114–118 (1996).
20. N. C. Meng, Expedition on the name of saltpeter in Han and Tang Dynasties, *Research of Natural Science History* (in Chinese), **2**, **2**, 97–111 (1983).
21. M. G. Shi, O. L. He and F. Z. Zhou, Study on the potash silicate glasses unearthed from Han tomb, *J. Chin. Ceram. Soc.* (in Chinese) **14**, **3**, 307–313 (1986).
22. K. H. Zhao, Probing to the origin of Chinese traditional glasses and the contribution of lead metallurgy to it, *Research of Natural Science History* (in Chinese) **2**, 145–156 (1991).
23. H. Q. Yuan, Glass making in the history of Chinese chemical technology, *Digest of 1957 Meeting of Chinese Chemical Society* (in Chinese), 80–81 (1957).
24. H. C. Beck, *Glass Before 1500 BC*. Ancient Egypt and the East (Corning, New York, 1962), pp. 83–85.
25. R. H. Brill and H. Shirahara, Lead isotope analysis of some Asian glasses, in *Proceedings of the 17th International Congress on Glass* (Chinese Ceramic Society, Beijing, 1995), Vol. 6, pp. 491–496.

Chapter 2 第二章

The Silk Road and Ancient Chinese Glass

Gan Fuxi

Shanghai Institute of Optics and Fine Mechanics,
Chinese Academy of Sciences,
Shanghai 201800, China

The name "Silk Road" was put forward by Baron Ferdinand von Richthofen in the late 19th century. It was used as a general expression for the transportation routes linking China with the Hellenistic Rome via the areas of Xiyu (the Western Regions). As is well known, silk was originally produced in China, and has a long history of more than 5000 years. It was introduced to Europe not later than the 4th century BC, according to a literary record by Ctesias, a Latin writer of that time. The Silk Road, in fact, is synonymous with the main artery connecting Asia with Europe for economic, political, cultural and technical contacts.[1] China is the center of the Silk Road in Asia, but not the terminal. It extended to peripheral regions, such as the Korean Peninsula, Japan and Southeast Asia. The Sea Silk Road is another route, linking China with the Mediterranean countries by the maritime route passing through Southeast Asia, India and West Asia. So the Silk Road was hugely expanded both in time and in space afterward.

The Silk Road played a very important role in cultural and technical exchange between China and foreign countries in antiquity. Investigation of the roles which it played in the exchange and spread of ancient glass products and technology is very significant. The influences of cultural and commercial contacts between the East and the West on the ancient Chinese glass products and the technology for making them can be seen from the development of ancient Chinese glasses.

This article deals with the formation and development of the ancient Silk Road, and its function in the flow of ancient glasses, as well as in promoting the development and spread of ancient Chinese glasses.

1. The Ancient Silk Road and Glass Exchange

The contacts between China and the Western world in protohistoric times, long before the Han Dynasty explored the Western Regions (Xiyu) and the formation of the Silk Road, have not been proved, due to the lack of physical evidence. But the practical cultural spread among nations and nationalities was always earlier than it appeared in the written evidence. Archeological materials and ancient legends can sometimes provide valuable information. *Mutianzi Zhuan* (*Biography of King Mu*) and *Shanhaijing* (*Book of Mountains and Seas*) are talelike ancient Chinese writings, but nowadays they seem to be more or less believable and have background. King Mu of the Zhou Dynasty made his travels westward in 989 BC and reached the Iranian plateau of Central Asia, west of the Chunlin Mountain (now the Permian plateau). Some people think that the so-called "boundless land" he reached is now the Kirgzistan grand steppe. His whole journey covered 35,000 *li* (1 *li* = 0.5 km). Silk is China's unique creation, first appearing 5000 years ago. According to the reports in foreign documents, exquisite beautiful silk appeared in the steppe of Central Asia, north of Tianshan Mountain, over 3000 years ago. Chinese silk had become the favored cloth of Hellenic top-class people from the 6th to the 3rd century BC. Therefore one may conceive that the contacts between China and the Western world started early, before 6th century BC. Now it can be conservatively estimated from scientific inferences that the early cultural intercourse between China and the West dated from the 11th century BC, i.e. between the Shang and the Zhou Dynastys in China.[2]

Zhangqian's traveling to the Western Regions was a magnificent undertaking, exploring a route for the East–West contacts. But long before his travels there had been ties between China and the West. And after his travels, some new transportation routes were explored between China and the outside world. Several years ago, UNESCO identified four main routes of the Silk Road: (1) the Northern (Steppe) Silk Road, (2) the Northwestern (Oasis) Silk Road, (3) the Southwestern (Buddhist) Silk Road and (4) the Southern (Sea) Silk Road. The following subsections will present the outlines of each Silk Road and the inflow of ancient glasses along these Silk Roads, focusing on the intercourse between China and the outside world during the pre-Qin and Qin–Han Dynasties. Figure 2.1 shows a schematic diagram of the four main routes of the Silk Road.

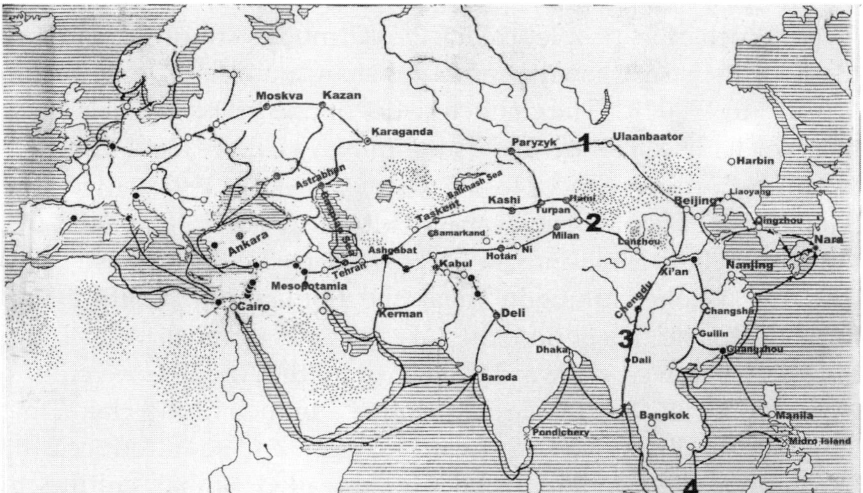

Fig. 2.1. The main routes of the Silk Road: (1) Steppe Silk Road, (2) Oasis Silk Road, (3) Buddhist Silk Road, (4) Sea Silk Road.

1.1. *The Northern (Steppe) Silk Route*

The steppe of Eurasia has a wide and plane topography. Barbarian nomadic tribes wandered along this grassland from place to place, with no fixed home, and thus played an important role in the contacts among ancient peoples. During the late Neolithic age (4000–3000 BC), the Indo-European peoples entered Central Asia in 3000 BC and moved westward to southern Russia, and even reached the central part of Europe. Also, they reached the Indian Subcontinent from Iran in 2000 BC.[3] It is of more concern whether the Indo-Europeans had marched eastward, entering Siberia, the Tarim basin and the Mongolian steppe. Recent archeological material has revealed that the Indo-European people really marched eastward. Figure 2.2 shows the nomadic tribes' movement during 3500–1500 BC. Scholarship has found the ancient Yuezhi (Yen Chin), Quici, Cheshi and Loulan peoples to be Tocharian people, because they all speak Indo-European languages and should belong to the European race. It was the earliest nationality, and they settled in the north and south of the Tianshan Mountain.[4] According to archeological excavation, the earliest culture of Tocharian people, called the Kirmuche culture, created in the area between the Altai and Tianshan mountains, dated from 2200 BC to 1900 BC. The characteristics of unearthed artifacts show that they had been influenced by the Yannaya culture (3600–2200 BC) of Indo-European people in Central Asia, also called the Shizhongmu ("stone graveyard") culture. Around 1800–1600 BC, the Kirmuche people moved southward to the Tarim basin, and thus the Xiaohe-Gumugou (near the Loulan ruin) culture was formed. During this period, the Ariya people, a branch of the Indo-Europeans, moved eastward to Tarim, and the Anderonova culture was introduced. The Qiang nationality, a branch of the Han–Tibet language family, moved westward to Tarim and so introduced the Hexi (Westward Yellow River, or so-called Shiba) culture to Xinjiang. Thus the Northern Tianshan culture, the New Tara culture (southern Tianshan) and the Niya Northern Bronze culture (Tarim basin) had formed 1500 BC.[5] This could be the earliest

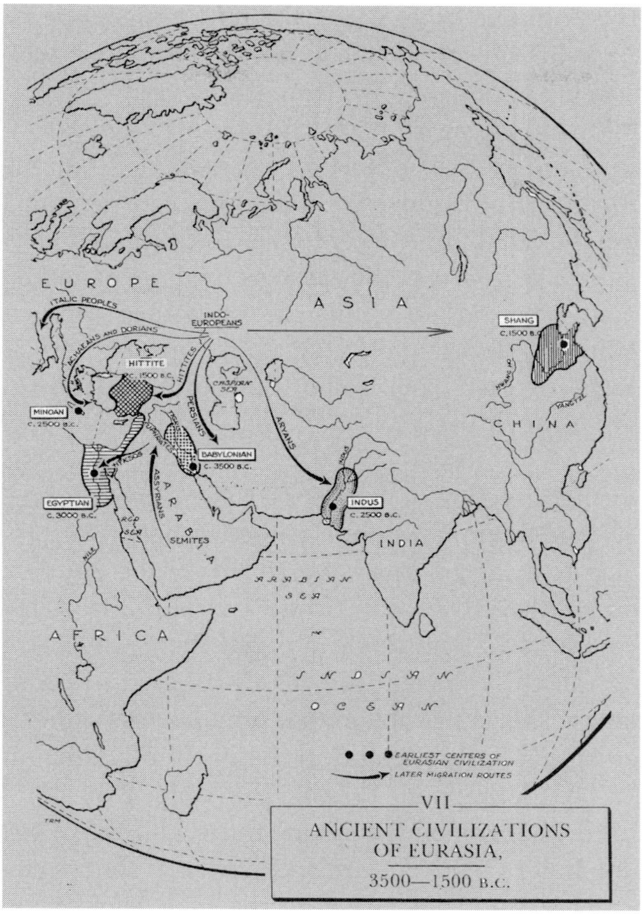

Fig. 2.2. Nomadic tribes' movement during 3500–1500 BC.*

* This map has been adapted from L. S. Stavrianos, *A Global History*, 4th edition (Prentice Hall, New Jersey, 1988). The red arrow line has been added by the author.

contact between China and the Western world, and the prototype of the Northern (Oasis) Silk Road in prehistory.

The northern steppe route of the Silk Road was closely related with the north and northwest of China, Altai, Mongolia and Outer Siberia areas. The nomadic tribes in China are mainly the Sai (Saka), Xiong Nu (Huns) and Xian Bei (Sienpi) nationalities.

Because the nomadic tribes along the northern steppe route wandered from year to year, very few tombs with some more funeral objects during back to 1000 BC were discovered. Only if they had settled down and engaged in agricultural activities was it possible to have ruins and relics available for excavation. A famous Pazyryk ancient frozen tomb was located at Uragan, Altai province, Russia, dating from the 6th to the 4th century BC and the 3rd to the 1st century BC, which period was the same as the Spring and Autumn to Warring States periods of China. Its unearthed artifacts include Chinese silk, jade, lacquer, etc., but no report about ancient glassware was discovered. This is the tomb of a chieftain of the Sai nationality. Pazyryk used to be an East–West trade center during that time.[6]

The northern section of the steppe route started from due north of China to the Mongolian steppe, extended northward to the outer Baikal Lake area, and turned westward to the southern Russia steppe and southward to Iran (route 1 in Fig. 2.1). A few ancient glasses were found along this route, but there was also no report concerning the early ancient glasses discovered in Mongolia and Siberian areas. This is a problem that remains for further investigation. Inner Mongolia is situated in the north of China and is covered with broad steppes. It was the only way for traveling southward to China along the steppe route, and was also a convergence of the Inner China culture and northern steppe culture. Not many ancient glass artifacts before the Han Dynasty have been found in Inner Mongolia. Photos 2.1 and 2.2 show a string of glass-like beads, agate and turquoise beads from a late Spring and Autumn period (~500 BC) cemetery of the Hun in Taohongbala, and a pendant set of the Han Dynasty (~200 BC) composed of glass-like beads and quartz beads unearthed at Zhungeerqi of Eerduosi city, Inner Mongolia.

In recent years, we have made systematic analyses of the chemical composition of the glass samples found in Inner Mongolia dating back to some of its major periods, using the nondestructive physical technique of PIXE.[7] Tables 2.1 and 2.2 give the condition and chemical composition of ancient glass samples unearthed in

Photo 2.1. String of glass beads, agate beads and turquoise from a late Spring and Autumn period of the Hun at Taohongbala, Inner Mongolia.

Photo 2.2. Pendant set of the Han Dynasty (~200 BC), composed of glass beads and quartz beads, unearthed at Zhungeerqi, Eerduosi city, Inner Mongolia.

Table 2.1. Condition of the ancient glass samples unearthed in the Inner Mongolia area.

Glass sample	Era	Unearthed locality	Description
NM02-2	Han Dynasty	Hu-Lun-Bei'er-Meng, Zha-Lai-Nuo-Er tomb	Green glass bead
NM03-2	Han Dynasty	Hu-Lun-Bei'er-Meng, Zha-Lai-Nuo-Er tomb	Yellow glass bead
WJ-09-a (blue section 1), WJ-09-b (blue section 2), WJ-09-c (green section)	Western Zhou	E-Ji-Na-Qi, Lvcheng	Green small glass bead, 5 mm in outside diameter, 2 mm in inside diameter
WJ-05	Northern Wei	Cha-You-Zhong-Qi, Qilangshan M6	Yellow glass bead, 17 mm in length, 7 mm in outside diameter, 3 mm in inside diameter
WJ-06-a (white glass bead) WJ-06-b (black section of black glass bead) WJ-06-c (white section of black glass bead)	Northern Wei	Cha-You-Zhong-Qi, Qilangshan	White and black glass beads

(Continued)

Table 2.1. *(Continued)*

Glass sample	Era	Unearthed locality	Description
WJ-07-a (white glass bead), WJ-07-b (blue glass bead)	Northern Wei	Cha-You-Zhong-Qi, Qilangshan M20	White and blue glass beads, 6 mm in outside diameter of white bead, 8 mm in inside diameter of black bead, 2 mm in inside diameter of both beads
WJ-01	Yuan Dynasty	Outside of Yuan-Shang-Du south wall	Relic of hexagonal rhombic glass bead, 18 mm in length, 13 mm in width
WJ-02	Yuan Dynasty	Outside of Yuan-Shang-Du south wall	Relic of hexagonal rhombic glass bead
WJ-03-a (white glass bead), WJ-03-b (black glass bead)	Northern Wei	Cha-You-Zhong-Qi, Qilangshan M20	White and black glass beads
WJ-08-b (white small glass bead)	Yuan Dynasty	Yuan-Shang-Du	White glass bead
WJ-10	Yuan Dynasty	Outside of Yuan-Shang-Du south wall	Blue rhombic glass bead, 16 mm in length, 9 mm in width
WJ-12	Yuan Dynasty	Outside of Yuan-Shang-Du south wall	Relic of blue plum blossom-shaped hairpin

Table 2.2. Analytic results on the early glasses unearthed in the Inner Mongolia area, determined by the PIXE technique.

	w/%															
No.	Al_2O_3	SiO_2	P_2O_5	SO_3	Cl	K_2O	CaO	TiO_2	Cr_2O_3	MnO	Fe_2O_3	CoO	NiO	CuO	ZnO	BaO	PbO
NM02-2	3.22	47.91	9.19	0.43		0.49	2.91		0.03	0.02	0.37	0.02		0.26	0.04	0.08	35.04
NM03-2	3.42	57.05	4.75			0.48	1.54				0.41	0.02		0.21	0.02	0.02	32.08
WJ-09-a	3.52	84.33	0.44	1.93	0.97	1.16	0.99				0.46			6.20			
WJ-09-b	2.64	89.55	0.69	1.00	0.96	1.04	0.84	0.06			0.29			2.95			
WJ-09-c	3.35	89.07		0.86	1.56	1.31	0.48				0.18			3.20			
WJ-05	6.66	74.72	0.85	1.42	1.38	5.05	7.00	0.29	0.04	0.10	2.36	0.03		0.12			
WJ-06-a	8.44	75.70	1.42	4.44	0.95	0.70	5.19	0.30		0.09	2.55	0.09		0.14			
WJ-06-b	11.54	70.39	1.39	1.18	0.97	2.95	6.37	0.31		0.92	3.95			0.03			
WJ-06-c	11.29	69.57	3.02	1.57	0.92	1.52	6.16	0.38		0.78	4.62	0.11	0.04	0.04			
WJ-07-a	3.73	72.41	2.04	2.40	2.54	1.44	12.15	0.24	0.03	0.09	2.48	0.03		0.41			
WJ-07-b	6.47	73.29	3.15	1.13	0.50	0.76	6.92	0.26	0.02	0.22	7.26		0.01	0.03			
WJ-03-a	4.25	82.90	0.12	1.05	1.27	2.75	6.36	0.11	0.03	0.07	1.11						
WJ-03-b	4.37	82.17	0.33	1.09	1.52	2.28	5.36	0.14		0.78	1.94			0.03			
WJ-02	1.93	78.04		0.88	0.58	5.03	8.60	0.09	0.03	0.09	4.66	0.05		0.04			
WJ-08-b	4.11	76.11	0.54	1.91	2.73	6.14	8.10	0.03			0.28			0.05			
WJ-10	8.43	62.20	0.48	1.58	1.76	15.38	6.09	0.12		0.05	2.01	0.05		1.87			
WJ-12	3.70	83.18		1.67	0.42	3.50	5.60	0.08		0.02	1.11	0.02		0.72			
WJ-01	4.86	57.44	0.16	2.03	0.67	18.65	7.03	0.09		0.09	4.81	0.11	0.02	3.33			0.70

the Inner Mongolia area. A faience sample (WJ-09) with typical Chinese characteristics, dating back to the Western Zhou Dynasty, was found. The glass samples (NM02, NM03) from the Han Dynasty mainly belong to the lead barium silicate system, and the glass samples (WJ-01, 02, 08, 10, 12) from the Yuan Dynasty belong to potash lime silicate glasses. Investigation and analysis have shown that these ancient glass wares came from Inner China and were influenced by the culture and technique of Inner China due to their close relationship. Several glass samples (WJ-66, 07) from the Northern Wei Dynasty (~600 AD) would have been imported after Zhangqian's travel to the West; they were soda lime silicate glasses (Na_2O could not detected by this measurement). Till now, the glass wares from the north (Outer Mongolia and Russian Siberia) along the steppe route before the Eastern Zhou Dynasty have not been found.

The southern section of the northern steppe route was relatively active from the pre-Qin Dynasty. It started from the Seven Rivers basin (Balkash Lake area), a part of the Ili River valley in Central Asia, passed through the Erchith River valley along the south of the Altai mountain, to the northwest steppe of Inner Mongolia, and then turned to the Hetao (Yellow River turn) area. It was said that King Mu of the Western Zhou Dynasty returned from his westward travels and could make his way just along this route. Comparatively, the Tarimu basin was sparsely populated and was blockaded during that period.

As mentioned above, the faience was unearthed at Ejinaqi in northwestern Inner Mongolia along this route, dating from the Western Zhou Dynasty. A glass eye bead necklace dating from the 6th to the 5th century BC was unearthed near the Kazakhstan area. Along the Ili River valley and the west of Altai Mountain, glass beads and ornaments were discovered, dating from the Spring and Autumn and Warring States period to the Western and Eastern Han Dynasties.[8] The above archeological evidence could reveal that the northern steppe route connecting the East and the West had been opened since then. The later steppe route, connecting eastward a branch of the Oasis Silk Road, starting from Hami and passing through the Barikun steppe and the Inner Mongolia steppe to Hetao, was more important for the transportation of jade and glass from the Xinjiang area. The soda lime silicate glass found in Inner China, dating back to the late Spring and Autumn and early Warring States periods, could have come from the southern section of the northern steppe route, as mentioned in Ref. 9; the glass eye beads unearthed from the Chu tomb and Gushihou Pile, Xujialin of Xichuan county, Henan province, are examples.

The Northern (Steppe) Silk Road after the Qin and Han Dynasties had been ruled by nationalities such as Xiongnu (Huns), Xianbei (Sienpi), Tujue (Turks) and Qidan (Khitan). It was not operated well in early times. But after the Northern Song Dynasty, the Liao and Jin Kingdoms opened the East–West route, and the interchange between the northeast of China and Central Asia and West Asia was operated along the Northern (Steppe) Silk Road through Mongolia. So a number of Islamic glasses had been unearthed in the northeast of China since the Liao and Jin Dynasties.[10, 11] Photos 2.3–2.5 show a glass plate, a glass cup and a glass bottle with Western styles unearthed from the tomb of the

Photo 2.3. Glass plate unearthed from the tomb of the Princess of Chen State of the Liao Dynasty in Inner Mongolia.

Photo 2.4. Glass plate unearthed from the tomb of the Princess of Chen State of the Liao Dynasty in Inner Mongolia.

Photo 2.5. Glass bottle of the Liao Kingdom unearthed from the tomb of the Princess of Chen State of the Liao Dynasty in Inner Mongolia.

Princess of Chen State of the Liao Dynasty in Inner Mongolia. The Northern (Steppe) Silk Road also extended to Korea and Japan through northeastern China, connecting with Central Asia and West Asia.[12] Particularly since the Tang and Song Dynasties, the contacts between China with the Lee and Chen Dynasties of Korea and the Heian and Kamakura periods of Japan had run along the so-called "Northeastern Silk Road," which streched from the Bohai Kingdom to Qingjin and Vladivostok, then across the sea to Japan; this was the "Japan Route." Alternatively, by crossing the Yalu River and the sea to Dengzhou, this was the "Yalu River Route."[13]

1.2. *The Northwestern (Oasis) Silk Road*

This route had been the main conduit between China and the rest of the world since the Han Dynasty explored the western regions. Xinjiang was the major part of the road, which comprised three subroutes; the south, north and new routes. Figure 2.3 shows the three subroutes. The south subroute (S) started at Dunhuang and went through Shanshan (now the northeast of Ruoqiang, Xinjiang), Yutian

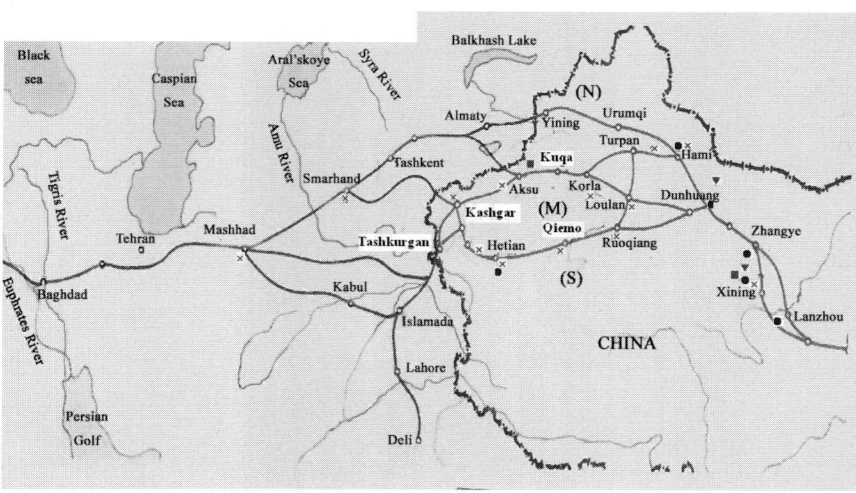

Fig. 2.3. Northwestern Silk Road and ancient Chinese glass distribution. ● PbO–BaO–SiO_2 glass × Na_2O–CaO–SiO_2 glass ▼ Alkali faience ■ K_2O–SiO_2 glass.

(now Hetian), Kashgar, etc., across the Chunlin (now the Pamir plateau) and into Great Yen Chin (now the middle part of the Amu River valley in Afghanistan) and Anxi (Persia, now Iran), extending westward to Tiaozhi (now the Persian Gulf) and Daqin (Roman Empire, now the east of the Mediterranean). The north subroute (M) started at Dunhuang and went through Cheshi (Gaochang, now Turfan of Xinjiang), Quci (now Kuga), Sule (now Aksu), etc., across Chunlin, and into Dawan (now Kyrgyzstan and the Falgana basin of Uzbekistan) and Kangju (now the middle of Syra River of Kazakhstan), extending southwestward through Anxi into Daqin. Later a new subroute (N) far northward was opened. It started at Yumengguan (Yumeng Pass), turned westward through Hengken, Turfan and Cheshi, then along the Ili river to Yining, off China to Almaty (Kazakhstan), further turning to Tashkent (Uzbekistan).

The oasis route took two directions after crossing the Pamir plateau: one was westward through Mashhad in Central Asia into Iran and West Asia; the other was southward and divided into two ways: (1) the Snow Mountain Way — across Tashkurgan and Tiegaishan, then westward along the upper reaches of the Penchihe River to the south of Bark, eastward across the Hindu Kush mountain to Kabul and Peshawar, and then into Punjab (India); (2) the Kapisa (Kasmira) Way — Kasmira (now Kashmir) is located in the lower reaches of the Kabul River, which passes through the valley of the snow mountain into Yutian (Hetian), south of Xinjiang. The route is stretched southwestward from Pishan of Xinjiang, along the upper reaches of the Yerqe River and across Xiandu of the Pamir plateau (now the southwest of Tashkulgan) into Kasmira, moving southward to the Wuyi mountain and into India.

The oasis route might have been a conduit linking the Chinese nationalities with Central and West Asia during the prehistoric times. Friedrich Engels pointed out in his book *Origin of Family, Private Ownership and States* that the Ariya people, a branch of the Indo-Europeans originating from Asia, who used to be an ancient nomadic tribe, settled in the grassland along the present Amu River and Syra River. In the middle of the 20th century BC, they moved to India and Iran, as well as to the Tarim basin and the Hexi

Corridor of China, respectively. Therefore, one may infer that ancient Chinese people during the Xia and Zhou Dynasties could have had some contact with the Ariya people. King Mu's (of the Zhou Dynasty) travel westward started at Yanmenguan (Yanmen Pass, Shanxi), and went westward through the Yanran mountain (Hangai mountain), Xining of Qinghai, the Chaidamu basin into the Tarimu basin, linking with the southern route of the oasis route; this was what the ancients called Qinghai Road.

As Xiyu ("western regions," now Xinjiang) was the major part of the oasis route within Chinese territory, quite a lot of ancient glass artifacts were excavated by the foreign explorers of the end of the 19th century and in the early 20th century, and by the archeological teams of the Xinjiang Institute of Cultural Relics and Archeology from the 1950s to the 1990s. Figure 2.4 shows the famous ancient glass unearthed sites in the Xinjiang Uygur Autonomous Region. The earliest glass beads dating back to the late Western Zhou Dynasty (~1000 BC) were found in the Kiziltur graveyard (photo 2.6). The Loulan, Milan and Niya ruins were the

Fig. 2.4. Famous ancient glass unearthing sites in the Xinjiang Uygur Autonomous Region.

Photo 2.6. Glass beads from the M26 tomb of the Western Zhou to the Spring and Autumn period at Kiziltur, Baicheng, Xinjiang.

Photo 2.7. String of glass beads and agate beads unearthed from the tomb of the Han Dynasty to the Jin Dynasty at the Niya site, Minfeng, Xinjiang.

most famous places for discovering ancient artifacts along the southern oasis route. The unearthed glass artifacts before the Han Dynasty (200 BC–200 AD) belong to different kinds of glass beads. Photos 2.7–2.10 show the glass strings and glass eye beads

Photo 2.8. Glass beads of the Warring States to the Han Dynasty, collected from the ancient Yuansha city site at Yutian, Xinjiang.

Photo 2.9. String of glass beads from the cemetery of the Warring States at Shangmiaoergou, Hami, Xinjiang.

Photo 2.10. String of glass eye beads and agate beads of the late Western Han Dynasty, from the No. 2 cemetery at Shanpula, Luopu, Xinjiang.

unearthed at the Niya site at Minfeng, the Yuansha city site at Yutian, the Shangmiaogu site at Hami and the Shanpula site at Luopu, respectively. After the Eastern Han Dynasty to Sui Dynasty (200–600 AD) period, some glass artifacts with a Western art style, such as glass cups (photos 2.11 and 2.12) and a glass goblet (photo 2.13), were excavated. Because these ancient glass artifacts have different provenances, they show with different glass chemical compositions.

Table 2.3 lists the lead barium silicate glass and potash silicate glass with ancient Chinese characteristics unearthed in the northwest of China. These glasses had spread westward to Hetian, Baicheng and Wensu of western Xinjiang since the Han Dynasty. To date, lead barium silicate glass and potash silicate glass have not been discovered in Central Asia. So the glasses with ancient Chinese characteristic had not been spread westward across

Photo 2.11. Glass cup of the Eastern Jin Dynasty, unearthed from the M49 tomb at Zhagunluke, Qiemo, Xinjiang.

Photo 2.12. Glass cup unearthed from No. 9 tomb at Yingpan, Weili, Xinjiang (Han Dynasty to Jin Dynasty).

Photo 2.13. Goblet of the Sui Dynasty, unearthed at the Senmusaimu Grotto at Kuche, Xinjiang.

Chunlin (Pamir plateau). Potash silicate glass and high lead silicate glass were unearthed at Pohrovka in southern Russia they dated from the 4th century BC to the 2nd century BC, but were considered to have come from China.[14] Most of the ancient glass artifacts along the oasis route after the Han Dynasty belong to the soda lime silicate glass system.

Reference 9 gives an introduction to the formation and development of the Pb–BaO–SiO_2, PbO–SiO_2 and K_2O–SiO_2 glass systems solely in Central China. These glass systems are dated more than 1000 years later than the ancient glasses in Mesopotamia and Egypt areas. During that millennium, contact and exchange between China and the West inevitably existed. The earliest glass in China was unearthed in the Xinjiang area, and was 500 years earlier than the ancient glass in Central China. Xinjiang is near to Central Asia, so a general cognition reached is that the Chinese silk introduced to the rest of the world represents the earliest cultural and technical exchanges between China and foreign countries, while the inflow of ancient glass technology into China would become the earliest physical evidence of these exchanges.

Table 2.3. Ancient glasses of the $PbO–BaO–SiO_2$, $K_2O–SiO_2$ and $Na_2O–CaO–SiO_2$ systems, unearthed in northwest China.

Type of glass	$PbO–BaO–SiO_2$	$K_2O–SiO_2$	$Na_2O–CaO–SiO_2$	Faience
Unearthed site and artifact	Glass ear pendant; Hantomb, Lanzhou, Gansu, GSU-2	Glass bead; later Western Han to Wang Mo period, Datong, Qinghai	Glass bead; Eastern Han tomb, Datong, Qinghai	Faience containing Pb and Ba; Warring States, Lixian, Gansu
	Glass bead; Han tomb, Wuwei, Gansu, GSU-3	Glass bead; tomb of Warring States, Baozidong, Weishu, Xinjiang	Glass bead; Kiziltur tomb of Spring and Autumn period, Baicheng, Xinjiang	Tooth-shaped bead; Eastern Han tomb, Datong, Qinghai
	Glass bead and ear pendant; Han tomb, Datong, Qinghai		Glass bead; tomb of Warring States, Zagonluk, Qianmo, Xinjiang	Small bead; tomb of Western Zhou, Ejinaqi, Inner Mongolia
	Glass ear pendant; Han tomb, Jiuquang, Gansu		Glass bead; Eastern Han tomb, Baozi, Xinjiang	

(*Continued*)

Table 2.3. *(Continued)*

Type of glass	$PbO–BaO–SiO_2$	$K_2O–SiO_2$	$Na_2O–CaO–SiO_2$	Faience
	Glass bead; tomb of Warring States, Hami, Xinjiang		Glass bead; Shanpula tomb of Warring States, Luo Pu, Hetian, Xinjiang	
	Glass eye bead; Han Dynasty, Hetian, Xinjiang		Fragment of glass cup; Wei-Jin Dynasties, Loulan, Rueqiang, Xinjiang	
			Glass tube; Tang Dynasty, Qunqimu, Keping, Xinjiang	
			Fragment of glassware; Tang Dynasty, Kumutula, Kuche, Xinjiang	
			Fragment of glassware; Tang Dynasty, Boxikeruemu, Sufu, Xinjiang	
			Fragment of glassware; Tang Dynasty Bugeiwuyilik temple, Meyu, Xinjiang	

One may learn from the study of ancient Xinjiang glasses that the earliest Western glass-making technique was introduced to China along the Oasis Silk Road, dating from the Western Zhou Dynasty to the Spring and Autumn period (1100–800 BC); an example is the glasses unearthed at Kiziltur, Xinjiang.[15,16] Table 2.4 lists the condition and chemical composition of the Kiziltur glass samples. These Kiziltur glass beads are very similar to the ancient West Asia glasses in their main chemical compositions. But glasses containing both high PbO and Sb_2O_3 have not been found in the West Asia and Central Asia of the same period. This is correlated with using local raw materials and bronze metallurgy. These glasses contain more bubbles, showing that the making technology was not so good. This just reveals that it was mainly the glass-making technology introduced from the West during that time rather than the glass products imported, because the learning and re-creation of a technology have to be step-by-step. The early local glass beads were made by using West Asia's glass-making technology, as well as local ores and slag from bronze melting, etc.

Analysis of the ancient glasses of various periods found in Xinjiang shows that the glass artifacts that came from the West during the Qin and Han Dynasties are few in number,[17] and include the ancient Egypt and Roman types of soda lime silicate glass with low K_2O, MgO, Al_2O_3 contents and that with high K_2O, MgO, Al_2O_3 contents of the Two Rivers (Tigris and Euphrates) valley type and Iran plateau types. The glass artifacts that came to Inner China along the Silk Road are even fewer, from which one may infer that the cultural and technical exchanges were not well operated during those years.

The inflow of glass-making technology from West Asia to China was mainly caused by the migration of nomadic tribes, rather than by the fixed production sites spreading step by step, because less glassware of the 3rd century BC has been discovered in Central Asia. From the migration of nomadic tribes and nationalities, we may obtain a more detailed background of the spread of ancient glass technology. During the years of Indo-Europeans entering central East and West Asia, there was a branch called the

Table 2.4. Condition and chemical composition of glass samples unearthed at Kiziltur village (Baicheng county) and E'min village (Tacheng Country), Xinjiang province.

A. Description of glass samples

Number of sample	Name of cemetery	Unearthed site	Description of sample	Date
XJ-1A	90BKKM36:6	Kiziltur cemetery	M26 glaucous glass bead	Western Zhou to Spring and Autumn period (1100–800 BC)
XJ-2A	90BKKM36:6		M26 buff glass bead	
XJ-2B	90BKKM36:6		M26 buff glass bead	
XJ-3A	90BKKM4:7		M4 corroded glass bead	
XJ-4A	90BKKM11		MII fragment of corroded glass bead	
XJ-30	90BKKM3:9		M3 fragment of corroded glass bead	
XJ-44	Bao-Zi-Dong M-41	E'min Tacheng village cemetery	M1 fragment of corroded glass bead	Spring and Autumn period (700–500 BC)

(Continued)

Table 2.4. *(Continued)*

B. Chemical composition.

Number of sample	Chemical composition of glass samples (mass percentage)													Measurement method
	SiO_2	Na_2O	CaO	MgO	K_2O	Al_2O_3	PbO	BaO	CuO	Fe_2O_3	TiO_2	MnO	Sb_2O_5	
XJ-1A	66.10	18.27	5.88	5.20	2.57	1.12	0.09	0.02	0.79	0.57	0.07	0.04		ICP-AES, I
XJ-2A	65.38	11.54	8.88	5.02	1.59	1.99	1.93	0.01	0.01	1.03	0.02	0.04	0.72	
XJ-2B	64.31	12.05	4.80	2.67	2.42	1.36	9.01	0.008	0.001	1.10	0.07	0.02	1.60	
XJ-3A	65.19	15.27	6.65	3.66	2.93	1.44	0.02	0.02	0.76	0.86	0.13	0.03		
XJ-4A	66.11	14.29	6.61	4.58	2.19	1.89	0.62	0.01	0.90	1.07	0.17	0.03	1.44	
XJ-30	75.44	9.08	7.74	3.35	1.51	1.43	0.02	0.005	0.56	0.34	0.11	0.02	0.03	
XJ-44	68.88	15.93	6.11	4.03	2.20	0.87	0.02	0.005	1.10	0.56	0.04	0.08	0.01	
West ancient glass	65.5	16.0	7.0	4.5	2.0	2.0								

Hurrian people who had settled in West Asia and had known how to make bronze, iron and glass by the 20th century BC. They established the Mitanni Kingdom in West Asia from the 15th century BC to the 14th century BC, and moved to Armenia by the 13th century BC.[18] Among them, one branch called the Scythian people came to China. These Persian–Scythian people are called the Sai people in the Chinese historical literature. Their footmarks covered the current Ili area of Xinjiang, the Seven Rivers valley along the north of Central Asia, Altai and the Mongolia steppe, etc. The Sai nomadic people played a pioneering role in spreading Chinese and Western cultures, including the glass culture and technique.[18]

The soda lime silicate glasses unearthed in Xinjiang and Central Asia are mostly dated after the Han Dynasty, and were made into glass artifacts by using the blowing method, and traded along the Oasis Silk Road. So there were a number of glass relics and ruins along this road.

The Hexi region, including Gansu province and part of Shaanxi province, was an important sector, connecting the northwestern region (now Xinjiang) with the ancient capitals of China, Xian and Luoyang. Table 2.5 shows the chemical composition of ancient glass samples unearthed in Gansu. It can be seen that the frit sample of the Warring States and the glass artifacts of the Han Dynasty belong to the lead barium silicate glass system, and the glass bottles of the Song Dynasty are high-lead-oxide containing silicate glass. Looking at shape, pattern and glass chemical composition, these glass samples are consistent with the glass artifacts unearthed in the central China of the same time. Therefore, the influence of the glass culture and technology from central China was dominant.

1.3 *The Southwestern (Buddhist) Silk Road*

The historical literature records that Zhangqian (100 BC) submitted a report to Emperor Hanwu after his travels to the Western Regions. In the report he said that he saw Gong bamboo sticks and fabrics of the Han Dynasty made in Sichuan when he was in the

Table 2.5. Chemical composition of early ancient Chinese glass samples from Gansu province.

Number of sample	Sample	Chemical composition (wt%)												Measurement method
		Na_2O	Al_2O_3	SiO_2	MgO	K_2O	PbO	BaO	CaO	CoO	MnO	Fe_2O_3	CuO	
GSU-1	Pale green frit tube, Warring states, Lixian, Gansu		3.27	51.84		0.17	28.89	8.46	1.30		P_2O_5 2.43	0.68	2.81	PIXE
GSU-2	Glass ear pendant, Han tomb, Lanzhou, Gansu		7.24	45.15		1.40	33.85	9.54	0.83			0.61	0.51	PIXE
GSU-3	Glass bead, Han tomb, Wuwei, Gansu		1.53	56.45		0.57	24.10	10.34	3.80		0.34	0.66	0.09	PIXE

(Continued)

Table 2.5. *(Continued)*

Number of sample	Sample	Chemical composition (wt%)												Measurement method
		Na_2O	Al_2O_3	SiO_2	MgO	K_2O	PbO	BaO	CaO	CoO	MnO	Fe_2O_3	CuO	
	Glass ear pendant, Jiuquan, Gansu	9.30	2.10	49.33	1.40	0.51	21.62	8.83	3.16	0.04	0.33	0.48	0.09	CA
	Squash-shaped glass bottle, Song Dynasty, Lingtai, Gansu	0.29	2.75	38.32	0.10	10.09	50.31		0.13		0.02	0.16	0.13	CA
	Squash-shaped glass bottle, Song Dynasty, Lingtai, Gansu	0.11	2.75	36.32	0.04	11.94	53.40		0.18		0.01	0.29		CA

Daxia Kingdom (Bactrian, now Afghanistan), and these goods were traded from the Shendu Kingdom (now India). So Emperor Hanwu wanted to open a "southwestern foreign route," anciently named the Shu–Shenduguo Route (Sichuan–India Route), now called the Southwestern Silk Road. This is regarded as a proposal for opening the Southwestern Silk Road.

However, Emperor Hanwu had not explored the southwestern route to India, although he had made efforts for 11 years. He tried to open the way by creating four lines from Sichuan, but all failed, though he was successful in opening a way from Yunnan and Guizhou. In fact, primary road is walked out by people; cultural intercourse between China and India along the Yunnan–Burma route had existed during the Neolithic age. According to the related historical literature available, the southwestern route (Sichuan–Yunnan–Burma–India) can be outlined like follows. The Sichuan–Yunnan section of this old road, as shown in Fig. 2.5, had two ways. One way[5] was called the Yak Route (also known as the Linguan Route or Xiyi Route in the Han Dynasty, and the Qingxiguan Route in the Tang Dynasty), which ran from Chengdu via Yaan, Yuexi and Xichang to Huili, turned to the southwest, across the Jinsha River to Panzhihua, then to Dayao of Yunnan and finally into the Dali area. The other way ran from Chendu along the Min River to Leshan, Yibin, then southward along the Wuchi Road of the Qin Dynasty (the old Chu Road, the Nanyi Road of the Han Dynasty, the Shimen Road and Yangke Road of the Sui–Tang Dynasties), via Gaoxian, then extended from Zhaotong, Qujin to Kunming, and finally into the Dali area.

The route from Dali to India through Burma comprised three ways, namely the Bonan Road and Yongchang Road of the Han Dynasty, and the Xier, Tianzhu Road of the Tang Dynasty.[19]

To investigate the Southwestern Silk Road, it is necessary now to pay attention to its extension northward from Chengdu. The Shu (Sichuan) people were a branch of the Qiang nationality and came from the north of China. Ancient Qiang people had settled around the Huangshui River valley, east of Qinghai, since very early times. Before 2000 BC, there was a route connecting the north and the

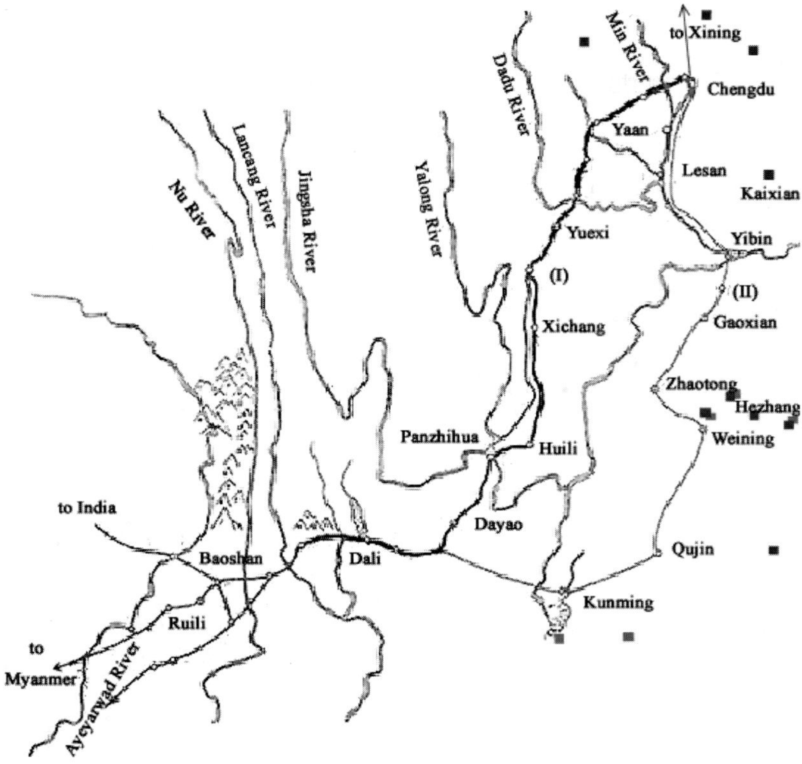

Fig. 2.5. Southwestern Silk Road and distribution of lead barium silicate and potash silicate glasses in the Han Dynasty. ■ $PbO–BaO–SiO_2$ glass ■ $K_2O–SiO_2$ glass.†

† This map has been adopted from Jiang Yuxiang's, *Study on the Ancient Southwestern Silk Road*, Vol. 2 (in Chinese); (Sichuan University, 1995), and translated into English by the author. The unearthing sites of Chinese ancient glasses have been added by the author.

south for commercial and cultural intercourse in the late Neolithic age, which started at the Huangshui River valley and ran via the Longwu River valley to the upper reaches of the Bailong River.[20] So, one could take this ancient route from Chengdu to Guangyuan, then across the Ming mountain along the lower reaches of the Bailong River northward to Xining. It was further connected to the Hexi Corridor of the northwestern (oasis) route. An alternative route ran northward along the same way as the current Baoji–Chengdu railway, i.e. along the Shu road to Baoji and then to Xian.

A lot of writings, both in China and abroad, have described the Southwestern Silk Road, but less archeological evidence was involved. The trade relationship between China and India was a kind of indirect exchange. In recent years, we have made more efforts to study ancient glasses in the southwest of China[21] in order to understand the distributions of ancient glasses along this road. From the analyzed results on the chemical compositions of the ancient glasses dating back to the Qin and Han Dynasties along this road, we learn that they are mainly of three types, i.e. the lead barium silicate, potash silicate and soda lime silicate glass systems, but mostly the first two types. Table 2.6 lists the unearthing sites and dates of these glass artifacts. We can see from the table that the ancient glass artifacts unearthed in Yunnan and Guizhou mostly belong to lead barium silicate glass and potash silicate glass, which should have come from central China. These ancient glass artifacts unearthed in the southwest were deeply influenced by the Inner China culture, particularly the Chu culture and technique, whatever their shapes or patterns. Photo 2.14 shows the faience tube of the Warring States, unearthed at Qianwei, Sichuan. Photos 2.15–2.16 show the inlaid glass eye beads of the Warring States to the Han Dynasty (300 BC–200 AD) unearthed in Chongqing, at Qingzhen, Guizhou, and at Jinming, Yunnan, respectively. The glass ritual disk (*bi*) and ear pendant are typical ancient Chinese artifacts, as shown in Photos 2.17 and 2.18; they were unearthed at the Baocheng railway site, Sichuan, and at Qingzhen, Guizhou respectively. The chemical compositions of these ancient glasses belong to the $BaO–PbO–SiO_2$ and $K_2O–SiO_2$ systems. Only a few of them belong to soda lime silicate glass of the Western types, which might have come from India via Burma. However, the Southwestern Silk Road was not suitable for transporting such fragile products as glassware due to the hardship of traversing so many high mountains and sharp peaks. It was also possible to transport them along the northwestern (oasis) route through Qinghai to the south of China.

Table 2.6. Ancient glasses of the K_2O–SiO_2, PbO–BaO–SiO_2 and Na_2O–CaO–SiO_2 systems, unearthed in southwest China.

Type of glass	K_2O–SiO_2 glass	PbO–BaO–SiO_2 glass	Na_2O–CaO–SiO_2 glass
Unearthed site and date	Western Han; Shizhaishan, Jinning, Yunnan	Eastern Han; Nanchun, Sichuan	Eastern Han; Maliuwan, Wanxian, Sichuan
	Warring States; Lijiashan, Jiangchuan, Yunnan	Warring States; Qingchuan, Sichuan	Six Dynasties; Baolun, Zhaohua, Sichuan
	Western Han; Zhongshuiliyuan, Weining, Guizhou	Warring States; Baxian and Kai Xian, Sichuan	Six Dynasties; Dongsunba, Baxian, Sichuan
	Western Han; Kele, Hezhang, Guizhou	Han Dynasty; high lead glass, Lixian, Sichuan	Six Dynasties; Changshan village, Zhangmin, Sichuan
	Han Dynasty; Qingzhen, Guizhou	Han Dynasty; Qingzhen, Guizhou	Han Dynasty; Shangsunjai-zhai, Datong, Qinghai
		Late Warring States; Hezhang, Guizhou	
		Han Dynasty; Weining, Guizhou	
		Eastern Han; Xinren, Guizhou	
		Eastern Han; Qianxi, Guizhou	
		Warring States; Baxian, Chongqing	
		Warring States; Kaixian, Chongqing	
		Eastern Han; Pingba, near Qingzhen, Guizhou	

Photo 2.14. Faience tube of the Warring States, unearthed at Qianwei Sichuan.

Photo 2.15. Glass eye beads from the tomb of the Western Han Dynasty in Chongqing.

As mentioned above, Xining was an important place, which connected the northwestern (oasis) route with the southwestern (Buddhist) route. Near Xining was a famous Shangsunjai Han cemetery (a group of tombs) located in Datong county,

Photo 2.16. Glass eye beads of the Western Han Dynasty, unearthed at Shizhaishan, Jinning, Yunnan.

Photo 2.17. Glass disk (*bi*) of the Warring States, unearthed in the southern area of the Baocheng Railway.

Photo 2.18. Blue glass ear pendants unearthed from the tomb of the Han Dynasty at Qingzhen, Guizhou.

Qinghai province.[22] Photos 2.19 and 2.20 show two ear pendants of the late Western Han Dynasty (~100 BC) unearthed at Datong county. One is a blue ear pendant with a typical Chinese style, and the other is a green ear pendant with a special shape (T shape). Table 2.7 lists the condition and chemical composition of glass samples unearthed from Han tombs. It can be seen that besides a sample of faience there are 11 glass samples, 5 of which belong to lead barium silicate glass, and they are all glass ear pendants with typical Chinese characteristics. Two glass samples are potash silicate glasses, and both are glass beads. The remaining four samples belong to the soda lime silicate system, which should be considered as being imported from the West. The different types of unearthed glass samples reflect the multiple provenances of ancient glass artifacts and the intercourse of glass culture and technology along the ancient Silk Road.

Photo 2.19. Blue ear pendant of the late Han Dynasty, unearthed at Datong country, Qinghai.

Photo 2.20. Green ear pendant with a T-shape of the late Han Dynasty, unearthed at Datong country Qinghai.

Table 2.7. Chemical composition of ancient glass samples of the Han Dynasty, unearthed in Qinghai province.

Number of sample	Condition of sample	Chemical composition (wt%)										Measurement method
		SiO_2	Al_2O_3	Fe_2O_3	PbO	BaO	CaO	MgO	K_2O	Na_2O	Others	
M128:6	Ear pendant, late Western Han	35.06	1.09	0.30	42.28	11.71	2.38	1.79		5.29	CuO 0.53	ICP
M130:1	Glass bead, late Western Han	77.78	3.98	1.97			0.55	0.29	14.16	0.34	CoO 0.12 MnO 0.25	AAS
M90:8	Ear pendant, early Eastern Han	39.18	0.38	0.22	37.26	15.79	0.45	0.12		4.62	CuO 0.13	ICP
M53:02	Ear pendant, late Eastern Han	54.32	0.85	0.63	17.12	11.13	3.65	0.95	1.69	6.27	CoO 0.04	ICP
M9:17	Ear pendant, late Eastern Han	45.85	1.66	0.14	27.25	12.93	0.77	2.16	0.34	9.31		EDX
M37:31	Glass bead, late Eastern Han	68.11	2.11	1.04			4.71	0.54	0.11	18.23	MnO 1.30	ICP

(Continued)

Table 2.7. *(Continued)*

Number of sample	Condition of sample	Chemical composition (wt%)										Measurement method
		SiO_2	Al_2O_3	Fe_2O_3	PbO	BaO	CaO	MgO	K_2O	Na_2O	Others	
M23:8	Glass bead, Eastern Han	64.49	7.28	0.93			5.90	2.24	0.70	17.6		EDX
M5:25-1	Glass bead, late Eastern Han	65.44	4.02	0.87			8.16	2.26	0.59	17.35		EDX
M5:25-2	Glass bead, late Eastern Han	57.38	8.59	0.19			2.62	5.28	0.84	23.48		EDX
M23:7	Tooth-type faience, Eastern Han	89.11	4.81	1.67				0.68	1.15	2.07		EDX
M53:02	Ear pendant, late Eastern Han	52.26	2.33	0.34	19.33	9.51	2.31	2.07	1.48	9.16	MnO 1.37	EDX
M130:1	Glass bead, late Western Han	78.05	4.39	1.83					13.78			EDX

1.4. *The Southern (Sea) Silk Road*

The Sea Silk Road was a maritime transportation line from ancient China to Western countries, and a great network connecting Europe, Asia and Africa. It started from the South China Sea, ran through the Indian Ocean, the Red Sea and the old canal of Egypt, into the Mediterranean Sea along the Nile River. This maritime road had two sections. Along the western section, ancient Egyptians, Hellenians, Babylonians, Phoenicians, Arabians and Indians had actively sailed on the Mediterranean Sea, Red Sea, Arabian Sea and Indian Ocean since 2000 BC. In early times they could only sail along the coast, but later they could go across the sea by sailing close to the monsoon wind. In the late first century AD a Greek writer from Alexandria wrote a book titled *Annals of a Voyage Along the Eliteria Sea*, which recorded in detail his navigation and what he saw along the Sea Silk Road (Eliteria Sea is now the Red Sea). In the eastern section, the road originated from the South China Sea and passed through the Indian Ocean, connecting with the Arabian Sea. This maritime route also connected with Southeast Asia, including the Philippines, Indonesia and Malaysia. It was explored earlier by the ancient countries along the west of the Indian Ocean, and Chinese people followed during the Han Dynasty. In the "Geographic Annals" section of *Han Shu* (*History of the Han Dynasty*), there are some records about the navigation line of sailing westward on the sea. Regarding the southern (sea) route, there have been many writings, in both modern and ancient times in China and elsewhere. Here, an introduction and a discussion focusing on the early exchange of ancient glasses and glass technology are presented.

The development of maritime transportation in ancient China started mainly in the Han Dynasty; by then Emperor Hanwu had conquered Vietnam and set up nine counties in the south of China, and ruled the coastal area of the South China Sea (now the south of Guangdong and Guangxi, Hainan Island and the northeast of Vietnam). The exchange activities with Western countries were carried out along the South China Sea route. Glass products were one of the main items of this cultural and commercial relationship. Ancient glasses have frequently been excavated along this route, but the places capable of making glasses were not many, and were mainly in India and China.

The Indian area (including the current India, Pakistan, Bangladesh, Sri Lanka, etc.) was the middle part of the ancient Sea Silk Road and the converging–diverging area of the East and the West. So it was a very important place. Reference 23 gives an introduction to the ancient glasses in this area. The earliest glasses unearthed in India, dating from the 7th century to the 6th century BC, belong to the soda lime silicate system having K_2O and a high content of Al_2O_3, similar to the compositions of the glasses from the Tigris River and Euphrates River valleys (the Two Rivers valley). Later, from 200 BC to 200 AD, the glass compositions paralleled to those of the ancient Egypt and Roman type glasses, i.e. soda lime silicate glass. So these glasses should have come from the Two Rivers valley and ancient Egypt and Rome along the sea route. India had experienced invasion by the Persian Empire and Alexander's empire, and also the trade relationship between the Roman Empire and India was well established. Consequently, the glass products in large amounts and the glass-making technology of West Asia came to India. Another special type of glass, potash silicate glass, was unearthed at Hastinapur, Arikamedu and Udaygiri; it dates back to the 3rd to 2nd century BC, but is less in quantity, while the unearthed glass dating back to the 1st century BC or later is much more in quantity. Arikamedu is commonly considered to be the earliest site of glass-making in India, from the 3rd century BC to the 10th century AD. Therefore the ancient potash silicate glasses were made in India and then spread elsewhere through the Sea Silk Road. The western terminal, which Chinese commercial envoys reached along the eastern section of the sea route, was the Yichengbu Kingdom (now Sri Lanka), where Mantai was also a glass-making center, from the 1st century AD. For Thailand, near to the Bengal Gulf, its Kuan Luk Pat was a possible site for making glasses from the 2nd century to the 6th century AD. The ancient glass-making technology in these areas might have been introduced from India.

The starting point of the Sea Silk Road in China in antiquity is now Hepu and Xuwen of Guangxi province, where the major ports of China from the Han Dynasty to the Six Dynasties were. Hepu was not only the site for the prefecture government, but also a flourishing harbor, and became one of the political, economic and cultural centers and metropolises of south China. In its southern suburb, there remains a cemetery of 1056 Han tombs. Large amounts of glass ornaments, including about a thousand glass beads, were excavated from the Han tombs, showing that they were rather popular at that time.[24] The chemical compositions of most unearthed glass objects belong to the potash silicate glass system; a few of them belong to lead barium silicate glass and high lead silicate glass. The majority of these glass objects are small ornaments, including beads, ear pendants and holed pendants, and all are glass objects with the Chinese characteristics, popular in central China, such as the Chu Kingdom (now Hubei), Wu Kingdom (now Jiangsu) and Yue Kingdom (now Zhejiang). So the glass objects of the Western and Eastern Han Dynasties excavated at Hepu and Xuwen should have been locally produced and used. Of course, they might have been exported overseas afterward. Photos 2.21–2.24 show a glass plaque, glass ring, ritual disk and tortoise-shaped glass ware of the Western Han Dynasty (~200 BC) unearthed in Guangdong and Guangxi, respectively. The glass finds dating from the Six Dynasties are mainly utensils and their chemical compositions belong to the soda lime silicate glass system (with some K_2O, MgO and Al_2O_3). It can be seen from their shapes and patterns that most of them were imported from the West and have mainly the Sasanian Dynasty style.[21,24] Photos 2.25–2.27 show a glass bowl, glass goblet and glass bottle of the Eastern Jin, Sui and Tang Dynasties (600–800AD) unearthed at Guangzhou of Guangdong, province and Hepu of the Guangxi region, respectively.

Guangzhou had been an export city of south China since the Eastern Han Dynasty. The unearthed glass objects from

Photo 2.21. Glass plaque of the Western Han Dynasty, unearthed from the Nanyue King's tomb, Guangzhou.

Photo 2.22. Glass ring of the Western Han Dynasty, unearthed at Hepu, Guangxi.

Photo 2.23. Glass disk (*bi*) of the Western Han Dynasty, unearthed at Hepu, Guangxi.

Photo 2.24. Tortoise-shaped glass ware of the Western Han Dynasty, unearthed from the Wenchang pagoda Heou, Guangxi.

Photo 2.25. Glass bowl unearthed from the tomb of the Eastern Jin Dynasty at Zhaoqing, Guangdong.

Guangdong show obvious features of the times and types. The glasses of the Warring States and Western Han Dynasty are mainly beads, *bi* disks, eye beads and ear pendants, showing the Chu Kingdom's style, such as those from the Southern Yue Kingdom tombs. They mostly belong to the lead barium silicate glass system; only a few are potash silicate glass, while the vessels, such as cup, bottle and bowl, etc., of the Eastern Han Dynasty and later, belong mainly to the soda lime silicate glass system, having the Western style. In general, the glasses of the Warring States period and Western Han Dynasty in south China were influenced by the cultures and techniques of central China, especially the Chu Kingdom, introduced from the northern part of China, and were also exported overseas from the ports of Guangdong and Guangxi; while since the Eastern Han Dynasty, the glass utensils had mainly been imported from the West, and then transported to the north from the ports of southern China. As the glass products were easy to break and there was difficulty in travelling along the land Silk Road, the Sea Silk Road played an even more important role.

Photo 2.26. Glass goblet unearthed from the tomb of the Sui Dynasty to the Tang Dynasty at Qingzhou, Guangxi.

Photo 2.27. Repaired glass bottle of the late Tang Dynasty, unearthed at Guangzhou, Guangdong.

2. The Ancient Silk Road Promoted the Development and Spread of Ancient Chinese Glass Technology

Cultural and technical exchange among various civilization centers and nationalities is a mutual relationship. This was also the case with the development and spread of ancient Chinese glass techniques. In early investigations of the exchange of Chinese glasses with the West, scholars put more emphasis on the comparison of their shapes, patterns and art, all of which were connected with cultural practice. So the glasses such as glass *bi* disks (ritual disks), ear pendants and *zhen* (ear cork) imitating jade ones were considered as being made in China. As for the inlaid glass beads that appeared in the Qin and Han Dynasties, a number of Chinese and foreign scholars compared their patterns with those of West Asia and Egypt, and thought that they possibly came from the West. Since the 1930s, scientific archeological study of glasses has achieved progress, so it is possible to investigate the development and spread of ancient Chinese glasses based on the different chemical compositions. Figure 2.6 shows the exchange of the ancient Chinese glass technology with places outside of China.

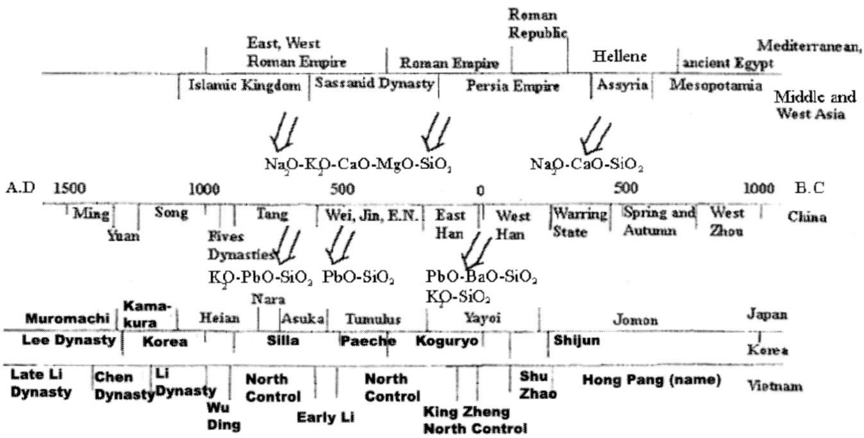

Fig. 2.6. Exchange of Chinese ancient glass technology with places outside of China.

2.1. *Import of glass products and techniques from the West*

As mentioned above, there had been contacts among the nomadic tribes along the primitive Silk Road (i.e. the north of the Euro-Asian steppes) since the first millennium BC, and the ancient Chinese glasses were also involved in the contacts and exchange. For example, the glass-making techniques and soda lime silicate glasses had spread to the Xinjiang area, although less in number and scale. We should have great esteem for Zhangqian's travels to the Western Regions as a pioneer in the opening of the Silk Road. The development of the Silk Road should be attributed to the appearance of four great empires in the 1st to the 5th century AD of the world's classic civilization age: the Eastern Han Dynasty of the East, the Roman Empire of west Europe and west Asia, the Persian Empire of central Asia (named Anxi in the Han Dynasty) and the Kushan Empire of south Asia. All of them were powerful at that time, which greatly facilitated the contacts among these empires along the Silk Road without any blockade, and promoted the exchange between China and the rest of the world.

The ancient Western glass-making technology was enhanced to a new level during the Roman Empire; particularly, the glass-blowing technique became popular, and at the same time the techniques of cameo glass, stained glass and twisted glass were developed. The world-renowned Roman glass techniques were transferred to the Persian Empire, and they also developed the glass-cutting technique during the Sasanian Dynasty (3rd to 5th century AD). Thus, a kind of very typical Persian culture glasses — the Sasanian glasses — appeared. These new glass-making techniques spread eastward through the Kushan Empire located in central and south Asia. The Kushan Empire was established by the Great Yen Chin people; they were oppressed by the Xiongnu (Hun) people of northern China during the early Western Han Dynasty, and moved to central and south Asia. So it was via the Great Yen Chin people that the Western glass techniques came to China. After the Eastern Han Dynasty, glass was one of the main articles produced in Daqin (the Roman Empire) and Anxi (the Persian Empire) and was brought into China by the Great Yen Chin people. Several records about this event can be found in the work *History of Later Han Dynasty*, in the section "Memoir of the Western Regions." Consequently, the ancient glassware production was enhanced in China. It can be seen from the unearthed ancient Chinese glassware that the earliest glass vessels made by the blowing technique and the mold-free forming technique date from the Wei, Jin, Southern and Northern Dynasties and the Sui Dynasty (3rd to 6th century AD); this shows a close relation to the opening of the Northern Silk Road. The earliest imported glass vessels in China include a Roman glass cup (from the Han tomb at Ganquan, Nanjing, Jiangsu, in the 1st century AD; photo 2.28), a barrel-shaped cutting glass cup (from the Langya King's tomb at Xiangshan, Nanjing, in the 3rd century AD) and a glass cup (from Mufushan, Nanjing, in the 4th century AD) all dating back to the Eastern Han and early

Photo 2.28. Roman glass cup unearthed from the No. 7 tomb at Xiangshan, Nanjing.

Photo 2.29. Duck-shaped utensil unearthed from the tomb of Feng Sufu of the Northern Yan Dynasty at Beipiao, Liaoning.

Southern and Northern Dynasties (1st to 3rd century AD). Some of them look like the imported Roman and Sasanian glass vessels in shape, such as a duck-shaped glass utensil unearthed from the tomb of Fengsufu at Beipiao, Liaoning (in the 5th century AD; photo 2.29), and a concave facial cut glass bowl of the Northern Zhou Dynasty (3rd century AD) unearthed from the tomb of Lixian in Ningxia (photo 2.30); they should belong to the Sasanian glasses in form.

During the flourishing Tang Dynasty in the 7th to 10th century AD, the contact and exchange between China and other countries became more frequent and extensive. While it was just the time when the Islamic religion and culture emerged in world history, at the same time when the Islamic glasses came out. Some large glass vessels were imported into China. Among them, the most famous one is an Islamic glass bottle uncovered from the underground palace of Famensi (9th century AD). These glass vessels display Islamic glass features in every respect, including the making technique, decorative artwork and patterns (photo 2.31 and 2.32).

The chemical compositions of these imported glasses all belong to the soda lime silicate glass system, with different contents of K_2O,

Photo 2.30. Concave facial cut glass bowl unearthed from tomb of Lixian of the Northern Zhou Dynasty at Guyuan Ningxia.

Photo 2.31. Blue glass plate of the Tang Dynasty, from the Famen Temple at Fufeng, Shanxi.

Photo 2.32. Glass bottle of the Tang Dynasty, from the Famen Temple at Fufeng, Shanxi.

MgO and Al_2O_3; this also reveals that different types of imported soda lime silicate glass are symbols of different times.[26] Table 2.8 shows the chemical composition of imported ancient glasses.

The glass vessels uncovered in northern China dating from the Northern Song Dynasty to the Liao Kingdom were imported Islamic glasses. One of the well-known glass unearthed sites was the Princess Tomb of the Chen State in Inner Mongolia. The uncovered glass wares are shown in photos 2.3–2.5; their chemical composition also belong to the soda lime silicate glass system.

Introduction of the Western blowing technique for glass making and soda lime silicate glass composition with high chemical stability raised the level of the glass-making technique of China. The glass vessels with the typical Chinese forms had been produced by using the chemical composition of the soda lime silicate system and the blowing method. The time from the Southern and Northern Dynasties to the Sui Dynasty was just the beginning of the glass-making technique introduced from Roman and Persia;

Table 2.8. Chemical composition of imported ancient glasses.

Unearthed place	Date	Glass artifact	Chemical composition (wt%)									Type
			SiO_2	Al_2O_3	Fe_2O_3	CaO	MgO	K_2O	Na_2O	CuO	MnO	
Jiangsu, China	Eastern Han	Twisted bowl	64.79	3.44	1.30	7.66	0.61	0.88	18.18	0.03	2.45	Roman glass
Nanjing, China	Eastern Jin	Pale yellow cup	67.70	3.43	0.58	6.05	0.94	0.45	19.23	0.02	1.63	
Feng Suofu tomb, Liaoning, China	Northern Yan	Pale green bowl	64.82	2.72	0.82	6.14	2.35	4.43	16.02	0.02	0.08	Sansania glass
Wu Lidun tomb Echeng, China	Western Jin	bowl	64.22	1.64		9.19	3.21	3.59	17.51	0.02	0.04	
Huafang tomb, Beijing, China	Western Jin	bowl	64.33	1.24	0.26	7.25	2.45	4.19	16.03			
Shigang tomb, Guangdong, China	Eastern Jin	bowl	65.0	2.0	1.0	9.0	2.0	4.0	17.0			
Famen temple, Shanxi, China	Tang	Blue bottle fragment	62.86	2.79	1.7	7.43	3.07	2.90	15.35	0.38		Islamic glass
Famen temple, Shanxi, China	Tang	Yellow bottle fragment	65.11	2.63	0.35	5.75	5.45	3.58	15.08		1.35	
Litai tomb, Hubei, China	Tang	Green amphorishes	61.58	1.66	0.69	6.27	6.43	3.53	17.86			

Photo 2.33. Glass cups of the Sui Dynasty, from the tomb of Li Jingxun at Xi'an, Shanxi (National Museum of Chinese History).

therefore, these types of glass vessels were few in number and not perfect in quality, such as a green flattened bottle and a narrow neck bottle uncovered from the Li Jingxun tomb at Xian, Shanxi (Sui Dynasty, 6th century AD; photo 2.33), and a glass alms bowl uncovered from the pagoda base at Dingxian, Hebei, dating from the Northern Wei Dynasty (photo 2.34). Another one worth mentioning is a thin-neck glass bottle uncovered from the Litai tomb at Yunxian, Hubei. Its shape is Chinese type, but the composition is soda lime silicate. So it should have been made inside China. The glass vessels from central China and south China were mostly made by the blowing technique and by using self-developed lead silicate glasses. We may see, therefore, that Chinese people had known well the foreign techniques and produced the glass artifacts with Chinese types.

From the Yuan Dynasty in the 13th century AD, European culture and technology, including the glass-making technique, were actively introduced to China by Catholic priests. Since the Qing Dynasty the major glass manufacturing site was the royal glassworks in the Imperial Palaces, Beijing. Emperor Kangxi (1662–1722)

Photo 2.34. Glass alms bowl unearthed from the pagoda base of the Northern Wei Dynasty at Dingxian, Hebei.

invited European glass-making technicians, such as the priest Kilian Stumpt, to the royal glassworks to fabricate glass artifacts with typical Chinese styles using European techniques. The most important glass techniques imported were as follows.

- Glass coloration was performed using metallic microparticles, such as silver (yellow), gold (red) and copper (red). The glass color could be gradually changed by the metallic microparticle's size, which was formed by different heat treatment of glass pieces (nowadays, it is called the quantum size effect). Photo 2.35 shows the red overlay on a white glass vase; the red glass layer was colored by gold (Au) microparticles. Photo 2.36 shows a "golden star" glass formed by larger copper (Cu) microparticles.
- Colored opaque glass artifacts were made using the glass phase separation (emulsion) effect. Photo 2.37 shows a "chicken oil yellow" glass vase created by codoping PbO and Sb_2O_3 as coloring and emulsion agents.
- A ribbon glass vase (photo 2.38) was made by the multicolor glass-twisting technique, which had been used since the Roman Empire period in Europe.

Photo 2.35. Glass vase with a long neck and a flower and bird design, and a red overlay on white glass beneath, of the Qing Dynasty (late 18th century).

Photo 2.36. Golden star glass calyx with a holly cock shape, of the Qianlong period of the Qing Dynasty.

Photo 2.37. Yellow glass vase with a long neck and a flower design, of the Qing Dynasty (late 18th century).

2.2. *The ancient Chinese glass and technique spread outside*

The neighboring countries nearest to China are Vietnam, Japan and Korea (peninsula). The intercourse between them can be traced back to the pre-Qin Dynasty and became frequent during the Qin and Han Dynasties. During the Han Dynasty, Japan and Korea sent their envoys to China, and China started to govern Vietnam (111 BC–938 AD).

We will now discuss the ancient Chinese style glasses, such as the lead barium silicate, potash silicate and high lead silicate glasses spreading to Japan, the Korean peninsula and Vietnam in earlier and later times.

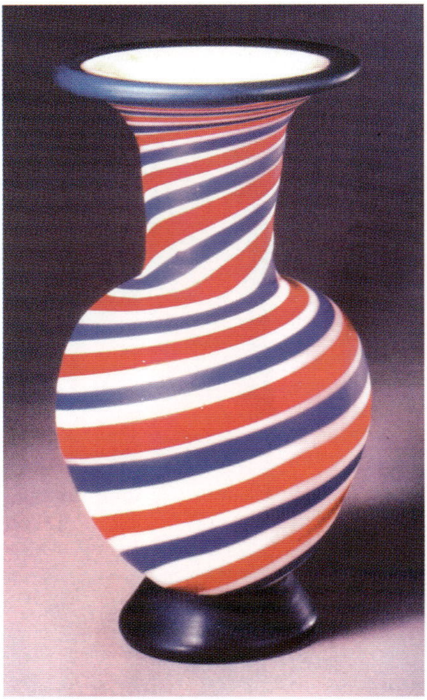

Photo 2.38. Ribbon glass vase of the Qianlong period of the Qing Dynasty.

(1) *Lead barium silicate glass ($BaO–PbO–SiO_2$)*

Lead barium silicate glass is the most typical ancient Chinese glass. The earliest glass discovered in Japan dates back to the late Yayoi period (2nd century BC to 3rd century AD). Chinese lead barium silicate glass came to Japan much earlier. This kind of glass was found in the tombs at Sugu-Okamoto, Fukuoka city and Tateiwa, Lizuka city, Kyushu, dating from the middle Yayoi period (100 BC–100 AD), corresponding to the late Han Dynasty (100 BC–100 AD).[27] The shapes and chemical compositions of the glass beads found in Japan are very similar to those of the glasses found in the Chu tombs of the Warring States period and Western Han Dynasty at Changsha, Hunan.

Export of ancient Chinese glasses to the Korean peninsula was done mainly along the land route. Glass beads made of lead barium silicate glass have been found there, dating from the first century BC to the third century AD. The glasses unearthed in Korea are dated earlier than those in Japan; this suggests that the glasses first went to Korea, and then to Japan by crossing the Tsushima Strait.[28]

A number of ancient glasses have been uncovered at Dong-han, Sa-huynh, Dong-Nai, etc., in Vietnam. The majority of them are glass beads, with a few of glass ear pendants and bracelets. The earliest glass dates back to the 3rd to 4th century BC. However, spectroscopic analysis has shown that the content of Al_2O_3 is higher, while the contents of Na_2O, K_2O and CaO are lower (<3%). So they should belong to frit and faience. Most of the glass artifacts uncovered in Vietnam date from the late Sa-Huynh Period (100 BC–100 AD).[29] Their chemical compositions are multiple, including the silicate glasses containing PbO and BaO. Therefore, the lead barium silicate glass of the Western Han Dynasty had gone to Vietnam since then.

As shown in Table 2.9, it can be seen from the above analysis that the lead barium silicate glass is all early ancient glass of Japan, Korea and Vietnam, and its appearance in these countries was later than in China; in addition, the shape and chemical composition of the glass artifacts are very similar to that of the Chinese lead barium silicate glassware. Definitely, they came from China.

(2) *High lead silicate glass ($PbO–SiO_2$) and potash lead silicate glass ($K_2O–PbO–SiO_2$)*

The high lead silicate glass artifacts in China appeared at a very early time. Most of them date from the Warring States period, some from the late Spring and Autumn period. The glass gradually evolved into potash lead silicate glass, and was produced in large quantities during the Six Dynasties and the Sui Dynasty. High lead silicate glasses appeared in the Korean peninsula, dating from the Shijun and Koguryo periods (100 BC–100 AD) and corresponding

Table 2.9. Chemical composition of ancient lead barium silicate glasses unearthed in Japan, Korea, Vietnam and China.

Unearthed place	Date	Glass artifact	Chemical composition (wt%)									
			SiO_2	Al_2O_3	K_2O	Na_2O	CaO	MgO	Fe_2O_3	PbO	BaO	CuO
Changsha, Hunan, China	400–200 BC	Glass ritual disk	36.57	0.46	0.1	3.72	2.1	0.21	0.15	44.71	10.1	0.02
Tahori, Korean peninsula	100 BC–300 AD	Green beads	39	0.43	0.06	3.35	3.69	0.4	0.16	37.5	14.12	0.84
Sugu-Okamoto, Fukuoka, Japan	100 BC–100 AD	Tubular beads	38	0.35	0.19	3.90	1.1	0.51	0.29	36.5	14	0.78
Tateiwa, Fukuoka, Japan	100 BC–100 AD	Tubular beads	41.2	0.46	0.25	6.82	0.42	0.27	0.06	35.72	11.43	
Vietnam	100 AD	Beads	Si 13.7	Mn 7.12	Ni 16.37	Cr 4.86	Ca 8.67		Fe 12.67	Pb 11.41	Ba 21.48	Cu 6.9

to the Eastern Han Dynasty of China. The potash lead silicate glass appeared in Korea during the Korean Three-Kingdom period (Koguryo, Paechke, and Silla; 4th century to 6th century AD). By then, there were more contacts between China and Korea, such as sending tributes to Chinese court and bringing back presents to Korea. Following such back and forth, the ancient glass wares were spread into Korea.[28]

The high lead silicate glass uncovered in Japan date from the 6th to the 7th century AD, and the potash lead silicate glass from the 12th to the 14th century AD, later than that in Korea, which may have come from China by the Sea Silk Road.[30] The potash lead silicate glass uncovered in Vietnam is dated even later — by 16th to 17th century AD.[29] These two kinds of glasses uncovered in China are dated earlier than those in the neighboring countries, so it can be accepted that they spread from China to elsewhere.

Table 2.10 gives a description of ancient high lead silicate and potash lead silicate glass samples (and their chemical composition) unearthed in Japan, Korea, Vietnam, India and China. It can be seen that the high lead silicate glass uncovered in China is not the earliest in the world. The same kind of glass was unearthed at Nidmrund in Mesopotamia, dating from the 6th century BC.[31] It was earlier than that in China, but less was found in later years. The glass containing PbO was also found in ancient India, dating from nearly the same time as the Chinese one. We have discussed the identification of the production sites of ancient glasses containing PbO by lead isotope analysis in Ref. 26

It can be concluded scientifically that Chinese glasses containing PbO, including lead barium silicate, high lead silicate and potash lead silicate glasses, were all made in Inner China, and then spread to peripheral regions later.

(3) *Potash silicate glass (K_2O–SiO_2)*

The origin of ancient potash silicate glass is a problem still in dispute within the glass archeological field. It is commonly recognized that this kind of glass had not been produced in the ancient

Table 2.10. Chemical composition of ancient high lead silicate and potash lead silicate glasses unearthed in China, Japan, Korea and Vietnam.

			Chemical composition (wt%)								
Unearthed place	**Glass artifact**	**Date**	**SiO_2**	**Al_2O_3**	**Fe_2O_3**	**CaO**	**MgO**	**PbO**	**K_2O**	**Na_2O**	**CuO**
Baicheng, Xinjiang, China	Glass beads	~800 BC	64.31	1.36	1.10	0.01	2.67	9.01	2.42	—	0.01 Sb_2O_5 1.60
Luoyang, Henan, China	Beads	400–300 BC	18.20					74.01		3.29	
Guangxi, China	Glass cup	600 AD	34.92	1.57				62.1		1.43	
Pingba, Machang, Guizhou, China	Glass beads	500–600 AD	49.38	2.61				35.52	7.48	3.69	0.63 Cl (0.66)
Pingba, Machang, Guizhou, China	Glass beads	500–600 AD	47.91	2.41	0.1		1.1	38.68	7.36	1.78	Cl (0.59)
Asukaik, Nara, Japan	Glass making relics	7th cent. AD	25.0	0.1	0.1	0.06		74.2	0.2	0.2	CuO 0.05
			32.5	0.1	0.1	0.02		66.7	0.2	0.2	0.05
Gukokri, Shell-mound, Korea	Green glass beads	6th–8th cent. AD	25.1	0.39	0.14	0.06	0.13	72.5	0.14	1.27	CuO 0.2
Lam son, Loc chall, Vietnam	Glass relics	16th–18th cent. AD	55–56			3.6–3.8		27–29	7.3–9.6	2.0–2.5	

area of the Tigris and Euphrates Rivers valley, as well as ancient Egypt and Rome. However, the potash silicate glass in the 2nd century BC was unearthed in India, and Arikamedu is considered to be its production site. From there the glasses were exported to Southeast Asia, Japan and Korea by maritime transportation.[12] Since the 1980s a lot of potash silicate glass beads of the Han Dynasty (200 BC–100 AD) have been unearthed at Hepu, Guangxi; the Chinese origin has attracted scholars' attention.[32,33] Recent glass scientific archeological information indicates that the earliest potash silicate glass beads uncovered in China date from the Warring States period to the Western and Eastern Han Dynasties; such glass beads were often buried together with lead barium silicate glass beads as funerary objects. It can be seen from Table 2.11 that the chemical compositions of the potash silicate glass uncovered in China, India, Japan, Korea and Vietnam are nearly the same. Also, glasses of this kind unearthed in China are the earliest and the most numerous. So it is quite possible that they were be exported from Hepu, Guangxi to elsewhere. However, the origin of potash silicate glass is to be further investigated by scientists and archeologists.

3. Conclusion

There were four routes of the Silk Roads in China from north to south. The route taken for importing glass artifacts was changeable with the political and geographical situations at that time. Generally speaking, from the Warring States period to the Western and Eastern Han Dynasties, the northern (oasis) route was the main channel for glass trading, and the majority of objects were glass beads and ornaments due to easier transportation. The lead barium silicate glass and high lead silicate glass beads and ornaments were produced in the Yangtze River valley with typical Chinese forms, and spread westward along this road. They reached at least the western side of the Xinjiang area. Whether lead barium glass is to be found in central Asia is worthy of our concern.

After the Eastern Han Dynasty, the Southern (Sea) route was opened up. Glass vessels, especially large breakables, were imported

Table 2.11. Chemical composition of ancient potash silicate glasses unearthed in Japan, Korea, Vietnam, India and China.

			Chemical composition (wt%)								
Unearthed place	**Date**	**Glass artifact**	**SiO_2**	**Al_2O_3**	**Fe_2O_3**	**CaO**	**MgO**	**K_2O**	**Na_2O**	**CuO**	**PbO**
Guangxi, China	200 BC–10 AD	Beads	81.2	2.69	0.65	1.0	0.49	12.16	0.79	0.36	0.3
Zhongyang Dong, Korea	100 AD	Blue beads	73.47	3.48	2.38	1.42	0.42	14.9	0.89	0.62	0.01 BaO 0.3
Pusan, Korea	300 AD	Blue beads	77.32	1.36	1.89	1.16	0.32	17.6	0.36	0.04	BaO 0.27
Japan	100 BC–200 AD	Beads	75.4	2.7	0.8	<1.5	<1.5	17.9	<1.5	MnO 1.0	—
Okayama, Japan	300–400 AD	Blue beads	76.94	4.40	0.83	0.38	0.17	14.7	0.62	1.4	0.05
Sa-huyth, Vietnam	100 BC–200 AD	Beads	Mostly	0.6–1.3	0.2–1.3	2.8–7	0.5–0.8	18–22	0.22	—	—
Akikamedu, India	100 AD or later	Beads	76–78	2–4	—	1–4	<1	13–19	<1	—	—

into China mainly along the maritime route, first to a port of the south coast and unloaded there, then transported to central China. The ancient Chinese glass artifacts made of lead barium silicate glass that spread to East Asia, Southeast Asia and India were also transported along the Southern (Sea) Route.

The Southwestern (Buddhist) Route was used earlier than the Northern (Oasis) and Southern (Sea) Routes; it was the main route tying southeastern China to India. There had been contacts since Zhangqian's exploration of the Western Regions during the Han Dynasty. Historical literary sources have some records about the Indian glass artifacts called *liuli*, *biliuli* etc. exported to China. From the historical geography aspect, there were two alternative routes connecting India and China in addition to the southwestern route. One ran westward from Kasmira to Hetian of Xinjiang; the other ran from the north of India through Tibet into Huangzhong (now the Xining region) of Qinghai along the current Xinjiang–Tibet road. To have more knowledge of the Silk Road, it is necessary to further study the early ancient glasses uncovered in India, Tibet, Qinghai, Yunnan, Sichuan, etc.

References

1. Boulnois Luce, *La Route de la Soie* (Paris, Arland, 1993).
2. F. C. He and M. Wan, *History of Ancient Cultural Exchange Between the East and West* (Commerce, Shanghai, 1998), in Chinese.
3. L. S. Stavrianos, *A Global History*, 4th edn. (Prentice Hall, New Jersey, 1988), pp. 58–66.
4. H. Bechman and A. H. Wang, translated from Chinese, *Archeology in Xinjiang* (Xinjiang People's Press, Urumqi, 1996).
5. M. C. Lin, Origin and movement of Tocharian (*in Chinese*), *J. Xi Yue Study* **3**, 581–586 (2003).
6. Z. L. Wang, *Introduction to the History of Central Asia* (Xinjiang People's Press, Urumqi, 2004), in Chinese, pp. 58–63.
7. F. Li, Q. H. Li, F. X. Gan *et al.*, Chemical composition analysis for some ancient Chinese glasses by proton-induced X-ray emission technique (*in Chinese*), *J. Chin. Ceram. Soc.* **33**, 581–586 (2005).

8. P. Zhang, Ancient glass technology in Northern and Northwestern China. In: F. X. Gan (ed.), *Development of Chinese Ancient Glass Technology* (Shanghai Science and Technology Publishers, 2004), in Chinese, pp. 166–182.
9. F. X. Gan, H. S. Cheng and Q. H. Li, Origin of Chinese ancient glasses — study on the earliest Chinese ancient glasses, *Science in China E: Technological Sciences* **49**, 701–713 (2006).
10. W. K. Ma, Islamic glasses unearthed from Liao tombs and pagoda (*in Chinese*), *Archaeology* **8**, 736–743 (1994).
11. Institute of Cultural Relics and Archaeology (Inner Mongolia), Brief report on excavation of Princess and her husband's tomb of Cheng Kingdom, Liao Dynasty (*in Chinese*), *Cultural Relics* **11**, 4–24 (1987).
12. I. S., Lee The Silk Road and ancient Korean glass, *Korea Culture* **14**(4), 27–30 (1993).
13. J. B. Hou and J. Lin, Discussion of the characteristics and practical meaning of "Northeastern Silk Road" (*in Chinese*), *Liaoning Silk* **4**, 27–30 (2000).
14. E. H. Mark and Y. Leonid, Chemical analysis of Sarmatian glass beads from Pokrovka, Russia, *J. Archeol. Sci.* **25**, 1239–1245 (1998).
15. F. X. Gan, Q. H. Li, D. H. Gu *et al.*, Study on early glass beads unearthed from Baicheng and Tacheng of Xinjiang, *J. Chin Ceram. Soc* (in Chinese), **31**(7), 663–668 (2003).
16. W. Qian, P. Zhang and Q. M. Li, Study on early glass beads unearthed from graveyard in Kiziltur, Xinjiang Province. In: F. B. Wan and E. Bamo (eds.), *Proceedings of the 5th International Conference of Chinese Minorities Science and Technology History* (Guangxi National Press, Nanning, 2001), in Chinese, pp. 138–145.
17. Q. H. Li, D. H. Gu and F. X. Gan, Chemical composition analyses of ancient glasses found in Xinjiang province, China, in *Proceedings of the XX International Congress on Glass* (Kyoto, 2003, O-15-004).
18. A. Engle, Glassmaking in China, *Reading in Glass History* **6–7**, 1–38 (1976).
19. Y. X. Jiang, *Study on Ancient Southwestern Silk Road*, Vol. 2. (Sichuan University Press, Chengdu, 1995), in Chinese, p. 1.
20. Y. T. Shi, *Brief History of Early Cultural Exchange Between East and West* (Commerce Press, Beijing 1998), in Chinese.

21. F. X. Gan (ed.), *Study on Ancient Glasses in Southern China — Proceedings of the 2002 Nanning Symposium on Ancient Glasses in Southern China* (Shanghai Scientific and Technical Publishers, 2003), in Chinese.
22. Qinghai Institute of Cultural Relics and Archaeology, *Shangsunjai Han-Jin Tombs* (Cultural Relics Press, Beijing 1993), in Chinese, pp. 250–254.
23. F. X. Gan, Development of ancient oriental glasses. In: F.X. Gan (ed.), *Development of Chinese Ancient Glass Technology* (Shanghai Scientific and Technical Publishers, 2005), in Chinese, pp. 52–59.
24. Q. S. Huang, Study on the ancient glassware discovered in Guangxi. In: F. X. Gan (ed.), *Study on Ancient Glasses in Southern China* (Shanghai Scientific and Technical Publishers, 2003), in Chinese pp. 10–20.
25. L. C. Qiu, Ancient glassware discovered in Guangdong. In: F. X. Gan (ed.), *Study on Ancient Glasses in Southern China* (Shanghai Scientific and Technical Publishers, 2003), in Chinese, pp. 21–25.
26. F. X. Gan, Chinese ancient glasses — the evidence of cultural and technical exchange between China and the outside world. In: F. X. Gan (ed.), *Development of Chinese Ancient Glass Technology* (Shanghai Scientific and Technical Publishers, 2005), in Chinese, pp. 242–249.
27. K. Yamasaki, The relation between Chinese ancient glass and Japanese ancient glass unearthed in Yayoi Period tombs (*in Chinese*). In: F. X. Gan (ed.), *Study on Chinese Ancient Glass — Proceedings of the 1984 International Symposium on Glass* (Chinese Architecture Press, Beijing, 1984), pp. 47–52.
28. Tuneo Yoshimizu, Junji Tanahashi. *Orient Glass* (Three Colors Press, Tokyo, 1977), in Japanese.
29. T. K. Nguyen, *Vietnamese Ancient Glass* (in Vietnamese).
30. T. Koezuka and K. Yamasaki, Investigation of some K_2O–PbO–SiO_2 glasses found in Japan — a historical survey, in *Proceedings of the 17th International Congress on Glass* (Chinese Ceramic Society, Beijing, 1995), Vol. 6, pp. 469–474.
31. E. R. Caley, *Analysis of Ancient Glasses* (Corning Museum of Glass, New York, 1962), pp. 83–85.

32. M. G. Shi, O. L. He and F. Z. Zhou, Study on several potash-silicate glasses from the Han tomb, *J. Chin. Ceram. Soc.* (in Chinese) **14**(3), 307–313 (1986).
33. M. G. Shi, O. L. He and F. Z. Zhou, Chemical composition of ancient glasses unearthed in China, in *Proceedings of the 15th International Congress on Glass* (Session: Archeology) (Leningrad, 1989), pp. 7–12.

Chapter 3

第三章

Opening Remarks and Setting the Stage: Lecture at the 2005 Shanghai International Workshop on the Archaeology of Glass Along the Silk Road

Robert H. Brill

The Corning Museum of Glass, Corning, New York, 14830, USA

This symposium is being held under the auspices of TC-17 (Archaeometry of Glass). It was organized by Prof. Gan Fuxi. On behalf of all the members of TC-17, I thank Prof. Gan for his leadership and for his efforts.

Since its beginnings in 1984, TC-17 has held eight meetings. Five of them have dealt with Asian glass — the glass of China, Japan, and Korea — as well as glass found in Central Asia, India, and Southeast Asia. We are pleased to continue that tradition with this symposium. The published proceedings of the previous meetings are included in the bibliography at the end of this paper. Those proceedings include important papers by Prof. Gan, An Jiayao, and many other notable Chinese authors.

The purpose of this symposium is to bring together archaeologists, historians, and archaeological scientists who have an interest in glass found along the Silk Road. We hope to learn from one another, to exchange ideas, and to plan for collaboration in the future. This is the first time that an international meeting has been held on this subject. Although the meeting may be small in size, we regard it as a truly significant event.

Because scientific investigations are so useful in studying ancient glass, it might be helpful to outline how such studies can be applied to Asian glass. Therefore, in order to set the stage for the papers that will follow I would like to review two techniques that have proven helpful in earlier research. These are chemical analysis and lead-isotope analysis. In addition, I will summarize the results of some of our museum's more recent research on glasses found in Xinjiang and elsewhere in Central Asia.

1. Chemical Analysis

In order to understand scientific research on glass of the Silk Road, it is necessary first to understand something about Chinese glass.

Working since 1979, we have done chemical analyses of more than 200 glass fragments or objects from China, Korea, and Japan. In addition, we have completed analyses of more than 220 glasses from India and Southeast Asia, and more than 100 from various sites in Central Asia. Of course, many valuable analyses have also been done elsewhere — especially in China and Japan. However, we shall concentrate here on the analyses done by our museum, only because they are what I am most familiar with.

Most of the glasses found in China dating from the Warring States Period and the Han Dynasties were probably also made in China. This is apparent from their forms, which, in most cases, are typical Chinese forms. For example, there are these pieces such as *bi*, cicadas, and sword terminals, many of which are turbid white in color. There are also many dark blue glasses (and glasses of other colors), like eye beads and assorted other small objects. Among the other finds is a curious little inlay that was found still inside the mold in which it was formed. A large majority of these glasses, whatever their color, are lead:barium:silica glasses. [Note: In the oral presentation of this lecture, numerous examples of objects were shown in slides.]

Some glasses of the lead–barium type might have been made in Japan and Korea, but none were ever made in the West. This chemical composition is unique to East Asia.

There are also some Asian glasses with very high lead contents that do not contain barium. These glasses are often somewhat later than the Han Dynasty.

Ancient glasses of the corresponding periods made in the West have very different chemical compositions. They are soda:lime:silica glasses. So it is often possible by chemical analysis to distinguish between early East Asian glasses and glasses made in the West.

There is another type of Asian glass besides the high-lead glasses. This was first discovered by a Chinese chemist — I believe it was Shi Meiguang — who analyzed some glass beads from Guangzhou. He found that the beads contained only two major oxides: potash and silica. Since then, we, too, have found many potash:silica glasses. Two "ear spools" and a group of dark blue beads are typical examples.

Although many such glasses have been found in China, potash:silica glasses have also been found elsewhere, as shown by the circles on the map in Fig. 3.1. They cover a broad geographical area, extending from Japan and Korea in the east, through China, to Thailand, Vietnam, and Indonesia, and all the way to southern India. The particular glasses we have studied range in date from as early as the 2nd century BC to about the 4th century AD. We do not know for sure where this chemical type of glass was made, but it was very likely made not only in China but elsewhere as well.

Following the Han Dynasty, glasses found in China and elsewhere in East Asia sometimes have compositions resembling the alkali-silicate glasses made in the West. Some of these were undoubtedly imported from the West. But others have compositions that, although they are similar, are nevertheless often distinguishable from Western glasses if one examines them carefully.

Next, we will look at glasses found along the Silk Road. The Silk Road was not a single road, but a group of several trade routes that connected East Asia with the Western world. A few years ago,

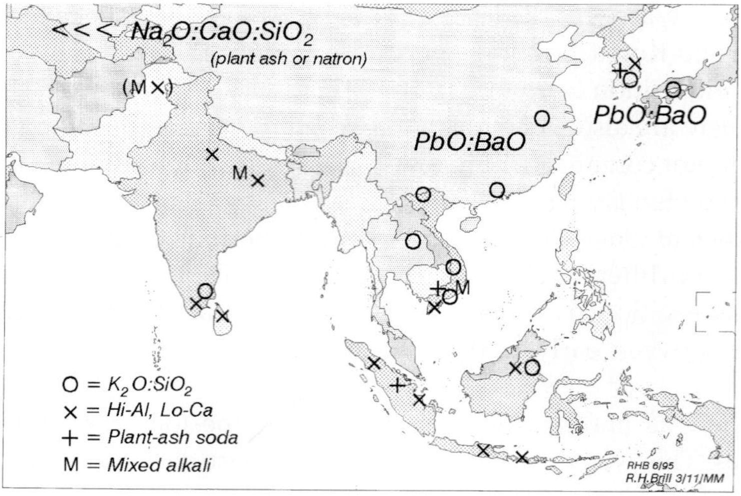

Fig. 3.1. Map showing distributions of various chemical families of Asian glasses.

Unesco identified four main routes, which they called the Steppe Route, the Desert Routes, the Maritime Route, and the Buddhist Route. Following the campaigns of Alexander the Great — which ended when he died in 323 BC — the earlier trade routes coalesced into what later became known as the Silk Road. One of the most important effects was that Greek culture became blended with Eastern cultures. Trade flourished between East and West, and all kinds of commodities, ideas, and ways of thinking passed back and forth during the centuries that followed.

Among the items of trade that traveled eastward were glass artifacts. Perhaps the most famous glass objects that passed eastward along the Silk Road are two faceted, hemispherical bowls that found their way to Japan. One is in the Tokyo National Museum and the second, nearly identical, bowl is in the Shōsō-in. Both are very well preserved. A similar bowl in The Corning Museum of Glass is very heavily weathered and has lost its transparency after centuries of burial in the soil. According to chemical analyses, all these bowls were probably made in Iran in the 4th to 6th centuries.

Returning to glasses uncovered in China, there is a little bowl in the National Museum of China that is probably known to all of you. It is believed to have been made in either Iraq or Iran, and is probably late Sasanian, dating perhaps from the 5th century. A small cup in The Corning Museum of Glass is very similar to it in color and has the same type of applied latticework decoration. The Corning piece is believed to have been made in Iraq.

Certain other glass vessels found in China are known that also have Western parallels. Most of them are probably familiar to you. Two objects from the Famensi were among those placed in the crypt in 874 AD, when it was sealed. They have parallels in both their shapes and decorations among glasses of about the same date that were found in Iran. One piece has an elaborate scratched decoration and the other is an uncommon type of luster ware.

Fragments of a dish with nearly identical scratched decoration were found at Nishapur in Iran. A group of luster-decorated glass fragments in Corning also came from Iran. They have the same type of colorful stained decoration as the bowl from the Famensi. It was produced by painting a design on the vessel using a paint that contained both silver and copper compounds. After the glass was fired a second time, in a reducing atmosphere, the dense yellow and orange decoration would have appeared. The Corning fragments of luster glass, as well as that with the scratched decoration, all have typical Islamic chemical compositions. A sample of a small yellowish bowl or lamp from the Famensi was analyzed with Mr Shi Meiguang. It also had a typical Islamic composition. By "typical Sasanian and Islamic compositions," we mean soda-lime glasses with K_2O and MgO levels of greater than about 1.5 weight percent. They were made with certain types of plant ashes as their source of soda. Roman and Hellenistic glasses were soda limes with somewhat lower levels of K_2O and MgO, having been made with natron.

However, there are other objects found in China that are puzzling. One is a well-known small, greenish glass cup in the National Museum of China. It has a large bubble in it and cut grooves near its rim. At first glance it looks as if it was Roman.

However, after more careful examination, it really does not quite look like a Roman vessel. But it does resemble very closely a bowl in a private collection in America. A chemical analysis of the second bowl showed that it has a potash:silica composition, so that bowl definitely was not made in the West — and it definitely is not Roman. Thus we have an example of an object whose origin is in question and have seen how a chemical analysis was used to rule out a Roman origin. It could have been made in China, or in India — or somewhere else on the map showing where potash:silica glasses were found.

Perhaps it would be helpful for some Chinese chemist to analyze the cup in Beijing to see if it also has a potash:silica composition. I was once told that it has a low specific gravity, which indicates that it is not a lead-containing glass and that it could possibly be a potash:silica glass.

2. Lead-Isotope Analysis

Another scientific method that is useful for studying ancient glasses is lead-isotope analysis. For this method, very tiny samples of any materials that contain lead (such as bronzes, pigments, glazes, or glasses) are analyzed in a mass spectrometer. The instrument generates numbers that are ratios of the various isotopes of the lead in the sample. There are several factors that must be carefully considered in interpreting the data, but for our purposes right now, the important thing is this. By plotting a graph of the isotope ratios for the artifacts, it is possible to separate them according to where the leads in them could have come from and where they could not have come from. Therefore, if used carefully, this method tells us something about where ancient artifacts were made.

Figure 3.2 summarizes the data we have collected for nearly 2000 artifacts and lead ores from all over the ancient world. Leads from Greece fall in the small ellipse marked "L." Leads from Europe are marked "E," leads from Mesopotamia "M," and leads from Egypt fall toward the left side of the graph.

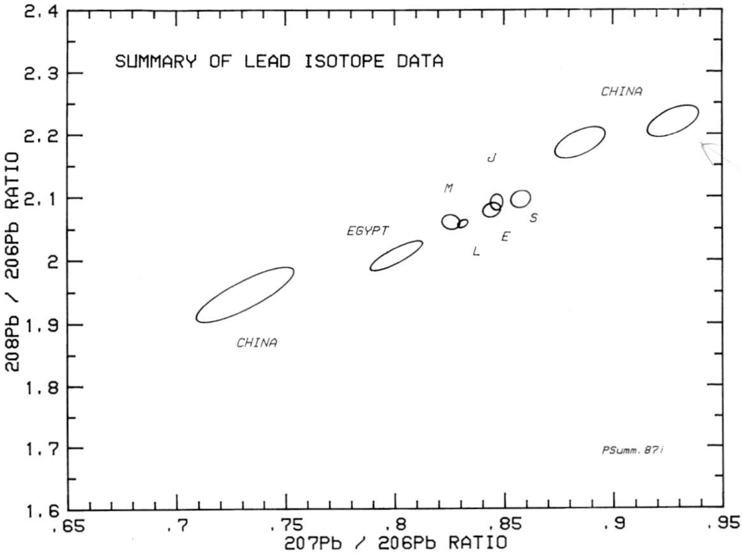

Fig. 3.2. Lead-isotope data for approximately 2000 ancient artifacts made of various materials from all over the ancient world. Many ores are also included. Note that leads in Chinese glasses are among the highest and lowest ratios found.

Since many early Chinese glasses contain lead oxide, over the years we analyzed about 100 ancient lead-containing glasses and related materials from Asia. The related materials were Chinese blue and Chinese purple artifacts and faience glazes.

Most of the leads from Chinese glasses fall in the two ellipses at the highest part of the graph. There are galena ores (lead ores) from China that match these ratios, verifying that the glasses were almost certainly made in China. Certain other Chinese glasses contain leads that fall in the lowest range of isotope ratios we have ever measured. However, there are other lead ores in China that match those low ratios, so these glasses were also made in China.

You will recall that some of the lead–barium glasses are made of white opaque glass. They include the *bi*, the cicada, and the sword terminal mentioned above. Perhaps they were made to imitate white jade. Nine of them have very low lead ratios and fall in a cluster of points at the left of the graph. On the other hand, most of the dark blue, green, and amber glasses we have analyzed — including the Chinese blue and Chinese purple artifacts — have very high ratios. We conclude that the two types of glasses were made of lead that came from different mining regions and that the white and blue glasses were, therefore, probably made in different places.

A group of Central Asian glasses lies intermediate between the high and low ratios. They apparently were made somewhere else. The potash:silica glasses have traces of lead in them. Analyses show that their traces of lead also have intermediate ratios. However, the traces of lead might have come in with the cobalt colorant. Therefore, the isotope ratios may be telling us some-thing about where the cobalt came from, instead of where the glass was made.

3. Some Glasses Excavated by Early Explorer–Archaeologists

Finally, I want to tell you about chemical analyses of some glasses from along the Silk Road. These samples are, indirectly, controversial. They came from small fragments of glass collected by Western explorer–archaeologists who traveled in Xinjiang a century ago. The fragments were brought back to various Western museums. With the cooperation of these museums, we removed very small samples of glass from some of these fragments. The samples, being very small, were consumed in the chemical analyses. The original fragments, which were themselves small and mostly nondescript, are still in the museums' collections.

The samples we analyzed came from fragments collected by men whose names are familiar to all of us: Sir Aurel Stein, Albert von Le Coq, Paul Pelliot, Sven Hedin, and Petr Koslov. We have also analyzed a number of samples of glass excavated at Begram and elsewhere throughout Afghanistan, as well as samples from Uzbekistan. All of the glasses from Afghanistan turned out to be soda:lime:silica glasses of typical Roman, Sasanian, Islamic, and Central Asian types. They have already been published elsewhere and will not be discussed further here. However, the samples from Uzbekistan are included in the discussion that follows.

The author is very grateful to the institutions and individuals who provided the samples for these analyses. They are listed in the "Acknowledgments" section. Many of the analyses (except those from Lou-Lan and the Stein collection) have been reported previously in the book *Chemical Analyses of Early Glasses*, published in 1999 (A-91 and A-92 in the bibliography). Brief sample descriptions are given in Volume 1 (pp. 144–149) of that work and the data are reported in Volume 2 (pp. 340–348).

Figures 3.3–3.6 illustrate some of the parent fragments from which the samples were removed. They are from: Qizil (the familiar type of faceted, hemispherical bowl); three other sites on the northern Desert Route (Duldur-Aqur near Turfan, Qoch Homa at Kucha, and Hazar-tam near Kashgar); Kucha Oasis; and Lou-Lan on the southern Desert Route.

The graphs in Figs. 3.7–3.12 show that there is a wide variability in the compositions of the glasses analyzed. The data are plotted as reduced compositions, i.e. the seven major and minor oxides used were normalized to 100.00%. In all, 60 glasses were analyzed. One proved to be a lead:silica glass, so it was not plotted in the graphs.

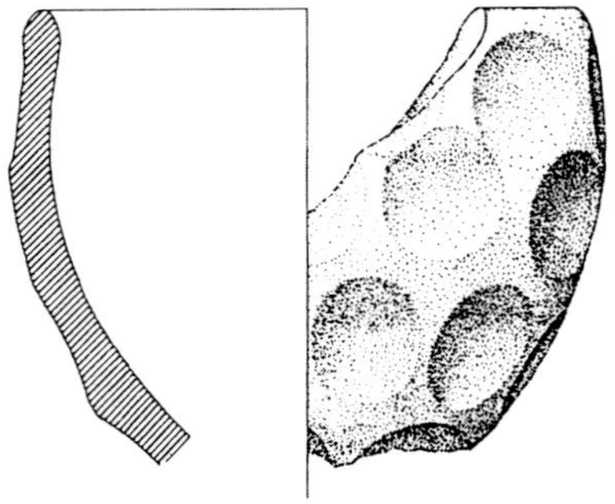

Fig. 3.3. Fragment of facet-cut hemispherical bowl from Qizil. CMG-6130. Drawing courtesy of Jens Kröger.

Fig. 3.4. Fragments excavated (*from top left*) at Duldur-Aqur (1), Qoch Homa (1), and Hazar-tam (4). CMG-6120, 6121, 6124, 6125, 6126, and 6128.

Fig. 3.5. Fragments excavated at Kucha Oasis (*top left*) and Apartak (3). CMG-6110, 6112, 6111, and 6113.

The graphs are very complicated and difficult to interpret. However, after examining them carefully, we concluded that the data could be summarized as shown in Tables 3.1 to 3.3. In the future, we will attempt to analyze the data by multivariate statistical methods to see what additional information can be gathered.

Fig. 3.6. Fragments excavated at Lou-Lan. CMG-6810–6827 series.

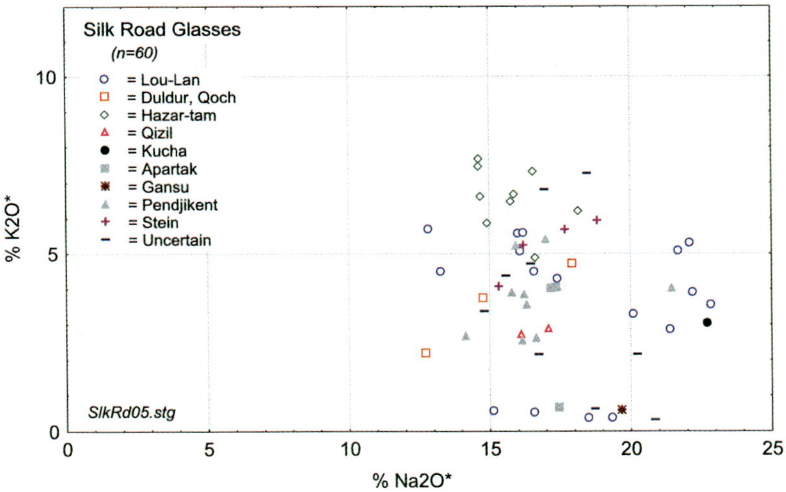

Fig. 3.7. K_2O^* versus Na_2O^* plot for 59 glasses from along the Silk Road. The asterisk indicates that the data have been normalized to 100.00% for seven major and minor oxides.

In several instances, glasses from certain sites were nearly identical, chemically, to glasses found at other sites. Whether this is just by chance or whether it indicates a relationship among the glasses will be investigated further.

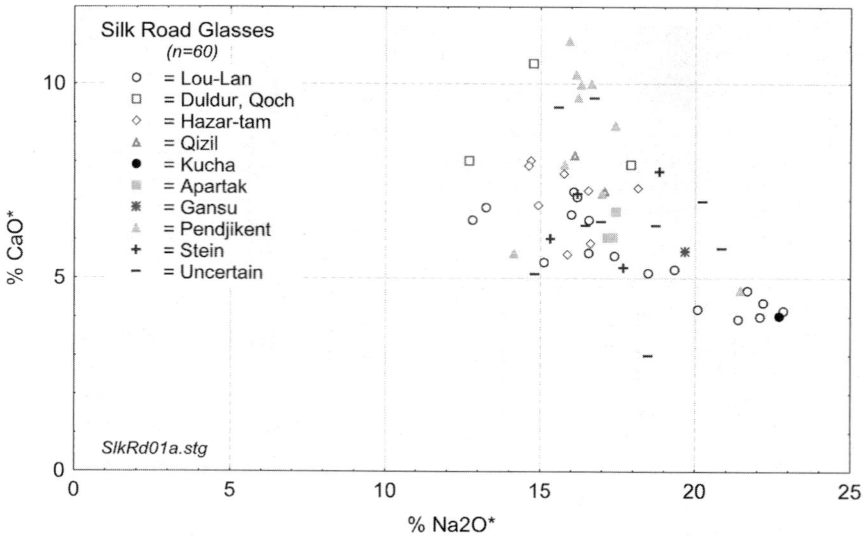

Fig. 3.8. CaO^* versus Na_2O^* plot.

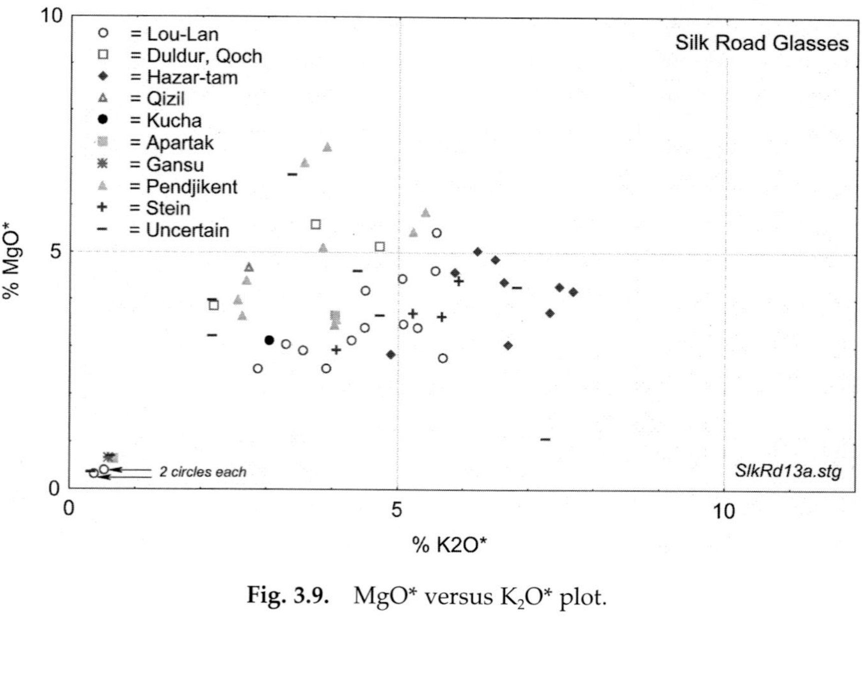

Fig. 3.9. MgO^* versus K_2O^* plot.

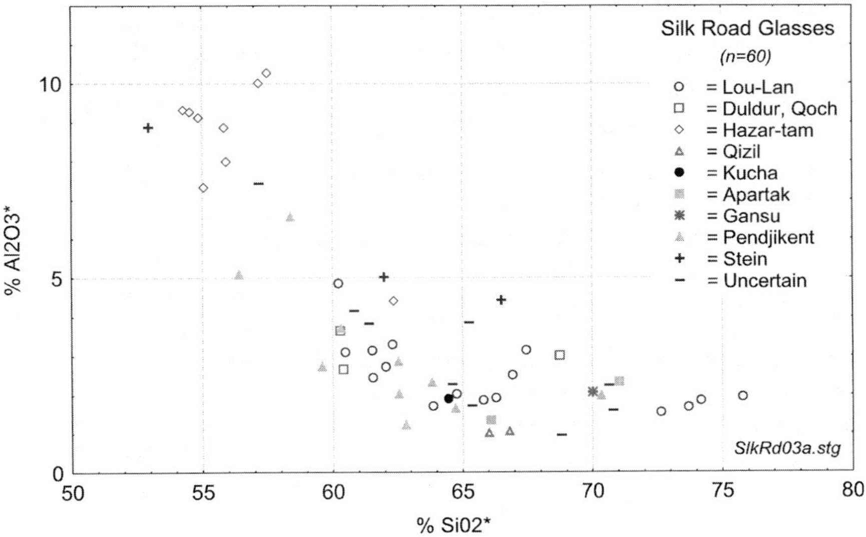

Fig. 3.10. Al_2O_3* versus SiO_2* plot.

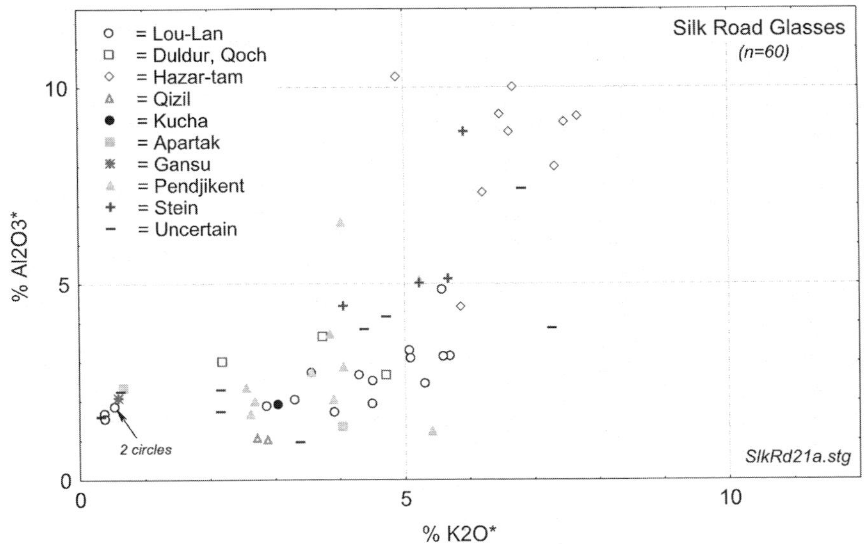

Fig. 3.11. Al_2O_3* versus K_2O* plot.

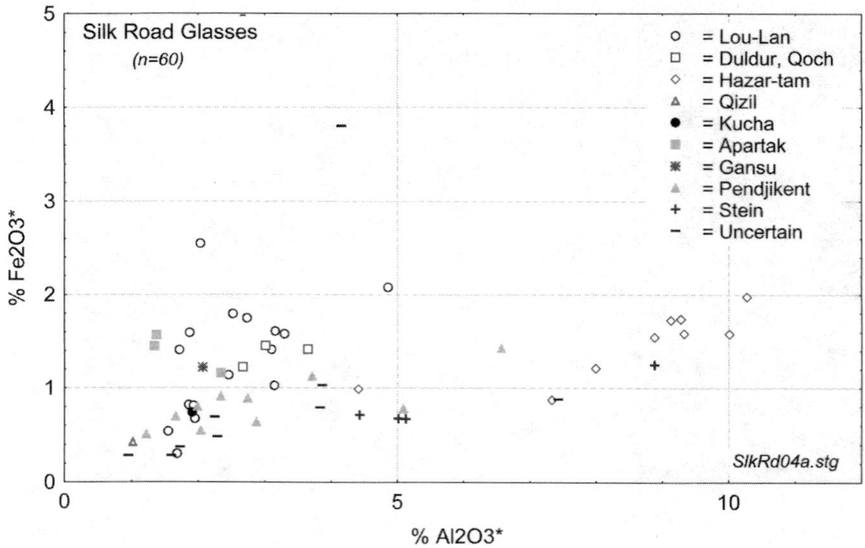

Fig. 3.12. Fe_2O_3* versus Al_2O_3* plot.

Table 3.1. Some chemical families of Silk Road glasses* (approximate limits).

$Na_2O:CaO:SiO_2$ (K_2O, MgO < 2%)	Natron type "Roman"
$Na_2O:CaO:SiO_2$ (K_2O, MgO > 2%)	Plant-ash type "Sas.-Islamic"
$Na_2O:CaO:SiO_2$ (K_2O > 4–4.5%)	"Central Asian"
$Na_2O:CaO:SiO_2$ (K_2O > 4–4.5%; Al_2O_3 > 5%)	"Central Asian, high Al_2O_3"
$Na_2O:CaO:SiO_2$ (K_2O > 4–4.5%; Al_2O_3 ~ 3–6%)	"Central Asian, moderate Al_2O_3"
(Na_2O, K_2O):CaO:SiO_2 (Na_2O ~ 10–16%; K_2O > 6%)	Borderline mixed alkali
PbO:SiO_2	Lead silicate

*In addition to PbO:BaO: SiO_2, PbO:SiO_2, and K_2O:SiO_2 types.

Of the total of 61 glasses analyzed:

- 8 were natron-based types typical of Hellenistic or Roman glass;
- 15 were plant-ash soda types typical of Sasanian and Islamic glass;
- 28 were plant-ash soda limes of various types that we believe also indicate Central Asian origins;

Table 3.2. Chemical families found for some Silk Road glasses.

Excavator	Location	"Roman" (natron type soda lime)	"Sas-Islamic" (plant-ash type soda lime)	Central Asian (high K_2O; soda lime)	Central Asian (high K_2O, high Al_2O_3; soda lime)	Central Asian (high K_2O, mod. Al_2O_3; soda lime)	Central Asian (probably mixed alkali)	No.
Historical samples								
S. Hedin (FME)	Lou-Lan	**4** (6814, 15, 16)	**4** (6820, 21, 24, 26)	**3** (6811, 19, 25)	**1** (6813)	**5** (812, 17, 18, 23, 27)		**17**
P. Pelliot (MNAA)	Duldur-Aqur		**1** (6119)	**1** (6120)				**2**
P. Pelliot (MNAA)	Qochhoma			**1** (6121)				**1**
P. Pelliot (MNAA)	Hazar-tam				**1** (6122)	**1** (6124)	7 (6213, 25, 26, 26a, 27, 28, 29)	**9**
A. von Le Coq (MIslK)	Qizil		**3** (6130, 31, 32)					**3**
P. Koslov (?) (HM)	Kucha		**1** (6110)					**1**

(Continued)

Table 3.2. *(Continued)*

Excavator	Location	"Roman" (natron type soda lime)	"Sas-Islamic" (plant-ash type soda lime)	Central Asian (high K_2O; soda lime)	Central Asian (high K_2O, high Al_2O_3; soda lime)	Central Asian (high K_2O, mod. Al_2O_3; soda lime)	Central Asian (probably mixed alkali)	No.
A. Stein (BM)	Hanguya Tati				**1** (8503)	**1** (8504)		**2**
A. Stein (BM)	Togujai				**1** (8505)			**1**
A. Stein (BM)	Kelpin				**1** (8506)			**1**
A. Stein (BM)	Uncertain	**2** (8500, 01)	**3** (8502, 09, 11)			**2** (8510b, 10w)	**2** (8507, 12)	**9**
Recent samples								
A. Abdurazakov	Apartak	**1** (6112)		**2** (6111, 13)				**3**
–	Xinjiang	**1** (7015)						**1**
B. Marshak	Pendjikent		**3** (6250, 57, 58)	**5** (6251, 54, 55, 59, 53?)	**2** (6252, 56)			**10**
Total		**8**	**15**	**12**	**7**	**9**	**9**	60

FME = Folkens Museum Etnografiska, Stockholm

MNAA = Musée National des Arts Asiatiques-Guimet, Paris

MIslK = Museum für Islamische Kunst, Berlin

HM = Hermitage Museum, St Petersburg

BM = British Museum, London

Table 3.3.1. Chemical analyses of Silk Road glasses.

	Lou Lan												
	6810	**6811**	**6812**	**6813**	**6814**	**6815**	**6816**	**6817**	**6818**	**6819**	**6820**	**6821**	**6822**
SiO_2 d													
SiO_2 a	74.98	65.62	60.55	59.22	77.30	76.08	75.48	66.00	60.40	64.94	62.33	67.02	28.87
Na_2O	18.81	13.00	15.63	15.74	20.57	16.99	15.06	12.55	15.90	16.24	22.94	21.79	0.54
CaO	5.22	6.68	7.02	6.53	5.56	5.79	5.39	6.36	6.96	6.36	4.18	4.01	0.07
K_2O	0.38	4.41	4.93	5.48	0.41	0.54	0.57	5.57	5.49	4.40	3.57	2.91	0.07
MgO	0.33	4.10	4.31	4.53	0.36	0.41	0.48	2.71	5.33	3.33	2.93	2.58	0.05
Al_2O_3	1.72	2.47	3.21	4.78	1.65	1.91	1.94	3.09	3.09	1.89	2.74	1.91	0.14
Fe_2O_3	0.14	0.79	0.69	0.92	0.26	0.38	0.30	0.71	0.45	0.36	0.79	0.73	0.06
TiO_2	0.074	0.11	0.10	0.11	0.11	0.071	0.050	0.082	0.073	0.033	0.19	0.11	0.007
MnO	0.00	1.33	1.28	1.22	0.09	0.05	0.08	0.56	0.10	0.30	0.08	0.12	0.09
CuO	0.12	0.078	0.074	0.074	0.021	0.13	0.17	0.12	0.070	0.025	1.80	1.39	0.11
CoO													
SnO_2													
Ag_2O													
PbO	0.17	0.080	0.023		0.091	0.031	0.22		0.093	0.040	0.034	0.047	76.29
BaO		0.032	0.050	0.048	0.033	0.044	0.063	0.026	0.11	0.041	0.083	0.038	
SrO													
Li_2O													

(Continued)

Table 3.3.1. *(Continued)*

	Lou Lan												
	6810	**6811**	**6812**	**6813**	**6814**	**6815**	**6816**	**6817**	**6818**	**6819**	**6820**	**6821**	**6822**
B_2O_3													
Cr_2O_3													
NiO													
ZnO													
Sb_2O_5													
V_2O_5													
P_2O_5		0.49	0.44	0.64	0.07	0.09	0.06	0.55	0.40	0.42	0.43	0.45	0.06
SO_3	0.19	0.32	0.22	0.26	0.28	0.13	0.25	0.22	0.30	0.29	0.34	0.28	0.00
Cl	1.36	0.44	0.56	0.36	1.07	1.06	1.09	0.46	0.48	0.42	1.39	1.32	0.61
As_2O_5													
Total	103.49	99.96	99.10	99.91	107.87	103.73	101.19	98.99	99.25	99.08	103.82	104.71	106.98
SiO_2*	73.82	67.60	62.85	60.93	72.85	74.51	76.07	68.05	61.87	66.59	62.65	66.39	
Na_2O*	18.52	13.39	16.23	16.19	19.39	16.64	15.18	12.94	16.29	16.65	23.06	21.59	
CaO*	5.14	6.89	7.29	6.72	5.24	5.67	5.43	6.55	7.13	6.52	4.20	3.97	
K_2O*	0.38	4.54	5.11	5.64	0.38	0.53	0.57	5.74	5.62	4.51	3.59	2.89	
MgO*	0.32	4.23	4.48	4.66	0.34	0.40	0.48	2.80	5.46	3.42	2.95	2.55	
Al_2O_3*	1.69	2.55	3.33	4.92	1.56	1.87	1.96	3.19	3.16	1.94	2.76	1.89	
Fe_2O_3*	0.14	0.82	0.72	0.95	0.24	0.37	0.30	0.73	0.46	0.37	0.79	0.72	
T*	100	100	100	100	100	100	100	100	100	100	100	100	

Table 3.3.2. Chemical analyses of Silk Road glasses (*cont.*).

	Lou-Lan					Duldur-Aqur		Qoch homa
	6823	**6824**	**6825**	**6826**	**6827**	**6119**	**6120**	**6121**
SiO_2 d						68.76	60.44	60.56
SiO_2 a	59.38	66.20	60.11	65.88	58.86			
Na_2O	21.33	20.53	16.87	22.90	21.16	12.49	17.45	14.31
CaO	4.60	4.29	5.40	4.50	3.83	7.89	7.72	10.20
K_2O	5.00	3.37	4.16	4.03	5.08	2.15	4.59	3.62
MgO	3.42	3.12	3.04	2.62	3.26	3.79	5.00	5.42
Al_2O_3	3.05	2.08	2.59	1.78	2.35	2.96	2.60	3.54
Fe_2O_3	1.39	1.17	2.15	0.65	1.09	1.43	1.19	1.37
TiO_2	0.18	0.087	0.18	0.12	0.13	0.13	0.090	0.13
MnO	0.15	0.16	0.07	0.00	0.198	0.04	0.05	0.05
CuO	0.33	0.70	1.59	1.67	1.91			
CoO								
SnO_2								
Ag_2O								
PbO	0.001	0.023			0.001	0.03	0.02	0.04
BaO						0.04	0.04	0.05
SrO								
Li_2O								
B_2O_3								

(*Continued*)

Table 3.3.2. *(Continued)*

	Lou-Lan					Duldur-Aqur		Qoch homa
	6823	6824	6825	6826	6827	6119	6120	6121
Cr_2O_3								
NiO								
ZnO								
Sb_2O_5								
V_2O_5								
P_2O_5	0.62	0.36	0.44	0.39	0.56	0.29	0.81	0.71
SO_3	0.19	0.15	0.14	0.48	0.070			
Cl		1.15	0.90	1.42				
As_2O_5								
Total	98.17	103.39	97.64	106.44	95.63	31.24	39.56	39.44
SiO_2*	60.49	65.70	63.73	64.36	61.55	69.13	61.06	61.16
Na_2O*	21.73	20.38	17.88	22.37	22.13	12.56	17.63	14.45
CaO*	4.69	4.25	5.73	4.40	4.01	7.93	7.80	10.30
K_2O*	5.09	3.34	4.41	3.94	5.31	2.16	4.64	3.66
MgO*	3.48	3.09	3.22	2.56	3.41	3.81	5.05	5.47
Al_2O_3*	3.11	2.06	2.75	1.73	2.46	2.98	2.63	3.58
Fe_2O_3*	1.42	1.16	2.28	0.64	1.14	1.44	1.20	1.38
T*	100	100	100	100	100	100	100	100

Table 3.3.3. Chemical analyses of Silk Road glasses (*cont.*).

	Hazar-tam or Saqal-tam								
	6124	**6122**	**6123**	**6125**	**6126**	**6126a**	**6127**	**6128**	**6129**
SiO_2 d	61.59	57.43	54.85	54.01	54.42	54.39			
SiO_2 a							54.24	55.34	56.41
Na_2O	14.65	16.31	14.15	15.32	14.15	14.14	17.88	16.38	15.67
CaO	6.75	5.78	7.74	7.48	7.65	7.67	7.21	7.18	5.54
K_2O	5.76	4.80	6.38	6.30	7.23	7.42	6.12	7.25	6.60
MgO	4.49	2.79	4.21	4.72	4.14	4.04	4.96	3.69	3.01
Al_2O_3	4.33	10.08	8.55	9.06	8.83	8.96	7.22	7.91	9.88
Fe_2O_3	0.97	1.94	1.49	1.54	1.67	1.68	0.86	1.20	1.56
TiO_2	0.060	0.15	0.11	0.11	0.11	0.12	0.050	0.090	0.12
MnO	0.75	0.06	0.05	0.86	0.95	0.98	0.08	0.87	0.06
CuO	0.010	0.010	1.87		0.030	0.010		0.010	0.010
CoO									
SnO_2									
Ag_2O					0.0005				
PbO	0.01	0.02	0.05		0.03	0.02		0.04	0.01
BaO	0.06	0.08	0.04	0.08	0.08	0.07	0.05	0.04	0.05
SrO					0.05				
Li_2O					0.001				
B_2O_3					0.01				

(*Continued*)

Table 3.3.3. *(Continued)*

	Hazar-tam or Saqal-tam								
	6124	6122	6123	6125	6126	6126a	6127	6128	6129
Cr_2O_3					0.01				
NiO					0.02				
ZnO					0.015				
Sb_2O_5					0.097				
V_2O_5									
P_2O_5	0.57	0.55	0.51	0.52	0.51	0.50	0.48	0.55	0.53
SO_3									
Cl									
As_2O_5									
Total	38.41	42.57	45.15	45.99	45.58	45.61	99.15	100.55	99.45
SiO_2*	62.50	57.93	56.33	54.87	55.48	55.33	55.07	55.93	57.17
Na_2O*	14.87	16.45	14.53	15.56	14.43	14.38	18.15	16.55	15.88
CaO*	6.85	5.83	7.95	7.60	7.80	7.80	7.32	7.26	5.61
K_2O*	5.85	4.84	6.55	6.40	7.37	7.55	6.21	7.33	6.69
MgO*	4.56	2.81	4.32	4.80	4.22	4.11	5.04	3.73	3.05
Al_2O_3*	4.39	10.17	8.78	9.20	9.00	9.11	7.33	7.99	10.01
Fe_2O_3*	0.98	1.96	1.53	1.56	1.70	1.71	0.87	1.21	1.58
T*	100	100	100	100	100	100	100	100	100

Table 3.3.4. Chemical analyses of Silk Road glasses (*cont.*).

	Qizil			Kucha Oasis	Apartak			Xinjiang
	6130	**6131**	**6132**	**6110**	**6111**	**6113**	**6112**	**7015**
SiO_2 d								68.70
SiO_2 a	65.51	65.93	66.15	63.84	65.15	65.03	67.83	
Na_2O	16.95	15.90	16.04	22.49	17.10	16.87	16.65	19.30
CaO	7.18	8.06	8.10	3.98	5.96	5.93	6.41	5.58
K_2O	2.86	2.68	2.66	3.01	3.98	3.95	0.64	0.58
MgO	5.29	4.62	4.59	3.10	3.61	3.58	0.61	0.65
Al_2O_3	1.01	1.05	1.06	1.89	1.33	1.36	2.23	2.03
Fe_2O_3	0.43	0.45	0.44	0.73	1.43	1.54	1.11	1.20
TiO_2								0.072
MnO	0.14	0.19	0.19	0.04	0.05	0.06	0.49	0.10
CuO		0.020		0.030	0.18	0.20	0.210	0.010
CoO								
SnO_2								
Ag_2O								0.005
PbO	0.07		0.03	0.04	0.16	0.19	0.33	0.21
BaO								0.05
SrO								
Li_2O								
B_2O_3								0.1

(*Continued*)

Table 3.3.4. *(Continued)*

	Qizil			Kucha Oasis	Apartak			Xinjiang
	6130	6131	6132	6110	6111	6113	6112	7015
Cr_2O_3								0.01
NiO								0.01
ZnO								0.093
Sb_2O_5	0.03	0.02	0.03	0.08	0.1	0.08	2.51	1.2
V_2O_5								0.01
P_2O_5	0.20	0.27	0.23	0.43	0.36	0.38	0.15	0.09
SO_3								
Cl								
As_2O_5								
Total	99.67	99.19	99.52	99.66	99.41	99.17	99.17	31.30
SiO_2*	66.02	66.81	66.79	64.46	66.10	66.18	71.04	70.07
Na_2O*	17.08	16.11	16.20	22.71	17.35	17.17	17.44	19.69
CaO*	7.24	8.17	8.18	4.02	6.05	6.04	6.71	5.69
K_2O*	2.88	2.72	2.69	3.04	4.04	4.02	0.67	0.59
MgO*	5.33	4.68	4.63	3.13	3.66	3.64	0.64	0.66
Al_2O_3*	1.02	1.06	1.07	1.91	1.35	1.38	2.34	2.07
Fe_2O_3*	0.43	0.46	0.44	0.74	1.45	1.57	1.16	1.22
T*	100	100	100	100	100	100	100	100

Table 3.3.5. Chemical analyses of Silk Road glasses (*cont.*).

	Pendjikent									
	6251	**6253**	**6255**	**6256**	**6259**	**6257**	**6250**	**6254**	**6252**	**6258**
SiO_2 d										
SiO_2 a	58.26	61.51	58.93	55.53	61.38	63.00	62.21	59.68	57.23	68.98
Na_2O	15.95	16.64	15.85	15.69	15.49	16.20	15.75	16.62	21.03	13.87
CaO	9.76	7.03	9.42	10.94	7.79	9.73	9.98	8.52	4.59	5.53
K_2O	3.48	5.30	3.76	5.15	3.83	2.55	2.49	3.87	3.95	2.63
MgO	6.75	5.74	4.99	5.36	7.10	3.55	3.88	3.41	3.39	4.31
Al_2O_3	2.68	1.20	3.63	5.01	2.00	1.62	2.28	2.74	6.44	1.95
Fe_2O_3	0.87	0.50	1.10	0.77	0.54	0.68	0.89	0.61	1.40	0.79
TiO_2	0.78	0.047	0.19	0.12	0.12	0.011	0.10	0.97	0.19	0.11
MnO	0.26	0.59	0.38	0.08	0.56	1.54	1.51	3.41	0.05	0.58
CuO	0.82									
CoO						0.0048	0.0095	0.013		
SnO_2	0.06	0.006	0.0046		0.0094					
Ag_2O										
PbO	0.2									
BaO		0.002	0.091	0.027	0.018	0.044			0.053	0.089
SrO										
Li_2O										
B_2O_3										

(*Continued*)

Table 3.3.5. *(Continued)*

	Pendjikent									
	6251	**6253**	**6255**	**6256**	**6259**	**6257**	**6250**	**6254**	**6252**	**6258**
Cr_2O_3										
NiO										
ZnO			0.026	0.019		0.017	0.002	0.018	0.009	0.017
Sb_2O_5	0.017	0.05	0.011	0.028	0.002	0.054		0.017	0.027	0.027
V_2O_5										
P_2O_5	0.60	0.64	0.60	0.64	0.33	0.37	0.44	0.56	0.49	0.29
SO_3	0.39	0.46	0.35	0.56	0.24	0.24	0.27	0.39	0.22	0.26
Cl	0.25	0.48	0.56	0.16	0.55	0.38	0.32	0.20	1.03	0.59
As_2O_5	0.02	0.01		0.00	0.02	0.00	0.00		0.00	
Total	101.15	100.21	99.89	100.09	99.98	99.99	100.13	101.03	100.10	100.02
SiO_2*	59.60	62.82	60.33	56.40	62.55	64.73	63.82	62.52	58.38	70.34
Na_2O*	16.32	16.99	16.23	15.94	15.79	16.64	16.16	17.41	21.45	14.14
CaO*	9.98	7.18	9.64	11.11	7.94	10.00	10.24	8.93	4.68	5.64
K_2O*	3.56	5.41	3.85	5.23	3.90	2.62	2.55	4.05	4.03	2.68
MgO*	6.91	5.86	5.11	5.44	7.24	3.65	3.98	3.57	3.46	4.40
Al_2O_3*	2.74	1.23	3.72	5.09	2.04	1.66	2.34	2.87	6.57	1.99
Fe_2O_3*	0.89	0.51	1.13	0.78	0.55	0.70	0.91	0.64	1.43	0.81
T*	100	100	100	100	100	100	100	100	100	100

Table 3.3.6. Chemical analyses of Silk Road glasses (*cont.*).

	Stein (uncertain)		Hanguya Tati			Togujai	Kelpin	Stein (uncertain)					
	8500	**8501**	**8502**	**8503**	**8504**	**8505**	**8506**	**8507**	**8509**	**8510b**	**8510w**	**8511**	**8512**
SiO_2 d													
SiO_2 a	71.27	71.62	63.89	52.60	66.16	61.40	61.58	65.34	69.32	61.25	56.53	64.21	57.04
Na_2O	18.89	21.10	20.01	18.71	15.25	16.05	17.58	18.50	14.93	16.58	14.35	16.44	16.92
CaO	6.41	5.83	6.89	7.70	5.99	7.11	5.24	3.01	5.15	6.40	8.65	9.46	6.44
K_2O	0.63	0.33	2.14	5.89	4.04	5.18	5.64	7.28	3.40	4.75	4.03	2.12	6.80
MgO	0.71	0.37	3.19	4.37	2.92	3.67	3.62	1.08	6.70	3.69	4.24	3.91	4.25
Al_2O_3	2.27	1.62	2.26	8.82	4.41	4.97	5.10	3.87	0.96	4.19	3.53	1.70	7.41
Fe_2O_3	0.7	0.29	0.48	1.24	0.71	0.67	0.67	1.03	0.29	3.83	0.73	0.37	0.88
TiO_2	0.14	0.070	0.080	0.10	0.16	0.080	0.17	0.17	0.080	0.16	0.15	0.070	0.11
MnO	0.08	0.02	1.43	0.91	0.04	0.71	0.18	0.06	0.31	0.02	0.13	2.23	0.15
CuO													
CoO													
SnO_2													
Ag_2O													
PbO													
BaO		0.05	0.03	0.06	0.11	0.05	0.13	0.05	0.10	0.06	0.06	0.10	0.12
SrO													
Li_2O													
B_2O_3													

(*Continued*)

Table 3.3.6. *(Continued)*

	Stein (uncertain)		Hanguya Tati			Togujai	Kelpin	Stein (uncertain)					
	8500	8501	8502	8503	8504	8505	8506	8507	8509	8510b	8510w	8511	8512
Cr_2O_3													
NiO													
ZnO													
Sb_2O_5													
V_2O_5													
P_2O_5	0.04	0.09	0.33	0.46	0.54	0.42	0.26	0.06	0.07	0.25	0.38	0.28	0.53
SO_3	0.17	0.30	0.25	0.17	0.10	0.34	0.21	0.05	0.33	0.39	0.31	0.14	0.37
Cl	1.06	1.31	0.78	0.50	0.88	0.45	0.63	0.95	0.42	0.51	0.51	0.50	0.70
As_2O_5													
Total	102.37	103.00	101.76	101.53	101.31	101.10	101.01	101.45	102.06	102.08	93.60	101.53	101.72
SiO_2*	70.65	70.80	64.63	52.95	66.51	61.99	61.93	65.27	68.80	60.83	61.41	65.38	57.19
Na_2O*	18.73	20.86	20.24	18.84	15.33	16.20	17.68	18.48	14.82	16.47	15.59	16.74	16.96
CaO*	6.35	5.76	6.97	7.75	6.02	7.18	5.27	3.01	5.11	6.36	9.40	9.63	6.46
K_2O*	0.62	0.33	2.16	5.93	4.06	5.23	5.67	7.27	3.37	4.72	4.38	2.16	6.82
MgO*	0.70	0.37	3.23	4.40	2.94	3.71	3.64	1.08	6.65	3.66	4.61	3.98	4.26
Al_2O_3*	2.25	1.60	2.29	8.88	4.43	5.02	5.13	3.87	0.95	4.16	3.83	1.73	7.43
Fe_2O_3*	0.69	0.29	0.49	1.25	0.71	0.68	0.67	1.03	0.29	3.80	0.79	0.38	0.88
T*	100	100	100	100	100	100	100	100	100	100	100	100	100

- 9 were mixed-alkali glasses that we believe indicate Central Asian origins;
- 1 was a lead:silica glass.

We hope that these graphs will be useful to Chinese scientists who are analyzing glasses from along the Silk Road. Perhaps they can serve as a guideline for the compositional classification of Silk Road glasses. Probably, however, they will have to be revised as other data are considered and new analyses are performed.

4. Conclusion

I apologize if this discussion does not seem entirely new to glass specialists in the audience. The intention was simply to set the scientific stage — a background, that is — for the discussions on more recent results that will follow.

I understand that much progress has been made in glass studies by Chinese archaeologists and laboratory scientists in recent years. As just one example, in the exhibition "China: Dawn of a Golden Age," which just closed at The Metropolitan Museum of Art in New York City, several beautiful and important pieces of glass were shown. This was the first opportunity many Western glass scholars had to learn about these objects. In the catalogue for that exhibition, An Jiayao wrote a fine essay bringing the subject up to date. Her essay is essential reading for anyone interested in the Silk Road.

I look forward to learning more today about how Chinese scholars and scientists view their glass. I sincerely hope that this exchange of ideas will continue and that it will raise opportunities for collaboration in the future. Thank you very much.

Acknowledgments

The author gratefully acknowledges the cooperation of the museums and individuals listed below in providing the samples used for analysis. The scholarly world owes a debt of gratitude to these individuals for the respect and concern they show for the materials entrusted to their care. They are: for the Sven Hedin samples, Folkens Museum Etnografiska, Stockholm (Håkan Wahlquist); for the Paul Pelliot samples, Musée National des Arts Asiatiques-Guimet, Paris (J. Giès, L. Feugère, F. Tissot); for the Albert von Le Coq samples, Museum für Islamische Kunst, Berlin (Jens Kröger); for the Petr Koslov sample, Hermitage Museum, St Petersburg (E. Lubo-Letnischenko); for the Aurel Stein samples, The British Museum (Carol Michaelson and Ian Freestone.) The samples from Apartak and Pendjikent were provided by Boris Marshak and Abdugani A. Abdurazakov. Electron microprobe analyses were performed by Philip M. Fenn and Colleen P. Stapleton. Shana Wilson and Jaci Saunders assisted in the preparation of the graphs and manuscript.

Bibliography

This is a list of the author's publications related to Asian glasses. The publications of the ICOG Congresses are marked with an asterisk. They contain many valuable references to the subject by other authors. Readers wishing to obtain reprints should refer to the reference numbers (A-29, etc.).

A-29 R. H. Brill, Appendix — chemical considerations, in: D. Blair, *A History of Glass in Japan* (Kodansha International and The Corning Museum of Glass, 1973), pp. 448–457.

A-44* R. H. Brill, S. S. C. Tong and D. Dohrenwend, Chemical analyses of some early Chinese glasses, in: F. X. Gan (ed.), *Research in Ancient Chinese Glasses — Proceedings of the International Symposium on Glass* (Beijing, 1984) (Chinese Building Industry Publications, 1986), pp. 15–35, in Chinese.

A-45* R. H. Brill and J. F. Wosinski, Physical properties of early Chinese glasses, in: F. X. Gan (ed.), *Research in Ancient Chinese Glasses — Proceedings of the International Symposium on Glass* (Beijing, 1984) (Chinese Building Industry Publications, 1986), pp. 10–14, in Chinese.

A-50* R. H. Brill, Chemical analyses of some early Indian glasses, in *Archaeometry of Glass — Proceedings of the Archaeometry Session of* *the XIVth International Congress on Glass* (New Delhi, 1986) (Indian Ceramic Society, Calcutta, 1987), Sec. 1, pp. 1–25.

A-51* E. E. McKinnon and R. H. Brill, Chemical analyses of some glasses from Sumatra, in *Archaeometry of Glass — Proceedings of the Archaeometry Session of the XIVth International Congress on Glass* (New Delhi, 1986) (Indian Ceramic Society, Calcutta, 1987), Sec. 2, pp. 1–14.

A-59 R. H. Brill, A possible derivation of the Chinese word *"boli,"* in *Digest — Proceedings of the 1988 Shanghai International Symposium on Glass* (Chinese Ceramic Society and Shanghai Association for Science and Technology, Shanghai, 1988), p. 51.

A-63* R. H. Brill, Thoughts on the glass of Central Asia with analyses of some glasses from Afghanistan, in *Proceedings of the XVth International Congress on Glass* (Leningrad, 1989; Archaeometry) (The International Commission on Glass, 1989), pp. 19–24.

A-64 R. H. Brill, S. S. C. Tong and Zhang Fukang, The chemical composition of a faience bead from China, *Journal of Glass Studies* **31**, 11–15 (1989).

A-65 R. H. Brill, Some thoughts on the origin of the Chinese word *"boli"* *Glass & Enamel* **18**, 1990, pp. 55–58, in Chinese.

A-67* R. H. Brill and J. H. Martin (eds.), *Scientific Research in Early Chinese Glass* (The Corning Museum of Glass, 1991). Contains many of the papers listed below.

A-68 R. H. Brill, Introduction, in: R. H. Brill and J. H. Martin (eds.), *Scientific Research in Early Chinese Glass* (The Corning Museum of Glass, 1991), pp. vii–ix.

A-70 R. H. Brill, S. S. C. Tong and D. Dohrenwend, Chemical analyses of some early Chinese glasses, in: R. H. Brill and J. H. Martin (eds.), *Scientific Research in Early Chinese Glass* (The Corning Museum of Glass, 1991), pp. 31–58.

A-71 P. M. Fenn, R. H. Brill and M. G. Shi, Addendum to Chapter 4, in: R. H. Brill and J. H. Martin (eds.), *Scientific Research in Early Chinese Glass* (The Corning Museum of Glass, 1991), pp. 59–64.

A-72 R. H. Brill and J. F. Wosinski, Physical properties of early Chinese glasses, in: R. H. Brill and J. H. Martin (eds.), *Scientific Research in Early Chinese Glass* (The Corning Museum of Glass, 1991), pp. 109–117.

A-73 R. H. Brill, Some thoughts on the origin of the Chinese word *"boli,"* *Silk Road Art and Archaeology* **2**, 1991/92, pp. 129–136.

A-74 R. H. Brill and P. M. Fenn, Glasswares in Famen Temple, in: *Selected Papers from the First International Symposium on the History and Culture of the Famen Temple* (1992), pp. 254–258 and 2 pp. of figs., in Chinese.

A-76 R. H. Brill, Scientific investigations of ancient Asian glass, *UNESCO Maritime Route of Silk Roads — Nara Symposium '91, Report*, (Mar. 1993), pp. 70–79.

A-79 I.-S. Lee in collaboration with R. H. Brill and P. M. Fenn, Chemical analyses of some ancient glasses from Korea, in *Annales du 12 e Congrès de l'Association Internationale pour l'Histoire du Verre* (Vienna, 1991) (The International Association for the History of Glass, Amsterdam, 1993), pp. 163–174.

A-80 R. H. Brill, The Dominick Labino Fund Lecture: Glass and glass-making in ancient China, and some other things from other places, *The Glass Art Soc. J.* 1993, pp. 56–69.

A-83* R. H. Brill, Scientific research in early Asian glass, in *Proceedings of the XVIIth International Congress on Glass* (Beijing, 1995) (International Academic Publishers, Beijing, 1995), Vol. 1, pp. 270–279.

A-84* R. H. Brill, P. M. Fenn and D. E. Lange, Chemical analyses of some Asian glasses, in *Proceedings of the XVIIth International Congress on Glass* (Beijing, 1995) (International Academic Publishers, Beijing, 1995), Vol. 6, pp. 463–468.

A-91 R. H. Brill, *Chemical Analyses of Early Glasses, Vol. 1: Catalogue of Samples* (The Corning Museum of Glass, 1999).

A-92 R. H. Brill, *Chemical Analyses of Early Glasses, Vol. 2: Tables of Analyses* (The Corning Museum of Glass, 1999).

A-97 R. H. Brill, Chemical analyses of some glasses from the collection of Simon Kwan, in *Early Chinese Glass* (The Chinese University of Hong Kong, Baofung Printing Co., 2001), pp. 448–471.

A-98 R. H. Brill, Some thoughts on the chemistry and technology of Islamic glass, in: D. B. Whitehouse and S. Carboni, *Glass of the Sultans* (The Metropolitan Museum of Art, 2001), pp. 25–45.

C-5 R. H. Brill, Lead and oxygen isotopes in ancient objects, *The Impact of the Natural Sciences on Archaeology* (The British Academy, London, 1970), pp. 143–164.

C-11 R. H. Brill, K. Yamasaki, I. Lynus Barnes, K. J. R. Rosman and M. Diaz, Lead isotopes in some Japanese and Chinese glasses, *Ars Orientalis*, v. 11, 1979, pp. 87–109.

C-13* I. Lynus Barnes, R. H. Brill and E. C. Deal, Lead isotope studies of early Chinese glasses, in: F. X. Gan (ed.), *Research in Ancient Chinese Glasses — Proceedings of the International Symposium on Glass* (Beijing, 1984) (Chinese Building Industry Publications, 1986), pp. 36–46, in Chinese.

C-17 R. H. Brill, R. D. Vocke, Jr, S. X. Wang and F. K. Zhang, A note on lead-isotope analyses of Faience beads from China, *Journal of Glass Studies*, v. 33, 1991, pp. 116–118.

C-18 R. H. Brill, I. Lynus Barnes and E. C. Joel, Lead isotope studies of early Chinese glasses, in: R. H. Brill and J. H. Martin (eds.), *Scientific Research in Early Chinese Glass* (The Corning Museum of Glass, 1991), pp. 65–83.

C-19 R. H. Brill, M. G. Shi, E. C. Joel and R. D. Vocke, Addendum to Chapter 5, in: R. H. Brill and J. H. Martin (eds.), *Scientific Research in Early Chinese Glass* (The Corning Museum of Glass, 1991), pp. 84–90.

C-20 R. H. Brill and M. Chen, A compilation of lead isotope ratios of some ores from China published by Chen Yuwei, Mao Cunxiao, and Zhu Bingquan, in: R. H. Brill and J. H. Martin (eds.), *Scientific Research in Early Chinese Glass* (The Corning Museum of Glass, 1991), pp. 167–180.

C-23 R. H. Brill, C. Felker-Dennis, H. Shirahata and E. C. Joel, Lead isotope analyses of some Chinese and Central Asian pigments, in: N. Agnew (ed.), *Conservation of Ancient Sites on the Silk Road — Proceedings of an International Conference on the Conservation of Grotto Sites* (Dunhuang, Oct. 1993) (The Getty Conservation Institute, Los Angeles, 1997), pp. 369–378.

C-24* R. H. Brill and H. Shirahata, Lead-isotope analyses of some Asian glasses, in *Proceedings of the XVIIth International Congress on Glass*

(Beijing, Oct. 1995) (International Academic Publishers, Beijing, 1995), Vol. 7, pp. 491–496.

C-25 R. H. Brill, C. Felker-Dennis, H. Shirahata and E. C. Joel, Lead isotope analyses of some Chinese and Central Asian pigments, *Sciences of Conservation and Archaeology* v. 12 No. 1, May 2000, pp. 55–62, Chinese transl. of C-23.

C-27* R. H. Brill and H. Shirahata, The Second Kazuo Yamasaki TC-17 Lecture on Asian Glass: Recent lead-isotope analyses of some Asian glasses. International Congress on Glass (Kyoto, Japan; September 29, 2004).

D-5 R. H. Brill, The early history of glass and glassmaking in the Western world. Unpublished outline of lectures delivered in China, Mar. 1982, 6 pp., in English.

D-6 R. H. Brill, The early history of glass and glassmaking in the Western world. Unpublished slide captions of 161 slides used for lectures in China, Mar. 1982, in English.

D-7 S. S. C. Tong, J. C. Y. Watt and R. H. Brill, The early history of glass and glassmaking in the Western world. Unpublished slide captions of 161 slides used for lectures in China, Mar. 1982, in Chinese.

D-8 S. S. C. Tong, James C. Y. Watt and R. H. Brill, A Chinese–English glossary of terms used in the history of glass and glassmaking. Unpublished glossary of 155 terms, Apr. 1982.

Sample Descriptions

Lou Lan, Xinjiang; various dates. Excavated by Sven Hedin, 1903. (H. Wahlquist, Folkens Museum Etnografiska, Stockholm, 9/24/96.) See: A. Conrady, Die Chinesischen Handschriften und Sonstigen Kleinfunde Sven Hedins in Lou-Lan, Stockholm, Generalstabens Litografiska Anstalt, 1920, pp. 173–175, Abt. 3, Tafel III.

6810 Base or wall fragment of thin-walled vessel. Colorless (sl. smoky), eroded overall. T. ~1.5 mm. 1903.26.259A, no. 34.

6811 Foot fragment of small vessel. Colorless, some w. scum. Steep rise in profile. 1903.26.260A.

6812 Foot fragment of small vessel. Colorless, traces of w. scum. Flatter rise than no. 6810. 1903.26.260B.

6813 Foot fragment of small vessel. Colorless with sl. olive tinge. Flat profile. 1903.26.260C.

6814 Vessel fragment, flat ground surfaces, "split" into two layers with sand trapped in crevice. Colorless (sl. smoky). 1903.26.260D.

6815 As above, with ground and polished(?) surfaces. Poss. from same object as no. 6814. 1903.26.260I.

6816 As above, poss. from same object as nos. 6814 and 6815. 1903.26.260J.

6817 Wall fragment of vessel, with "nipt diamond waies" effect. Colorless, eroded. T. varies between 1 and 2 mm. 1903.26.260E.

6818 Neck fragment of vessel. P. green, eroded. 1903.26.260G.

6819 Wall(?) fragment of v. thin walled vessel. Completely colorless. Poss. with threaded design; raised portions eroded, wall well preserved. 1903.26.260K.

6820 P. blue transp. chip. 1903.26.253A.

6821 P. blue transp. chip. 1903.26.253B.

6822 Colorless chip. 1903.26.253C.

6823 Colorless (sl. smoky) chip. 1903.26.253D.

6824 Green transl. chip; filled with minute bubbles. 1903.26.253E.

6825 Green transl. chip; minute elongated bubbles. 1903.26.253F.

6826 P. blue transp. chip. 1903.26.253G.

6827 P. blue transp. chip. 1903.26.257A.

Duldur-Aqur (nr. Kucha), Xinjiang; prob. Sasanian. Excavated by P. Pelliot, 1906. (J. Giès, L. Feugère, F. Tissot, Musée National des Arts Asiatiques-Guimet, Paris.) See: M. Hallade and S. Gaulier, Douldour-Aquor et Soubachi: Mission Pelliot, IV, Paris, 1982, pp. 288–292.

6119 Vessel with facet cutting; Sasanian type. Green, little or no w., but eroded. MG 23736 (P. 595).

6120 Vessel with large applied prunt. Green, little or no w., but eroded. MG 23737 (P. 647).

Qoch homa (Kucha District), Xinjiang; poss. Islamic. Excavated by P. Pelliot, 1906. (As above.)

6121 Rim of large pattern-molded vessel. Green, little or no w., but eroded. MG 24048 (P. 460).

Hazar-tam or Saqal-tam (nr. Khan-Oi, Kashgar), Xinjiang; prob. Roman or Sasanian. Excavated by P. Pelliot, 1906. (As above.)

6122 Rim of a thin-walled vessel. Aqua, little or no w., but eroded. Bubbly. (P. 885, 25/9/1906.)
6123 Rim of a thin-walled vessel. Bl. green, little or no w., but eroded. Bubbly. Unnumbered.
6124 Wall of a thin-walled, pattern-molded object. P. pink, lightly w., but eroded. Unnumbered.
6125 Vessel (or lamp?) with applied and trailed prunt. P. olive, eroded.
6126 Wall fragment of vessel with threaded decoration. P. olive with orangy-amber threads. Bubbly. Little or no w., but eroded.
6126a As above, amber glass.
6127 Rim(?) fragment of three hollow threads. Olive, moderately w.
6128 Top of an ear ornament(?) or stem. Orangy-amber, streaky, little or no w., but eroded. MG 23806 (P. 889).
6129 Fragment of an ear ornament(?) or stem. Green, lightly w. and eroded.

Qizil ("Ming-Oi"), Xinjiang; 4th–5th c. (M. Yaldiz, Museum für Indische Kunst, Berlin, courtesy J. Kröger, Museum für Indische Kunst, Berlin.) See: A. von Le Coq, Buried Treasures of Chinese Turkestan, translated by A. Barwell, London, George Allen and Unwin Ltd, 1926. Available in reprint, with introduction by P. Hopkirk, from Oxford University Press, 1985.

6130 Hemispherical bowl with facet-cut disks; Sasanian type. Colorless, heavily w. MIslK III 7101. Excavated by A. von Le Coq, Third Turfan Exped., 1905–07. Found in workshop.
6131 Rim fragment, sliver showing parts of two top registers of facet-cut disks; Sasanian type. Colorless, moderately w. MIslK III 7686a.

Excavated by A. von Le Coq, Fourth Turfan Exped., 1913–14. Found in New Cave under cave of swordbearers, near cave with casettes.

6132 Rim fragment, showing top register of facet-cut disks; Sasanian type. Colorless, moderately w. MIslK III 7686b. From same location as above.

Kucha Oasis, Xinjiang, 6th–7th c. Given to RHB by E. Lubo-Lesnichenko, Leningrad, 7/5/89.

6110 Fragment of medium-sized bead, "nutshell" shape, possibly with shallow longitudinal grooving. Colorless, bubbly, moderately w. From a string of beads excavated by Oldenburg (or Koslov?) in 1905. (Now on exhibition in the Hermitage Museum.)

Apartak, Uzbekistan; 4th–5th c. Given to RHB by A. Abdurazakov, Samarkand, 7/13/89.

6111 Small round bead. Med. blue transp., moderately w. Mound no. 3.

6112 Fragments of small thin-walled, cylindrical bead. Blue transp., heavily w.

6113 Small ellipsoidal bead. Aqua, v. heavily w., little glass remains.

Xinjiang; Sasanian(?). (8/14/98.)

7015 Chip of small cup with facet-like cutting. Colorless, w. scum. Reminiscent of piece excavated by Sir Aurel Stein. (Same as Pb-3449.)

Pendjikent, Tajikistan; 8th–9th c. (B. Marshak and V. Raspopova, HM, 4/13/93.) See: B. L. Marshak and V. I. Raspopova, "A Hunting Scene from Panjikent," Bulletin of the Asian Institute, v. 4, 1990, pp. 77–94. See also the same authors' forthcoming book on the site.

6250 Flask with trailed and pincered decoration on sides and molded and/or trailed face on flattened side. Colorless (or p. amber?) with black w. crust. H. ~10 cm. No. 55-I-21.

6251 Fragment of bowl with vertical ribs. P. grn. blue transp. No. 70-XXIV-19; 71-XXIV-7/17; Fig. V-6.

6252 Base of bottle with trailed zigzag decoration around side. Yellowish amber(?) with black w. scum. No. 57-III-12.

6253 Base of bottle with horizontal spiraling, threaded decoration. Colorless, v. heavily w. No. 79-XXIII-8; Fig. XV-10.

6254 Base of bottle with large zigzags of applied decoration. Colorless, with black w. crust. No. 82-XXV-4.

6255 Shoulder and neck fragment of bottle with pincered spiraling, threaded decoration. P. grn. aqua, heavily w. No. 71-XXIV-8; Fig. XIV 3.

6256 Neck fragment of bottle with applied ring decoration. Aqua, with iri. No. 64-XIX-2.

6257 Base fragment of bottle(?) with heavy, applied, zigzag decoration. Colorless, with black w. crust. No. 57-III-11.

6258 Fragment of "Sasanian cut bowl." Colorless, no apparent w. on sample. No. 75-XXVIII-4; Fig. I-1.

6259 Fragment of unidentified type. Colorless, with black w. crust. No. 81-XXV-3.

These samples were among the small finds excavated by Sir Aurel Stein in 1902–07. The parent fragments are in the Stein collections of the British Museum. At the request of R. H. B., the samples were taken and provided by Dr Ian Freestone in collaboration with Dr Carol Michaelson. Received on 4/24/00.

8500 Site uncertain. Fragment of cut vessel. Colorless, lightly w. or eroded. MAS 576; slide no. 26b.

8501 Site uncertain. Fragment of cut vessel. Colorless, lightly w. or eroded. MAS 697; slide no. 26a.

8502 Site uncertain. Fragment of cut vessel. Purple, lightly w. or eroded.

8503 Hanguya Tati. Fragment of vessel. Amber, lightly w. or eroded. 1907, 11-11.240; slide no. 14.

8504 Hanguya Tati. Fragment of molded(?) vessel. Bl. aqua, lightly w. or eroded. 1907, 11-11.243; slide no. 14.

8505 Togujai. Fragment of vessel. Colorless, lightly w. or eroded. 1907, 11-11.14; slide no. 36.

8506 Kelpin. Glass bead. Yellow opq., moderately w. or eroded. Longitudinal section flow lines suggest the bead was formed by pincering a drawn tube. MAS 11163; slide no. 19. (Same as Pb-3408.)

8507 Site uncertain. One of two small pendants (poss. hollow?). Dk. blue, little or no w. MAS 1009; slide no. 27.

8508 Niya. Finial(?) with millefiori inlays. Thin strips of yellow opq. and red opq. on one flat side; body consists of amber colored phase with mineral inclusions. N XIViii 0035; slide no. 31.

8509 Site uncertain. Small glass bead, cylindrical with collar on one end. Aqua, little or no w. MAS 141; slide no. 33a.

8510 Site uncertain. Small bead, cylindrical. Black with white stripe, lightly w. MAS 147; slide 33b.

8511 Site uncertain. Molded glass seal(?). Greenish, lightly w. 1902, 12-20.159; slide no. 34a.

8512 Site uncertain. Glass stem(?). Green, lightly w. 1902, 12-20.162; slide no. 34b.

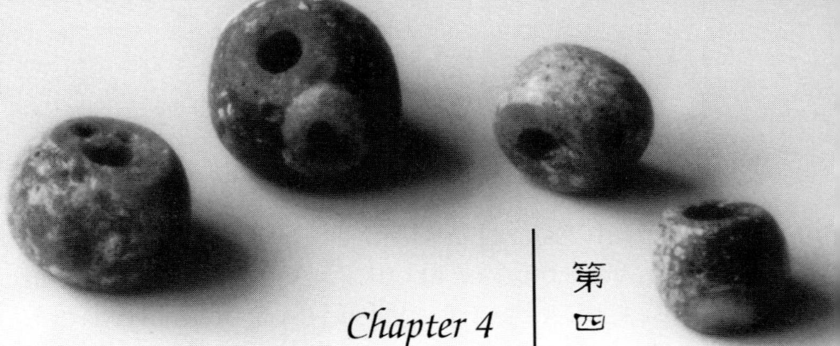

Chapter 4 第四章

The Second Kazuo Yamasaki TC-17 Lecture on Asian Glass: Recent Lead-Isotope Analyses of Some Asian Glasses with Remarks on Strontium-Isotope Analyses

Robert H. Brill

The Corning Museum of Glass, Corning, New York, 14830, USA

Hiroshi Shirahata

Muroran Institute of Technology, Muroran 050, Japan; ret.

1. Introduction

This paper is dedicated to Prof. Kazuo Yamasaki, a pioneer in the scientific investigation of archaeological artifacts and works of art, especially those found in Japan. Both of the authors of this paper have benefited from long friendships and professional collaboration with Prof. Yamasaki. Professor Yamasaki is a founding member, and a member emeritus, of TC-17 (Archaeometry of Glass), and was the first President of The Blair Society.

At previous meetings of the International Commission on Glass (the International Symposium in Beijing [1984],[1] and the Glass Congresses in New Delhi [1986],[2] Leningrad [1989],[3] and Beijing [1995][4]), the authors and their colleagues reported chemical analyses and lead-isotope analyses of numerous ancient glasses and related materials from East Asia, South Asia, Southeast Asia, and Central Asia. The lead-isotope data reported in those studies involve a total of about 75 Asian glasses. For archaeometric purposes, the objective of lead-isotope analysis is to classify artifacts according to which have similar isotopic compositions and which have differing compositions. This in itself can be very useful. Beyond that, comparisons with galena ores provide evidence which, under favorable conditions, can suggest possible geographical origins of the artifacts themselves. The method is independent of the chemical histories of the materials (providing no lead has been introduced from external sources). The analyses require the sacrifice of only minute samples.

As with all archaeometric methods, this one has its limitations. Leads from different mining regions sometimes have similar isotope ratios (the overlapping effect) and recycling of old metal can yield isotope ratios intermediate between those of the starting leads (the mixing effect). Nonetheless, lead-isotope analyses have been applied usefully to a wide variety of ancient artifacts of widely differing sources and dates.[5] The method has been notably successful in the study of Asian glasses. Most of the Asian glasses analyzed have been of Chinese origin, judging from their typologies and sources. But numerous glasses found in Japan, Korea, the Indian subcontinent, and Southeast Asia have also been analyzed.[6]

In many cases, the ratios found for Chinese glasses match those of galena ores occurring in China. Because these ratios are usually distinctly different from those of leads in glasses known to have been made in the West, lead-isotope analyses are especially useful for distinguishing between glasses made in Asia and those found in Asia, but imported from the West. Similarly, the leads in certain Korean and Japanese artifacts match Korean and Japanese ores.[7]

2. Results of Analyses

Data are reported here for 48 additional Asian glasses. The samples are described briefly in Table 4.1. Many of the samples came from objects donated for study by Dr Simon Kwan of

Table 4.1. Lead-isotope ratios of some Chinese glasses.

Pb no.; description	$^{208}Pb/^{206}Pb$	$^{207}Pb/^{206}Pb$	$^{204}Pb/^{206}Pb$	CMG anal. no. and notes
3450 cicada, turbid green	2.16373	0.88147	0.056779	(6720); PbO:BaO
3451 cicada, turbid white	1.91654	0.722056	0.044795	(6721); PbO:BaO
3452 cicada, turbid white	2.02613	0.778176	0.04908	(6722); PbO:BaO
2357 molded inlay, dk. blue	2.21554	0.92751	0.060931	(6700); PbO:BaO
2358 molded inlay, dk. blue	2.21176	0.926789	0.061043	(6701); PbO:BaO
3459 biconical bead, lt. blue	2.21565	0.933844	0.061641	(6734); PbO:BaO
3461 eye bead, dk. green	2.24659	0.93793	0.061660	(6742); PbO:BaO
3462 eye bead, w. eye	2.22106	0.92369	0.060288	(6745); PbO:BaO
3460 eye bead (fritted), w. eye	2.08902	0.84862	0.054454	(6737); PbO:BaO
3465 beads, em. green	2.18567	0.87955	0.056696	(6772); PbO:BaO
3475 eye bead (fritted), blue	1.94522	0.79704	0.050872	(6736); PbO:BaO
3453 plaque, turbid white	1.91599	0.714972	0.044370	(6723); PbO:BaO
3454 ear spool, blue	2.03340	0.82613	0.052375	(6724); $K_2O:SiO_2$. Some Th
3455 ear spool, blue	1.72134	0.69439	0.041411	(6725); $K_2O:SiO_2$. High Th
3472 bead, purple	2.10616	0.85947	0.054624	(6769); $K_2O:SiO_2$
3466 mellon bead, yellow transp.	2.09466	0.84562	0.053737	(6773); PbO:BaO
3471 bead, dk. blue	2.11643	0.86558	0.054849	(6768); PbO:BaO
3473 bead, lt. blue	2.15766	0.875709	0.056192	(6770); $K_2O:SiO_2$

(*Continued*)

Table 4.1. (*Continued*)

Pb no.; description	$^{208}Pb/^{206}Pb$	$^{207}Pb/^{206}Pb$	$^{204}Pb/^{206}Pb$	CMG anal. no. and notes
3465 bead (?)	1.90732	0.71112	0.04409	(6772?); PbO:BaO
3465 bead, bl. green	2.18567	0.87955	0.05670	(6772); PbO:BaO
3463 cubic eye bead, turbid white	2.21480	0.927401	0.06084	(6764); PbO:BaO
3464 cubic eye bead, lt. blue	2.19070	0.89309	0.05755	(6765); PbO:BaO
3448 *bi.* CMG (Strauss), turbid	2.11389	0.807448	0.051060	PbO:BaO (?)
2342 small bead, dk. blue	2.10800	0.86101	0.054696	$K_2O:SiO_2$, CoO
2343 small bead, dk. blue	2.11007	0.86142	0.054648	$K_2O:SiO_2$, CoO
2347 small bead, dk. blue	2.11053	0.86153	0.054612	$K_2O:SiO_2$, CoO
2345 small bead, dk. blue	2.11211	0.86403	0.054768	$K_2O:SiO_2$, CoO
2346 small bead, dk. blue	2.10949	0.86338	0.054837	$K_2O:SiO_2$, CoO
2344 small bead, dk. blue	2.11011	0.86351	0.054819	$K_2O:SiO_2$, CoO
2348 small bead, dk. blue	2.10609	0.8604	0.054714	$K_2O:SiO_2$, CoO
2349 small bead, dk. blue	2.10621	0.86077	0.054747	$K_2O:SiO_2$, CoO
2350 small bead, dk. blue	2.10983	0.86325	0.054813	$K_2O:SiO_2$, CoO
2351 small bead, dk. blue	2.11089	0.86347	0.054789	$K_2O:SiO_2$, CoO
2359 Qing bowl (crizzl.)	2.11070	0.86032	0.055182	(5874) Not anal.

(*Continued*)

Table 4.1. *(Continued)*

Pb no.; description	$^{208}Pb/^{206}Pb$	$^{207}Pb/^{206}Pb$	$^{204}Pb/^{206}Pb$	CMG anal. no. and notes
3474 Six Dyn./Tang bracelet, blue	2.08025	0.839244	0.053427	(6775); natron, 1.21% PbO, CoO
3467 Ming hairpin, white opq.	2.11222	0.857137	0.05480	(6780); $K_2O:PbO:SiO_2$+Sn
3457 faience bead, blue body	2.18959	0.87516	0.056468	(6729); PbO:BaO
3458 faience bead, buff body	2.18609	0.88296	0.057107	(6730); PbO:BaO (?)
3456 hexaganol rod, grn. blue	2.18524	0.879669	0.056767	(6727); PbO:BaO
3468 Chinese purple rod	2.19918	0.888459	0.05741	(6782); PbO:BaO
3469 Chinese purple rod	2.11542	0.86148	0.055565	
3470 bead, blue, date uncertain	2.10035	0.855198	0.05473	(6783); 10.7% PbO
3447 horse, modern	2.15152	0.873556	0.056453	(6767); natron, 0.23% PbO, known forgery
3404 vessel, green, Lou-Lan	2.07926	0.80351	0.050958	(6825); soda lime, high K_2O, 1.8% PbO
3405 vessel, colorless, Lou-Lan	2.10688	0.84203	0.052890	(6822); $PbO:SiO_2$, 56.4% PbO
3406 vessel, green, Lou-Lan	2.11299	0.86405	0.054747	(6824); soda lime, high K_2O, 1.2% PbO
3407 vessel, colorless, Lou-Lan	2.08919	0.84715	0.054083	(6816); natron; 0.35% PbO
3449 cup, colorless, Xinjiang	2.09250	0.84719	0.053665	(7015); natron+Sb; 0.21% PbO

Hong Kong. Chemical analyses of most of those samples have been published previously in a catalogue of Dr Kwan's glass collection.[8] Most of the other samples have also been analyzed chemically.[9] The samples of glasses from Central Asia were provided by Dr Håkan Wahlquist of the National Museum of Ethnography, Stockholm. They are from the Sven Hedin Collection. The few remaining samples came from objects in the collection of The Corning Museum of Glass. (Donors of the samples reported in our earlier studies are acknowledged in the corresponding publications.)

The analyses were performed at the Muroran Institute of Technology by one of the authors (H.S.) and his colleagues. A Finnigan MAT 262 surface-ionization, solid-source mass spectrometer was used. The data are reported in Table 4.1. Figures 4.1 and 4.2 show the data for the samples in this study separated according to lower and higher isotope ranges. Because of a peculiarity in the software used to plot the data, some of the points that lie close to one another are obscured in the graphs. To improve legibility the points have been drawn larger than they should be.

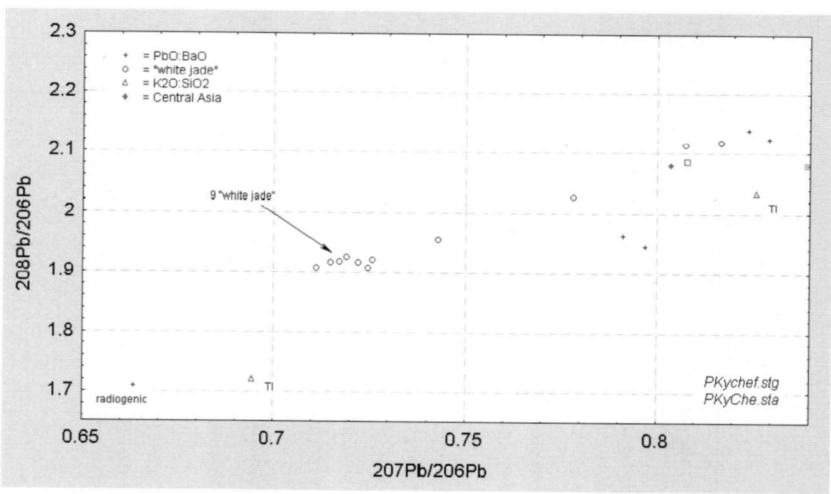

Fig. 4.1. Lead-isotope ratios for Asian glasses (lower range of values).

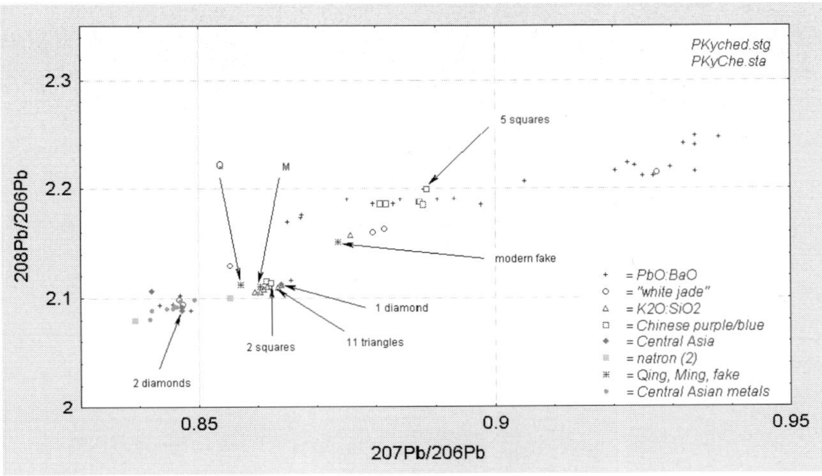

Fig. 4.2. Lead-isotope ratios for Asian glasses, Chinese purple pigments, and Central Asian metals (higher range of ratios).

3. Discussion of Results

As was known from earlier analyses, the range of isotope ratios of leads in Chinese glasses extends from the uppermost extreme to the lowermost extreme of the leads in some 2000 artifacts from all the other locations we have studied; see Fig. 4.3. The ellipses in this figure denoting ranges for Chinese glasses have been published previously, as cited above.[10] The wide range is a consequence of the complex geology of China. (For comparison, leads from the ancient mines at Laurion in Greece would fit easily within the range enclosed by two ticks on the horizontal axes of the graphs shown here.) This is an advantage for studying Chinese artifacts because it increases the chances of uncovering meaningful differences among them, as well as differences between them and artifacts made elsewhere.

Some white translucent glasses. There are nine glasses in the prominent cluster of points in the lower isotope range of Fig. 4.1. All are Han Dynasty, white translucent (or opaque) $PbO:BaO:SiO_2$ glasses. They might have been made to imitate white jade. Although there

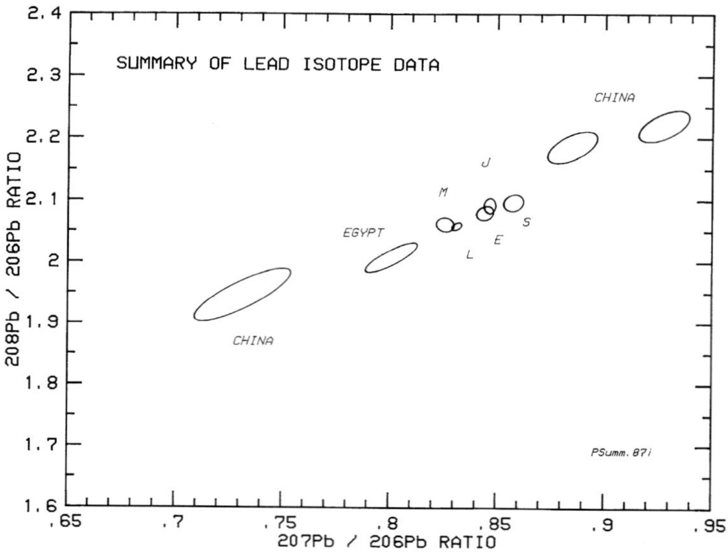

Fig. 4.3. Summary of lead isotope data. There are three main groupings of Chinese glasses, but some (not shown here) also overlap the upper range of European artifacts. More than 85% of all the leads analyzed fall between 0.82 and 0.87 on the horizontal scale, so overlapping often occurs there. L = Laurion; E = Europe; S = Spain; M = Mesopotamia; J = Japan.

are other similarly colored glasses that have higher (and more variable) ratios, we believe these particular nine glasses were made in the same region as one another, possibly even at the same workshop. They include two *bi* disks, a cicada, a sword terminal, a small ornamental plaque, and glass plaques of the sort used to make "burial suits." Close by lies the lead from an ancient glass animal. Leads with similarly low ratios have been reported for Shang Dynasty bronze objects that are believed to have been made in southwest China.[11] These bronzes are several centuries older than the glasses, which reminds us that lead-isotope data tell us something directly about where objects might have been made, not about when they were made.

Some K_2O:SiO_2 glasses. An interesting family of glasses having potash:silica compositions (K_2O:SiO_2), has come to light recently.[12]

Numerous examples have been found in China (said to date back to the Han Dynasty) and in Japan and Korea. Others have been excavated in India, Thailand, Indonesia, and Vietnam. Most of the objects are beads, but they do include a few vessels and pieces of cullet. Additional examples will undoubtedly continue to appear. Overall, the glasses date from about the 1st–2nd centuries BC, until perhaps the 4th century. Some are colored with cobalt while others have the natural aqua color produced by iron impurities. The glasses are remarkably durable, considering that they lack a stabilizer such as lime, but on a microscopic level they often show signs of incipient crizzling that can be developed into severe crizzling by moderate heating.

Figure 4.2 shows the ratios for traces of lead from 14 dark blue examples of such glasses. They span a considerable range. There is a large cluster of 11 points near the center of the graph, but that concentration of points is somewhat misleading, because 10 of those glasses were from a single cache of blue beads. Only one glass in that cluster came from a different source: a purple bead, which also contained some cobalt. These ratios are not a close match for any of the leads in typical heavily leaded, Han Dynasty, $PbO:BaO:SiO_2$ glasses. Assuming that the lead in the potash:silica glasses might have been introduced with the cobalt colorant, the isotope ratios would then be telling us more about the source of the cobalt than about the sources of the glasses themselves. If that is so, the cobalt in those glasses apparently came from a source that had different lead ratios than the lead in the heavily leaded glasses. Only one $K_2O:SiO_2$ bead — no. Pb-3473, which is said to date from the Warring States Period — has lead that matches some of the heavily leaded glasses. It is worth noting, too, that all these beads contain manganese, which is sometimes viewed as an indication that any accompanying cobalt came from China.

Two of the potash:silica glasses are especially interesting (nos. Pb-3454 and 3455). Both pieces are "ear spools," but of slightly differing shapes.[13] They have very similar chemical compositions and are colored with cobalt. Each contains about 0.1% PbO, which could have come in with the cobalt. They have markedly different lead isotope ratios. No. Pb-3454, in particular, has very low $^{207}Pb/^{206}Pb$ and $^{208}Pb/^{206}Pb$ ratios, characterizing it as being radiogenic.[14] Also, both glasses contain detectable levels of thallium, which is unusual. This may indicate that the cobalt colorant — or some batch material — came from southwest China, possibly in the vicinity of Lanmuchang (Xinren county, Guizhou), where thallium mineral deposits occur.

Some Central Asian finds. Five glasses analyzed in this study were excavated in Xinjiang. They include four vessel glass fragments from Lou-Lan that were collected by Sven Hedin just after the turn of the last century. The parent fragments were nondescript and it was not possible to determine what the original vessels were like, so it is not even possible to guess where they might have come from. However, judging from their chemical analyses, it appears almost certain that one of the glasses (Pb-3405, a colorless $PbO:SiO_2$ glass) was of Asian origin. Assuming that the piece was ancient, such compositions are not found among Western glasses. Two other glasses are soda:limes made with alkali derived from plant ashes. They have rather high potassium levels — in excess of 4% K_2O — so it is quite likely that they, too, were made in Central Asia.[15] The fourth glass from Lou-Lan is evidently a natron-based glass. The three alkali:silicate glasses all contain low-to-minor levels of lead, so lead isotope analyses were carried out on them, as well as on the lead:silicate glass. Another fragment from a cup excavated in Xinjiang[16] is a natron-based glass containing a low level of lead, so it was also analyzed. It contains an additive level of antimony, so it could be Roman or Middle Eastern.

The lead from one of these glasses found in Central Asia is plotted in Fig. 4.1. The leads from the other four glasses are plotted in Fig. 4.2, which also shows leads from a group of copper–alloy artifacts from Afghanistan.[17] The results for some of the glasses are in good agreement with those of the metals, which is consistent with their probable Central Asian origins (as based on their chemical compositions.). Leads from a few later glasses of unquestioned Chinese origin also fall close to them. Unfortunately, however, these isotope ratios are in a range that also includes some artifacts from European and other provenances, so overlapping is a complication here. Thus, the results for the two natron-based glasses are ambiguous. On the other hand, we are inclined to believe that the results for the two glasses with the high levels of potash are consistent with their presumed Central Asian origins, because they do not overlap the Western artifacts. The point for the lead:silicate glass falls well above the trend for the Western glasses, and we see it, tentatively, as a match for the Central Asian metals. Sample Pb-3474 is puzzling. It is from a thin, weathered, crudely made bracelet that is reminiscent of Indian bangles. But, unlike Indian glasses, it has a natron-based composition (and contains MnO), so it could well be Roman. It has lead ratios that place it where some Western, Central Asian, and Chinese leads overlap. One wonders if it could have been made somewhere in Asia of recycled Roman glass.

Chinese blue and Chinese purple. Seven of the objects studied here (and previously) are made of Chinese purple, and one is made of Chinese blue.[18] As far as we know, these remarkable synthetic materials are unique to China. They owe their color to the presence, respectively, of two compounds: copper:barium:disilicate (CuO:BaO:$2SiO_2$) and copper:barium:tetrasilicate (CuO:BaO:$4SiO_2$). The most intriguing aspect of Chinese blue and Chinese purple is that they are chemical analogues of the familiar synthetic material Egyptian blue that was widely used throughout Egypt and the Near East from the second millennium BC through Roman times (and possibly even later). Egyptian blue consists largely of the compound copper:calcium:tetrasilicate (CuO:CaO:$4SiO_2$). In the two Chinese analogues, barium can be seen as having replaced the calcium, somewhat as lead and barium had replaced the alkali and lime in Western glasses. There could very well be historical and technological connections between these synthetic materials that were made thousands of miles away from each other. However, unlike their Western counterparts, the Chinese-made materials ordinarily contain minor-to-major levels of lead.

The Chinese purple artifacts analyzed in this study, along with other examples reported previously, all contain leads that are isotopic matches for leads found in contemporaneous Han Dynasty and Warring States Period leaded glasses. The glasses and pigments are certainly connected somehow — probably both technologically and geographically.

Having completed chemical analyses, lead-isotope analyses, and (recently) some strontium-isotope analyses of Chinese blue and Chinese purple — and having successfully synthesized each of them in the laboratory — we plan to continue these lines of research.

There can be no question that the $PbO{:}BaO{:}SiO_2$ glasses found in China were made in China, as were, apparently, some such glasses found elsewhere in East Asia. Similarly, there is no doubt that Chinese blue and Chinese purple were also made in China — and possibly only in China. However, the circumstances that prompted the manufacture of these materials, both the glasses and the pigments, are still not really well understood. This returns us to that most intriguing of all questions about Asian glass — a question we have asked since 1979: Was Asian glass the result of an independent discovery (most likely in China), or could it have been triggered by some sort of contact with the West? Perhaps we shall never know for sure.

4. Strontium-Isotope Analyses

As this book is going to press, we have just published a progress report on a project recently undertaken by The Corning Museum of Glass and the University of North Carolina.[19] This project consists of a survey of strontium-isotope analyses of 325 samples of historical glasses and related materials (such as faience, Egyptian blue, Chinese purple, and batch ingredients). The analyses were performed by Prof. Paul D. Fullagar. The samples came from widespread locations and span 2600 years of glass-making history. The $^{87}Sr/^{86}Sr$ ratios of traces of strontium in the glasses vary according to the geochemical nature of the sources from which the batch materials were obtained. Therefore, they are useful for classifying the glasses. Previously, strontium-isotope analyses have been applied mainly to the study of skeletal remains, but now a few other laboratories are also conducting studies of glasses.

The preliminary findings of our survey are promising. They suggest that strontium-isotope analyses are a valuable — and independent — supplement to chemical analyses for classifying glasses according to their geographical origins. For example, there are small, but significant, differences among Egyptian and Mesopotamian glasses; pronounced differences between natron-based and plant-ash soda glasses; and differences among mediaeval stained glasses made in various parts of Europe. The data are especially useful for comparing individual artifacts with one another when there is a need to decide whether they are somehow related or not.

Certain glasses made in Kopia (and elsewhere in India) have the highest ratios of all, in keeping with the geological sources of the batch materials from which they are thought to have been melted. The results for a series of potash-silica glasses found at various Asian sites suggest that even though their chemical compositions are similar, they were probably made in different places. For example, three potash-silica glasses made in Japan have much lower ratios than the rest of the potash-silica glasses analyzed so far. On the other hand, three potash-silica glasses found in China have very high ratios and are entirely different than those excavated in Thailand, Vietnam, and Korea.

The ratios for seven $PbO:BaO:SiO_2$ glasses form a tight cluster of relatively low ratios — slightly lower than that of modern seawater. They appear to have been made of very similar batch materials and are in fact similar to three examples of Han Dynasty Chinese purple bars and two pieces of faience excavated in China. These results suggest to us that all of these artifacts could have been made in the same workshop. That would be consistent with the findings of our lead-isotope analyses.

The strontium-isotope analyses for 12 glasses excavated in Central Asia indicate that a "Central Asian range" of intermediate ratios may be emerging (based on 8 samples), while 2 glasses with chemical compositions suspected to be "Western" turned out to have lower ratios. One matched the ratios of Roman natron-based glasses while the other matched Islamic and Sasanian plant-ash-based soda glasses. Consequently, the strontium-isotope data are fully consistent with the chemical compositional data. The results for two other glasses from Uzbekistan were indeterminate. We plan to continue this research on additional Asian glasses.

Acknowledgments

The authors thank Shana Wilson and Jacolyn Saunders of The Corning Museum of Glass for their assistance in preparing the graphs for this paper.

References

1. R. H. Brill and J. H. Martin (eds.), *Scientific Research in Early Chinese Glass* (The Corning Museum of Glass, 1991); see esp. R. H. Brill, I. Lynus Barnes and E. Joel, Lead isotope studies of early Chinese glasses, *ibid.*
2. R. H. Brill, Chemical analyses of some early Indian glasses, in *Archaeometry of Glass — Proceedings of the Archaeometry Session of the XIVth International Congress on Glass* (New Delhi, 1986) (Indian Ceramic Society, Calcutta, 1987), Sec. 1, pp. 1–25; E. E. McKinnon and R. H. Brill, Chemical analyses of some glasses from Sumatra, *ibid.*, Sec. 2, pp. 1–14.
3. R. H. Brill, Thoughts on the glass of Central Asia with analyses of some glasses from Afghanistan, in *Proceedings of the XVth International Congress on Glass* (Leningrad, 1989; Archaeometry) (International Commission on Glass, 1989), pp. 19–24.
4. R. H. Brill, Scientific research in early Asian glass, in *Proceedings of the XVIIth International Congress on Glass* (Beijing, Oct. 1995) (International Academic Publishers, Beijing, 1995), Vol. 1, pp. 270–279; R. H. Brill, P. M. Fenn and D. E. Lange, Chemical analyses of some Asian glasses, *ibid.*, 1995, Vol. 6, pp. 463–468; R. H. Brill and H. Shirahata,

Lead-isotope analyses of some Asian glasses, *ibid.*, 1995, Vol. 7, pp. 491–496.

5. For example, our own analyses alone have included over 2000 samples of metallic leads, bronzes, brasses, silvers, gold, glasses, glazes, faience, kohls, white leads and other pigments — from such diverse artifacts as coins, stained glass window caming, oil paintings, wall paintings, mosaics, net sinkers, opus sectile work, etc.
6. R. H. Brill, K. Yamasaki, I. Lynus Barnes, K. J. R. Rosman and M. Diaz, Lead isotopes in some Japanese and Chinese glasses, in *Ars Orientalis*, Vol. 11, 1979, pp. 87–109; R. H. Brill, R. D. Vocke, Wang Shixiong and Zhang Fukang, A note on lead-isotope analyses of faience beads from China, *Journal of Glass Studies* **33**, 116–118 (1991); K. Yamasaki and M. Murozumi, Similarities between ancient Chinese glasses and glasses excavated in Japanese tombs, in: R. Brill and J. Martin, ref. cited in note 1; T. Koezuka and K. Yamasaki, Chemical compositions of ancient glasses found in Japan — a historical survey, in *Proceedings of the XVIIth International Congress on Glass* (Beijing, 1995), Vol. 6, pp. 469–474; M. G. Shi and F. Z. Zhou, Some glasses unearthed from a tomb of the Warring States Period, *ibid.*, Vol. 6, pp. 503–506.
7. R. Brill, K. Yamasaki *et al.*, cited in note 6.
8. R. H. Brill, Chemical analyses of some glasses from the collection of Simon Kwan, in: S. Kwan, *Early Chinese Glass* (The Chinese University of Hong Kong, Baofung Printing Co., 2001), pp. 448–471.
9. R. H. Brill, *Chemical Analyses of Early Glasses, Vol. 1: Catalogue of Samples* and *Vol. 2: The Tables* (The Corning Museum of Glass, 1999).
10. Although we have included here only the $^{207}Pb/^{206}Pb$ and $^{208}Pb/^{206}Pb$ plots, we also routinely plot the ^{204}Pb data. They are not shown here because they do not usually provide any classification information that is not revealed by the other plots. To improve legibility the points on the graphs shown here are plotted much larger than they should be.
11. I. L. Barnes, W. T. Chase and E. C. Joel, in: R. W. Bagley, *Shang Ritual Bronzes in the Arthur M. Sackler Collection* (1987, App. II), pp. 558–560. See also: Z. Y. Jin, A reassertion that the high-radiogenic lead in Shang bronzes originated in southwestern China (announced for the SEAA Conference, June 2004, Daejeon).

12. F. X. Gan (ed.), *Study on Ancient Glasses in Southern China*, 2002 (in Chinese), Huang Qishan, pp. 10–20 and Li Qinghui *et al.*, pp. 76–84; R. Brill, E. McKinnon and R. Brill, two refs. cited in note 2; R. H. Brill, Scientific investigations of ancient Asian glass, *UNESCO Maritime Route of Silk Roads, Nara Symposium' 91*, Report, Mar. 1993, pp. 70–79; R. H. Brill, Chemical analysis of some glasses from Jenné-Jeno, in: S. K. McIntosh (ed.), *Excavations at Jenné-Jeno, Hambarketolo, and Kaniana (Inland Niger Delta, Mali), the 1981 Season* (University of California Press, Berkeley, 1994), Chap. 5, pp. 252–256, and Fig. 38; R. H. Brill, ref. cited in note 4.
13. R. Brill in S. Kwan, cited in note 8, pp. 455–456, and p. 466.
14. See refs. cited in note 11.
15. R. Brill, 1989, cited in note 3.
16. R. Brill, 1999, cited in note 9, Vol. 1, p. 148 and Vol. 2, p. 344 (CMG 7015).
17. R. H. Brill, C. Felker-Dennis, H. Shirahata and E. C. Joel, Lead isotope analyses of some Chinese and Central Asian pigments, in: N. Agnew (ed.), *Conservation of Ancient Sites on the Silk Road — Proceedings of an International Conference on the Conservation of Grotto Sites* (Dunhuang, Oct. 1993) (Getty Conservation Institute, Los Angeles, 1997), pp. 369–378. For faience beads, see: R. Brill, R. Vocke *et al.*, cited in note 6.
18. R. Brill, S. Tong and D. Dohrenwend, 1991, pp. 36–39 and 43–47 in book cited in note 1; P. Fenn, R. Brill and M. Shi, 1991, pp. 59–64, *ibid.*; R. Brill, I. L. Barnes and E. Joel, 1991, pp. 71–79, *ibid.*; R. Brill *et al.*, 1991, pp. 84–89, *ibid.*; R. H. Brill, The Dominick Labino Fund Lecture: Glass and glassmaking in ancient China, and some other things from other places, *The Glass Art Society Journal* 1993, pp. 56–69; E. W. FitzHugh and L. A. Zycherman, *Studies in Conservation* 28, p. 15, 1983; R. Brill, cited in note 9, Vol. 1, pp. 206–207 and Vol. 2, pp. 466–472.
19. R. H. Brill and P. D. Fullagar, *Strontium-Isotope Analyses of Some Historical Glasses and Related Materials: A Progress Report*, AIHV-17 (Association Internationale pour l'Histoire du Verre, Antwerp; Sep. 5, 2006), in press.

Chapter 5 第五章

Glass and Bead Trade on the Asian Sea

Insook Lee

Busan Museum, Korea. 210 UN street, Nam-gu,
Busan, 608-812, Korea

1. The Silk Road – Maritime Trade Route in the Early Christian Era

(1) What is the Silk Road? A network of transportation between East and West
(2) The Silk Road: Oasis Route, Steppe Route and Sea Route
(3) The Maritime Trade Route: India and Rome, Southeast Asia and China
(4) Extension of sea trade to the Far East
(5) Trade on the South China Sea: evidence of the early sea trade

We know that the so-called "Silk Road" developed from ancient times as a product of cultural contacts between the East and the West. By looking back at the history of the Silk Road, we can deepen our understanding of different cultures and relationships among the peoples of different regions.

The name "Silk Road" in modern usage arose from the fascination with cultural diffusion, particularly in the 19th century in Germany and England. It was first used by Baron Ferdinand von Richthofen, a German geologist, traveler and economic historian.

In a paper published in 1877, he coined the term *"Seidenstrassen"* ("Silk roads") in referring to a Central Asian land bridge between China and Europe. He conceived of Central Asia as a subcontinent — a region that not only connected distant civilizations but also provided a source of cultural creativity in its own right. The Silk Road developed because the goods traded were valuable and useful, worth the trouble of transporting them over great distances.

The Silk Road spanned the Asian continent and represented a form of global economy when the known world was smaller but more difficult to traverse than nowadays. It was actually a worldwide network of thousands of miles of land and sea transportation routes for trade traversing regions of Asia, connecting markets and centers of cultural production in China, India, Central Asia, Iran and the Middle East, and extending to those in Europe, Korea, Japan, Southeast Asia and Africa. From these roads there were many terrestrial and maritime extensions, eastward from China to Korea, its old capital Geongju, and across the East Sea to Nara, Japan.

Routes turned northward from China to Mongolia, and southward from China to Burma, into what is now Bengal, and southward from Central Asia through Afghanistan, the Buddhist site of Bamiyan, the mountain passes into Kashmir, Pakistan and India; and northward from the Iranian Plateau through the Caucasus mountain regions.

Silk routes alternatively ran southward along the Persian Gulf, then northward through Turkey to Istanbul, and across the Mediterranean into the Balkans or to Venice. From these points, the network extended still further, to the coastal towns of South India and along the east coast of Africa. Just as it is not one Silk Road, or one historical period or product, it is not one story that conveys the essence of the Silk Road. Scholars working on the Silk Road have found a variety of stories to tell.

In fact, we need to understand that the Silk Road is not one but three main, widely separated routes. These are the Oasis Route, the Steppe Route and the Sea Route. My talk will concern the southern sea route trade, which is generally considered to have begun developing in earlier times than the northern land routes.

For the study of the Sea Route, it is necessary to understand the archeology and history of the South and Southeast Asian Sea coast area. Many archeological finds from Southeast Asia that can be dated from at least the middle of the first millennium BC were probably imported from India or farther to the west. By the late first millennium BC, Southeast Asia was part of a world trading system linking the civilizations of the Mediterranean Basin and Han China.

Before 500 AD, however, there are still very few material items of undoubted Western manufacture found in Southeast Asia in contexts which suggest that they reached there before the beginning of the Christian Era. But the acceleration of archeological surveys and excavations in Southeast Asia, particularly in Thailand and Vietnam, over the past 30 years has produced quite a number of items that can help us to extend the physical evidence for regular exchange systems spanning the Bay of Bengal back to about the middle of the first millennium BC.

This trade was, of course, closely related to what has come to be called the "Southern Silk Road," that great maritime network, an alternative to the northern desert route, which originated at the ports of southern China, passed along the Vietnamese coast, around Cape Ca Mau into the Gulf of Thailand, into western Indonesia, through the Straits of Malacca, with routes spanning the Malay Peninsula, then across the Bay of Bengal to India and the Mediterranean. This multibranched network of trade carried Chinese influence south, into Southeast Asia. Ideas, values and material objects from the west crossed the seas to take root in indigenous cultures, and occasionally to be deposited in the ground as evidence for later archeologists. And we should not overlook the countercurrent of Southeast Asian materials, technologies and values carried to the north and west. The surviving archeological evidence is meager indeed, since the bulk of the items of trade would have been perishable items such as spices, dye woods and marine products, but it should be remembered that Southeast Asia, with its extended coastline and numerous great rivers, was above all a maritime region, its peoples as much at home on the waters as on the land. All the surviving ancient water craft of Asia came from southern Thailand, Malaya and the Philippines.

The history of Chinese trade by sea dates far back and began no later than the overland trade. In China, Guangzhou of Guangdong province and Hepu of the Guangxi Zhuang Autonomous Region are well known as ancient ports. Archeological evidence of shipbuilding and foreign objects like Western types of glassware and Persian silver coins and vessels have been discovered in these regions.

Fragments of a Roman mosaic glass bowl from an Eastern Han Tomb (67 AD) at Ganjiang, Jiangsu province, are one of the earliest examples of Roman glass found in China. Also, glassware from the Roman and Sasanian Empires and the Islamic period, as well as silver coins and glazed pottery from Persia, were discovered at Fuzhou of Fujian province, Ruian of Zhejiang province, Yangzhou and Nanjing of Jiangsu province, Wuwei of Anhui province and Hanyang, Echeng and Anlu of Hubei province.

The glass cup and bowl with faceted cut decoration from the Roman and Sasanian Empires found in Nanjing, and Chungcheng, Jiangsu province, dating from the fourth to the fifth century, have a shape and surface decorations similar to bowls found in Korea and Japan — from the Great Tomb at Geongju in Korea and Tomb 126 of Niizawasenzuka in Japan.

The South China Sea, the main conduit between China and the rest of the world in antiquity and the hub of the world's maritime trade in more recent centuries, is an important geographic entity.

An important point is the question of when maritime trade actually began on the South China Sea. The birth of long-distance maritime routes which served both transport and trade purposes was no later than the late 3rd century BC. The region of Lingnan underwent development from the time of Qin rule to the period of the Nanyue Kingdom in the early Han Dynasty. Panyu (the old name for Guangzhou), first as the seat of the Nanhai prefecture, then as the capital of the Nanyue Kingdom, was the political, military, economic and cultural center of Lingnan. Situated at the mouth of a river, it was a natural port with an extensive hinterland. It developed its maritime transport and commerce, becoming one of the metropolises of ancient China, and the only one that could claim to have maritime trade. Imported items, shipbuilding sites and the depiction of ships on objects excavated in Guangzhou in archeological contexts of this period testify to this achievement.

According to archeologists and anthropologists, a relationship existed in prehistoric times between the primitive peoples who lived on the mainland of China and the racial groups which emerged on the islands in Southeast Asia and even in Australia. As for historic times, there are literary records of the early Yue people of Lingnan sending tributes to the Shang and Zhou court 3000 years ago. This amounted to a kind of indirect or direct commercial relationship, testifying to the maritime transport and commercial activities carried out by the early inhabitants of the Lingnan region.

We can find some written evidence of maritime contacts in Chinese literary sources. In the "Biographies of Wealthy Merchants" section of *Shiji* (*Historical Records*), there is the statement "Panyu was another metropolis, a place where pearls and jade, rhinoceros horns, tortoiseshells, fruits and fabrics were available," which testifies to the fact that Guangzhou in Han times was already one of the economically flourishing metropolises of China.

The "Geography" section of *Han Shu* (*History of the Han Dynasty*) refers to the South China Sea route, attesting to China's close maritime links with Southeast Asia, South India and Sri Lanka in Han times.

The "Territories of the Western World" section of *Hou Han Shu* (*History of the Later Han Dynasty*) records the exchange of envoys between the Roman Empire (specifically, Roman Emperor Antoninus) and Han China.

From the Qin and Han periods, Panyu (Guangzhou) became the starting point of the maritime trade route over the South China Sea and held its position for 2000 years. It owed this position to its superior geographical location, large economic hinterland, plentiful material resources, fine craft production and advanced shipbuilding technology. In the reign of Han Wudi (140–87 BC), the government sent envoys to trade with Southeast Asian and Indian Ocean countries. During the period from the Three Kingdoms to the Southern Dynasties, a new route from Guangzhou to Hainan Island was set up.

There is, by and large, a consensus of opinion as to how the route was developed from period to period. It is generally accepted that in Qin and Han times, because vessels were small and could only sail close to the coast, the farthest they reached was the east coast of the Indian peninsula. There is also concurrence on the subsequent expansion of the route: that by the Three Kingdoms period and the Western and Eastern Jin Dynasties, vessels could have reached the Persian Gulf and the Red Sea; and that by the Tang Dynasty, the route had extended to Africa and Europe. It was in the Southern Dynasties period that silver was imported into China by the maritime route, and Lingnan became a special zone of international trade where gold and silver were used as standard currency.

The period of the Southern Dynasties was a time of incursions by northern tribes and protracted wars in north China, which brought economic hardship to that part of the country. The relative stability enjoyed by Guangdong was conducive to further development of maritime trade and to East–West exchange.

Silk, the most highly prized commodity in the Mediterranean world, was the principal export of the Lingnan region. Spices, gold and silver wares, precious stones and glass wares were imported from overseas to Guangzhou by the maritime route. A glass bowl, gilt bronze cups and strings of etched beads are good examples of imported goods unearthed from Han Dynasty and Southern Dynasties tombs.

Many Buddhist monks arrived at Guangzhou on trading ships or embarked at Guangzhou on their westward voyage. The first to arrive at Guangzhou by sea was the famous Indian monk Jiva in the year 306. The Eastern Jin Chinese monk Faxian had taken the land Silk Route to India in quest of sutras, but returned by the Maritime Silk Route some years later. In the Southern Dynasties period, a succession of Buddhist teachers from Central India, including Gunarata, came to Guangzhou on proselytizing missions.

In the process of preaching, these monks introduced Indian philosophy, literature, medicine, painting, sculpture and architecture.

China was under the rule of the Southern Dynasties and Northern Dynasties in the early fifth century. In those days, the southern part of China was under the control of the Eastern Jin (or Dong Jin) Dynasty and Southern Dynasties, while the north was ruled by the Northern Wei Dynasty, which gradually united the nomadic states called "Wuhu Shiliuguo" ("Sixteen Kingdoms"). The capitals of the Han Dynasty, such as Changan and Luoyang, had already lost their glory, and in the Hexi Corridor along the Silk Road various small local governments started to gain power and occasionally the transportation route was blocked.

While the former capital of the Northern Wei Dynasty, Pingcheng (Datong of Shanxi province today), was being constructed, the transportation route between the Northern Wei Dynasty and the territory of Central Asia was not dependent on the Hexi Corridor, but on the route heading east from Yiwu (Hami of today's Ejinaqi in Inner Mongolia); the route reached Pingcheng, crossing the Mongolian Steppes. Even from Pingcheng, the route was extended farther east, to Yingzhou (Chaoyang of Liaoning province today), Liaodong and the Korean peninsula, then over to Japan by sea. This was really a great route, crossing over the northern steppe of China. However, in ancient documents, not much is mentioned about this route. Today, thanks to recent archeological finds, the existence of the "Steppe Silk Road" has been proven.

Many imported artifacts which might be related to this trade route were found at the archeological sites in China. Sasanian gilded silver plates were unearthed from the Fenghetu Tomb, of the first year of Zhengshi (504 AD) of the Northern Wei Dynasty, at Datong, Shanxi province; gilded bronze and silver stem cups in mixed Sasanian and Byzantine styles, gilded silver bowls with carved decoration, and a cut glass bowl were excavated at the Northern Wei ruins and from a tomb located in the southern suburb of Datong city. Two Sasanian silver coins were discovered at Mangshan, Luoyang, Henan province; Byzantine gold coins were found at Zanhuang, Hebei province and Luoyang, Henan province; a silver ewer with a decoration of Hu-people-like heads and a gilded silver plate were found at Aohanqi, Chifeng, in Inner Mongolia.

Five pieces of Roman glass were found at the Fengsufu tomb of the seventh year of Taiping (415 AD) of the Northern Yan Dynasty, (a kingdom in northeast of China), at Beipiao, Liaoning province. Among them the most famous is a greenish glass vessel in the shape of a duck and other types of Roman glass cups that can be compared to parallels from Silla Dynasty tombs in Geongju, southern Korea.

The "Steppe Silk Road" in northern China was a route long used by nomads when they moved around here and there. This has been proven by the archeological evidence mentioned above. In accordance with the change of time and environment, the trade routes between different regions might have undergone some change.

From the seventh century onward, many Korean and Japanese students and monks traveled to Sui and Tang China. They came to have direct contact with western culture in Changan, the starting point of the overland Silk Road, so what they introduced to Korea and Japan later was Tang culture mixed with Western culture. From Changan eastward, they eventually took ships from one of the following ports: Yangzhou, Mingzhou and Dengzhou (today's Penglai) in Shandong. Then they went back to Japan via Silla on the Korean peninsula. In this way, the Silk Road was extended to the east and finally to Japan.

We know that early in Japan, the inflow of civilization and culture was made mostly over the Korean peninsula, where we have found in Geongju, the capital of the ancient Silla Dynasty, quite a number of relics and artifacts that clearly reflect the cultural exchanges between this kingdom and West Asia.

We also know that the Shosoin, the ancient royal repository at Nara, has many artifacts and products from ancient Persia and elsewhere in West Asia, such as glassware, musical instruments, patterns and designs.

The "Silk roads" that ran through Central Asia could be said to have stretched farther beyond the sea to Japan through China and the Korean peninsula. The Central Asian elements in Korean and Japanese civilizations could have been introduced mostly over the land route through China, but some probably came by the maritime route. Such cultural inflows and exchanges in ancient times were the outcome of the active movements of men and materials among the nations of China, Korea and Japan from the early Christian Era and later, especially during the seventh to ninth centuries.

For 2000 years the Maritime Silk Route was a shipping route by which goods traveled between East and West. It also served as a two-way route for things which we call culture, such as ideas, knowledge and religion.

2. Bead Trade Along Asia's Maritime Trade Route

(1) Beads — an important item of trade in the ancient world
(2) Development of sea trade in East Asia
(3) Scientific analysis of beads
(4) Glass bead manufacturing sites
(5) Global stretch of the bead trade

In Harappa, the ancient Indus Valley site, archeologists found seashells, lapis lazuli, carnelian and other beads that indicate contact with other major urban centers in Arabia, Mesopotamia, Baluchistan, Central Asia, and possibly even China. For them, the Silk Road reaches far back, to somewhere around 2500–3000 BC.

The same land and sea routes that may have carried ancient silk also carried beads as trade items. Following the beads is a way of ascertaining cultural contact and understanding the growth of various centers of civilization.

The western sector of the Asian maritime bead trade was opened before 2000 BC, with Harapans bringing lapis lazuli and carnelian to Mesopotamia and the sea Arabs trading between them. Coral was perhaps exported to India this early. Despite the antiquity of this trade, only in the last few centuries BC did it link the whole area under study. Commerce increased dramatically in the third and second centuries BC after the establishment of powerful empires around the Mediterranean and in Persia, India and China. The Romans, Parthians, Maurya and Han built vigorous, wealthy and ambitious imperial units. Each produced surplus goods, and held a high status. With both surplus and demand, trade expanded dramatically.

The Asian maritime trade has operated continuously since then, despite occasional changes among the major participants. Politics, economics, technology and ecological adjustments affected the movement of goods and people as the trade waxed and waned.

My discussion on the evidence of Asia's maritime trade starts with Indo-Roman trade. By the early part of the Christian Era, Indo-Roman trade routes had brought together the previously rather disparate Southeast Asian exchange systems, linking them in a vast network stretching from Western Europe, via the Mediterranean Basin, the Persian Gulf and the Red Sea, to India, Southeast Asia and China. This period saw the first appearance of what we can recognize as a "world system" of trade linking the great metropolitan centers of the Mediterranean, India and China via their peripheral regions, which generally had less internally integrated political and economic structures. Indo-Roman commercial undertakings seem to have been highly organized and are quite well documented in classical writings dating from the 2nd century AD.

The great expansion of Southeast Asian, and particularly of island–mainland exchange, that is evident in later prehistory is closely connected with this Indo-Roman commerce and can be explained in part, at least, by rising demand for exotic and prestigious items of consumption and adornment in the sophisticated urban civilizations of the Mediterranean Basin, India and, of course, China; that "splendid and trifling" trade in spices, perfumes, precious stones and pearls, silks and muslin, tortoiseshell, ivory and rhinoceros horn, dyes and so on. There is sufficient detail for some historians to have been able to develop a comprehensive and, on the whole, convincing structure for the trade between India and the Roman world as it existed at the beginning of the Christian Era.

However, eastward from India, the data, both historical and archeological, become much more sparse.

Two Roman gold coins dating back to the 2nd century AD have been found at Oc Eo, an important maritime trade site in southern Vietnam. The gold coin of the Roman Emperor Antoninus Pius (138–161 AD) and that of his successor, Marcus Aurelius (161–180 AD), were found at Oc Eo. In addition, inscribed gemstones, rings, medallions and statuary from India and Mediterranean seals have been discovered at Oc Eo and other locations in Vietnam. Since then quite a few other finds have been made or recognized and these are enough to permit us to argue for at least indirect exchange links between Rome and Southeast Asia. Some other archeological finds from Southeast Asia bearing on trade with the West include a copper coin of the Western Roman Emperor Victorinus (268–70 AD) minted at Cologne and found at U-Thong in western Thailand. It is preserved in the National Museum there.

The site of Khlong Thom (also known as Khuan Lukpad, "Bead Mound") in Krabi province, southern Thailand, has become famous for its rich collection of glass and semiprecious stone beads that seem to be related to the west of Southeast Asia. Although maritime bead trade between India and Southeast Asia was already established by the 4th century BC, this trade became increasingly important with the increased wealth generated by the demand for luxury goods in the Roman Empire.

As we saw above, ancient societies were not isolated, but were rather closely connected and related by external trade and contacts with neighboring regions through Asia's maritime routes. In this context, it is necessary to examine the so-called Indo-Pacific bead, which is defined as a small monochrome drawn bead that was frequently unearthed in East Asia and actively traded among ancient societies. During the early first millennium, Indo-Pacific bead trade flourished and extended far to the east, for example to southern Korea and Kyushu, Japan. This is attested to by many excavated examples and also by Chinese written sources of the period describing the highly favored bead tradition of the Three Han people of Korea.

Studying the composition of glass is another way to trace diffusion routes and technology transfer. Ancient glass was produced to combine pure sand or quartz pebbles with some type of flux to lower the melting temperature. Sodium and potassium oxides were the most commonly used fluxes. Potassium glass and the Asian type of soda-lime glass were both present in ancient Korea and Japan, though they were probably made in Southeast Asia or India. Chinese coins at Iron Age sites confirm that southern Korea was part of a large "interaction sphere" by the first century BC, and we can consider the introduction of potassium glass and soda glass beads as part of this interaction. Indeed, some traditions of southern Korea suggest a connection to the south by sea. I mentioned above the connection already established between India and the West, as well as within Southeast Asia. I pointed out the similarities between glass beads found in Korea and those found in Thailand as well as India or Indonesia. We know that southern Korea was the site of extensive iron production in the first centuries BC and AD, and that it was the center of an important trade network. The beads may have been one article of exchange for iron.

Also, we can find more evidence of the trade by examining some special types of glass beads — such as mutisalah beads (sealing-wax red opaque glass beads), gold-foil glass beads, mosaic eye beads, and multifaceted, cornerless cube glass beads, which seem to have been mostly manufactured somewhere in India or Southeast Asia. These were also rather common in Korea from about the 1st century to 500. Among them, the most famous and exciting example is the one mosaic face bead found in Geongju, Korea, which is presently presumed to be an Indonesian Jatim bead. Thus, we cannot deny that there had been a close relationship between Southeast Asia and Far East during the early Christian era.

By using typological comparisons, specialists experienced in handling artifacts like glass or metal objects are sometimes able to distinguish between objects imported from distant places and those made close by. But such comparisons, more often than might be expected, may prove inadequate for this task. There are all too many cases where objects simply do not fall unambiguously into one category or the other. In such cases, chemical analyses, lead isotope analyses or other laboratory studies may indicate whether or not a glass was imported — and sometimes they can even be used for dating glass finds.

Consider the small aqua-colored glass cup excavated from an Eastern Han Dynasty tomb in Guangxi. Various experts have examined the cup but have arrived at differing opinions as to its origin. Some scholars might believe it to be an imported Roman glass, but Dr Robert H. Brill is inclined to see it as differing in subtle ways from Western types of glass that it superficially resembles. A recent test showed that some of these cups are potash-silica glasses made by molding and grinding, so this extremely important object is thought to have been made somewhere in India, Southeast Asia or East Asia (possibly in China).

One of the most significant recent events in the study of Chinese glass was the discovery of several examples of Han Dynasty potash-silica glasses. Scholars have not yet decided whether these glasses were made in China alone or elsewhere in Asia also. The potash glasses include some typical Chinese forms, but in addition there are beads which could just as well have been made somewhere other than China. It will be interesting to learn where that technology originated. One thing, however, is certain: these glasses were not made in the West.

We shall add a remark on one of the most fascinating aspects of the study of Asian glass: that of the glass from ancient Korean, Silla tombs and the glass in Shosoin, Japan. One question remaining about these glass cups and bowls is just how they made their way to Korea and Japan from their place(s) of manufacture. Although some parts of the journey must have been by sea, the initial and longest legs could have been across the steppe and desert routes, where fragments of other such glass have been found. We know many examples of facet-cut Sasanian glasses uncovered in Iraq and Iran, and some excavated in China. The star of the Shosoin glass is the marvelous dark blue cup with ringed decorations. Close parallels have been excavated in Korea, but questions still surround the origins of these glasses. Several authorities see them as imports from the West, while others see them as having been made in Asia, because (to some individuals) the glasses do not really look precisely like their proposed Western parallels.

Close examinations and scientific analyses of glass beads from Asia suggest that there were several major bead-making sites. Analysis of ancient glasses from archeological sites can shed light on ancient cultural exchange networks.

The Indo-Pacific beads mentioned above were widely traded within Southeast and East Asia and even further, to the African sea coast. They are abundant and are the best-preserved trade items from the early first millennium. Special attention should be paid to numerous discoveries of these tiny glass beads in East Asia in terms of cultural contacts between different regions. The southern part of the Korean peninsula, yielding various kinds of glass beads, should be specially considered as an important place in the ancient East Asian bead trade network. In this context, ancient Korean society can be thought as one of the centers for Indo-Pacific beads traded in East Asia by the sea route. Then, some questions are raised and need to be resolved as to where — the exact sites — these beads were made, and when they were brought to the regions. Scientific investigations of these beads can give us clear answers to these questions by comparing the data from the different regions. According to the recent research on Indian and Southeast Asian glass beads, about ten bead-manufacturing sites have been actually identified in India and South Asia, so far. Those sites are thought to be Arikamedu, Karaikadu in India, Mantai in Sri Lanka, Khlong Thom, Takua Pa in Thailand, Oc Eo in Vietnam, Kuala Selinsing, Sungai Mas in Malaysia, and Vijaya in Indonesia. Arikamedu, Karaikadu, Mantai, Khlong Thom, Oc Eo and Selingsing are especially noteworthy sites, as they are considered to belong to the early first millennium AD. Among them, Arikamedu is the most famous as an original site. Arikamedu, Khlong Thom and Oc Eo must be examined relative to the beads traded to the Far East, i.e. to China, Korea and Japan. From my previous scientific researches on the beads, it is presumed that many different types of glass beads with various shapes and compositions were actively traded and imported into East Asian countries, especially the southern part of the Korean peninsula from India, Indonesia, Thailand and Vietnam, during the early Christian Era. Patterns in the bead trade along the Asian maritime routes never changed abruptly; the process took longer through the ages.

Closer investigations and further discussions of the origin and the places of manufacture of each type of beads are needed in the future. Bead research on this trade adds another tool to the kit of those interested in getting a fuller understanding of our shared history.

There seems to have been rather active trade, most likely by sea, between East and Southeast Asian countries. This trade might have reached the southeastern Chinese sea coast, by the Indonesian sea route: from India to Malaysia, Indonesia, Thailand, the Philippines and then China. Early sea trade routes could also have continued further east, to the southern coast of the Korean peninsula. Gaya, an early confederated kingdom in the modern South Geongsang province, Korea, with a highly developed iron culture, seems to have played an important role in this sea trade, exchanging iron for such exotic goods as precious jewelry, tortoiseshell, ivory and glass objects accompanied by new technology and fashion.

On the other hand, many glass artifacts from South Korea have been proven to be typical late Roman glass types that spread throughout the Roman territory. Their distribution could have spread mainly into the northern steppes from the manufacturing sites in the Middle East. These types of Roman glass seem to have been traded via the Silk Road through Central Asia until they finally arrived in the Silla kingdom of Korea. Taking into account the evidence of glass vessels found along the southeastern sea coast of China during the Han and Chin periods, Roman glass vessels may have been first introduced into Korea by sea in early times. But, afterward, it is somewhat strange that late Roman glass vessels are rarely found in China or Southeast Asia, despite the abundance of such vessels found in Geongju, Korea.

Glass products, which may have been one of the most important trade items between East and West, and the knowledge of glass-making techniques of the West, could have reached the Far East via the famed Silk Road. The Sea Route and the Steppe Route were the most probable direct links of glass transport to East Asia.

More detailed investigations of glass beads will provide a clearer map for the influence or contacts among different peoples in East Asia and can deepen our knowledge of the cultural exchange in East Asia.

References

1. C. G. Seligman and H. C. Beck, Far Eastern glass, *Bulletin of Museum of Far Eastern Antiquities* (1938).
2. Robert H. Brill and J. M. Wampler, Isotope studies of ancient lead, *American Journal of Archaeology* **69** (1965).
3. Robert H. Brill, K. Yamasaki *et al.*, Lead isotopes in some Japanese and Chinese glasses, *Art Orientalis* (1979).
4. Robert H. Brill, Scientific investigation of ancient Asian glass. Unesco Silk Road – Maritime Route seminar (Nara, 1991).
5. Robert H. Brill and J. H. Martin, *Scientific Research in Early Chinese Glass* (The Corning Museum of Glass, 1991).
6. I.-S. Lee, R. H. Brill and P. M. Fenn, Chemical analyses of some ancient glasses from Korea, in *Annales of 12 congres de l'Association Internationale pour l'Histoire du Verre* (Vienna, 1991) (Amsterdam, 1993).
7. I.-S. Lee, Ancient glass trade in Korea, Korean material culture, in *Papers of the British Association for Korean Studies*, Vol. 5 (London, 1994).
8. I.-S. Lee, Early glass in Korean archaeological sites, Korean archeology and Korean exodus, *Korean and Korean American Studies Bulletin* **8**(1/2) 1997.
9. I.-S. Lee, *The Silk Road and Korean Ancient Glass, Korean Culture* (Los Angeles, 1993).

10. I.-S. Lee, *Ancient Jade Decorative Objects of Korea, East Asian Jade: Symbol of Excellence* (The Chinese University of Hong Kong, 1998).
11. I.-S. Lee and M. T. Wypyski, Comparison of prehistoric glass beads from Korea and Thailand, *Man and Environment* **27**(1) (2002).
12. I.-S. Lee, A study on ancient lead glass from Korea, in *Proceedings of the 17th International Congress on Glass* (Beijing, 1995).
13. J. W. Lankton, I.-S. Lee and J. D. Allen, Javanese beads in late fifth to early sixth-century Korean (Silla) tombs, in *Annals of the 16th Congress of the Association of History of Glass* (London 2003).
14. M. G. Shi, O. L. He and F. Z. Zhou, Investigations of some Chinese potash glasses excavated from tombs of the Han Dynasty, *Journal of the Chinese Silicate Society*, **14** (1986).
15. M. Stern, *Early Export Beyond the Empire, Roman Glass: Two Centuries of Art and Invention* (The Society of Antiquaries of London, 1991).
16. K. K. Basa, *Early Glass Beads in Thailand* (Southeast Asia Archeology, 1988).
17. K. K. Basa, I. Glover and J. Henderson, The relationship between early Southeast Asian and Indian glass, *Indo-Pacific Prehistory Association Bulletin* **10** (1991).
18. P. Francis Jr, Beadmaking at Aricamedu and beyond, *World Archeology* **23**(1) (1991).
19. J. Henderson, *The Scientific Analysis of Ancient Glass and Its Archaeological Interpretation, Scientific Analysis in Archaeology and Its Interpretation*, Oxford University Committee for Archaeology Monograph, Vol. 19 (1989).
20. I.-S. Lee, Silk Road trade and Roman glass from Korea, *Central Asian Studies* **6** (2001), The Korean Association for Central Asian Studies.

Chapter 6 | 第六章

Characteristics of Early Glasses in Ancient Korea with Respect to Asia's Maritime Bead Trade

Insook Lee

Busan Museum, Korea

210 UN Street, Busan, 608-812, Korea

1. Introduction

A great number and variety of glass beads have been found at numerous Iron Age sites in Korea. The types vary not only in shape and color, but also in chemical composition. Glass objects made with different components indirectly indicate their different places of manufacture and their origins.

It is presumed that every archeological site in South Korea yields varieties of glass beads that were not made in Korea, but could be considered as traded, imported ones. Most of them can be identified as so-called Indo-Pacific beads.

Indo-Pacific beads, which are defined as small monochrome drawn beads, were widely traded within South and East Asia and even farther, to the African coast. They are abundant and are the best-preserved trade items from the early first millennium. Special attention should be paid to the numerous discoveries of these tiny glass beads in South Korea in terms of cultural contacts between different regions. In this context, ancient Korean society can be considered as one of the centers for Indo-Pacific beads traded in East Asia.

From previous research, it has been proven that ancient Korean glasses can be classified into such major compositional categories as lead glass, potash glass, soda glass and mixed-alkali glass. Distinct variations in composition have been observed with the passage of time.

The transition of these kinds of glass can be briefly described as follows. Lead glass (with barium) and potash glass preceded soda glass. Soda glass became an important composition of glass beads during the Three Kingdoms period. Afterward, mixed-alkali glasses were found. High-lead glasses became dominant in the Unified Silla Dynasty.

Questions are raised and need to be resolved as to where — the exact sites — these beads were made, and when they were brought to Korea. Recent scientific investigations of these beads can give us clear answers to these questions by comparing the data from the different regions. Several meaningful issues on the study of early Korean glass should be studied in future research on East Asian glass.

2. Discussion

2.1. *Lead glass*

In Korea, two different kinds of lead glass have been identified: lead glass with barium and high-lead glass without barium. High-lead glasses are divided into two groups: the 60–70% lead-content group and the 30–40% lead-content group.

The earliest glasses in Korea have been identified as the lead-barium-silica type. The blue tubular beads from the stone chamber tomb in Hapsongri, dated back to the 2nd century BC, were proven to be closely related to contemporaneous Chinese glass in their chemical composition and lead isotope ratios, as well. This means that the earliest type of glass in Korea was of Chinese origin. Also, considering the iron assemblage accompanying the earliest glass beads, the initial introduction of glass into Korea apparently coincided with the arrival of the iron culture from China. These earliest glass objects in Korea are thought to have been manufactured with Chinese materials.

Next, Korea started to use a high-lead composition glass without barium around the 4th–5th centuries. Afterward, this type of glass became prevalent during the Unified Silla Dynasty, contemporaneous with the Tang Dynasty in China. Many glass artifacts (like *sarira* reliquary bottles) discovered in Korea were made with this kind of lead glass. Clay crucibles and clay molds for glassmaking were found and tested, and these objects are clear evidence for the making of lead glass in Korea. Considering the earlier dates of the appearance of high-lead glass in Korea, questions on where and when this type of lead glass was made still remain to be answered.

2.2. *Potash glass*

Utilization of glass material in ancient Korea began with the appearance of colorful beads mainly made with potash glass. These were first introduced around the 1st century AD, later than the introduction of lead-barium glass. What is the true identity of this type of potash glass and where did it originate?

According to recent research on Indian and Southeast Asian glass beads, about ten bead-manufacturing sites have so far been identified in India and South Asia. These are thought to be Arikamedu and Karaikadu in India, Mantai in Sri Lanka, Khlong Thom and Takua Pa in Thailand, Oc Eo in Vietnam, Kuala Selinsing and Sungai Mas in Malaysia, and Vijaya in Indonesia. Arikamedu, Karaikadu, Mantai, Khlong Thom, Oc Eo and Kuala Selingsing are especially noteworthy sites, as they are considered to belong to the early first millennium AD. Among them, Arikamedu is the most famous as an original site. Arikamedu, Khlong Thom and Oc Eo must be examined relative to the beads traded to Korea.

In my previous works, I assumed that this type of potash glass (10–20% of K_2O) was derived from Han China, as Chinese potash glass of this period is almost the same in its composition. But we do not have any direct information about the place(s) where this type of glass was made. The analysis of the Indo-Pacific glass beads from Arikamedu raises more speculation about the potash glass.

One especially interesting fact is that Korean potash glass in dark blue or purplish blue contains more manganese (MnO) on average than others. Peter Francis Jr noted that the translucent dark blue glass beads at Arikamedu and Karaikadu have elevated amounts of manganese (about 1.5%) in a potassium glass, which, with small amounts of cobalt, yields the color. Also, he said *wad* containing cobalt and *wad* not containing cobalt were used for dark blue and violet glass, respectively. If this is true, it very much suggests a viewpoint for the interpretation of Korean ancient glass. Considering the large numbers of dark blue glass objects found in Korea (the most dominant color of glass is dark blue), this can be a most significant clue to ancient cultural contact between the regions.

2.3. *Soda glass*

Research on the comparison of glass beads from Korea and Thailand (Khlong Thom) gives new evidence of a possible close relationship between the eastern Indian Ocean region and the Iron Age, pre- and proto-Three Kingdoms period in Korea.

One opaque yellow glass bead from Korea that was excavated from the Misari site near Seoul, dated 1st century, was analyzed and found to be similar to a yellow bead from Khlong Thom, especially in the presence of lead-tin yellow. This kind of opaque yellow drawn bead has a soda glass composition with low levels of magnesium and calcium, a very high level of aluminum and a small amount of lead. This appears to contain the yellow colorant-opacifier lead-tin yellow. The other soda glasses in opaque red from both sites have similar compositions.

2.4. *Indonesian Jatim glass beads in Korea*

Jatim beads are known for their strong association with far-east Java (*Jawa Timur* in the Indonesian language), although their rare appearance at sites ranging from Berenenike, Egypt, to Japan suggests widespread, if limited, trade during the first millennium. Dating for the Jatim beads has been difficult, due to the paucity of examples from scientific excavations.

Four Jatim beads found in Silla Kingdom tombs in Geongju, dating from the late 5th to the early 6th century, provide the earliest well-established dating for Jatim beads. This finding implies a trade of Jatim beads to Korea and further suggests that the famous mosaic glass bead with images of faces, birds and trees, found in the vicinity of a large mound tomb traditionally associated with King Michu, is consistent with a possible Indonesian origin as well.

Indonesian beads found in Korea clearly suggest direct contacts between these regions at early dates. Archeologically, this signifies an absolute clue to influence from the southern sea to the Korean society at the foundation of an early state in Korea.

2.5. *Coil beads*

Many glass coil beads (wound beads) in translucent or transparent blue or colorless glass are known from the archeological sites in Korea. Coil beads have generally been understood to be Chinese products. They flooded the market as Indo-Pacific beads were disappearing in the 12th century. However, the presence of coil beads at early Korean sites should be taken into account. The idea that most coil beads are made of lead glass must be reconsidered.

2.6. *Cornerless cube beads, gold-foil glass beads and melon beads*

Black or dark blue cornerless cube glass beads that are considered by-products of Indo-Pacific beads have been found in Korea.

Gold-foil (or silver-foil) glass beads with West Asian or Egyptian origins are also important glass items found at Korean archeological sites.

Many blue melon beads have been found as well.

Although these special shapes of glass beads are rarely found in other regions of East Asia, they are frequently uncovered at Korean sites. Each shape should be examined in detail, with a careful investigation of the origins and places of manufacture. More attention and further research are needed in this area.

3. Final Remarks

As mentioned above, in terms of Asia's maritime bead trade, special attention should be paid to the numerous discoveries of glass beads at archeological sites in the southern Korean peninsula.

The smallest beads may well require the most work to uncover their story. The rewards, however, are potentially very great.

Ancient Korea was not an isolated kingdom, but was closely connected by external trade and contacts with neighboring regions through Asia's maritime routes. We have to recognize that ancient Korea was one of the largest markets for the Indo-Pacific bead trade during the early first millennium, as attested to by excavated objects and by Chinese written sources of the periods describing the highly favored bead tradition of the Three Han people of Korea.

More detailed investigations of glass beads should provide a clearer map for the contacts among different peoples in East Asia. This topic seems to be the most crucial one to be examined and to be resolved in the study of Korean archeology. There is still much more work to be done.

References

1. Robert H. Brill, Scientific investigation of ancient Asian glass. Unesco Silk Road — Maritime Route Seminar (Nara, 1991).
2. I.-S. Lee, R. H. Brill and P. M. Fenn, Chemical analyses of some ancient glasses from Korea, *Annals of the 12th Congress of the International Association of History of Glass* (Vienna, 1991).

3. K. K. Basa, I. Glover and J. Henderson, The relationship between early Southeast Asian and Indian glass, *Indo-pacific Prehistory Association Bulletin* **10** (1991).
4. I.-S. Lee, *The Silk Road and Korean Ancient Glass, Korean Culture* (Los Angeles, 1993) (Winter, 1993).
5. I.-S. Lee, Ancient glass trade in Korea, *Papers of British Association for Korean Studies*, Vol. 5 (London, 1994).
6. I.-S. Lee, A study on ancient lead glasses from Korea, in *Proceedings of the 17th International Congress on Glass* (Beijing, 1995).
7. T. Koezuka and K. Yamasaki, Chemical composition of ancient glasses found in Japan, in *Proceedings of the 17th International Congress on Glass* (Beijing, 1995).
8. I.-S. Lee, Early glass in Korean archeological sites, *Korean and Korean American Studies Bulletin* **8**(1/2) (1997).
9. S. Gupta, New analyses of Indo-Pacific beads and glass waste from Arikamedu, India, *BSTN* **35**, 8–9 (2000).
10. K.-H. Kim, *A Study of Archeological Chemistry on Ancient Glasses Found in Korea* (Choongang University, Seoul, 2001), in Korean.
11. I.-S Lee, *Silk Road Trade and Roman Glass from Korea, Central Asian Studies* (The Korean Association for Central Asian Studies, 2001).
12. I.-S. Lee and M. T. Wypyski, Comparison of prehistoric glass beads from Korea and Thailand, *Man and Environment* **XXVII**(1) (2002).
13. Peter Francis Jr, *Bead Trade: 300 BC to the Present* (University of Hawaii Press, 2002).
14. J. W. Lankton, I.-S. Lee and J. D. Allen, Indonesian glass beads in early sixth century Korean tombs, in *Annals of the 17th Congress of the International Association of History of Glass* (London, 2003).

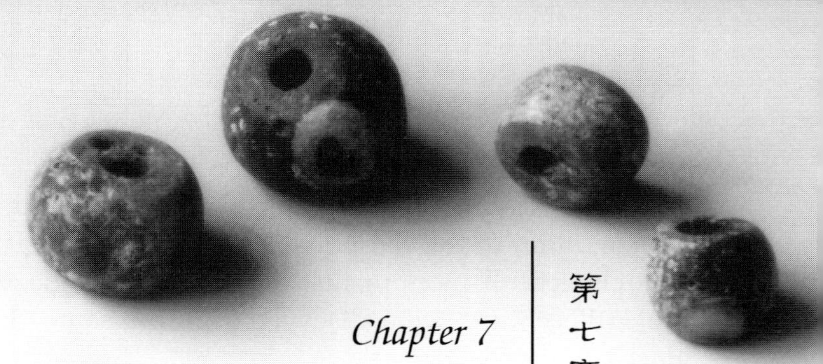

Chapter 7 第七章

Ancient Lead-Silicate Glasses and Glazes of Central Asia

Abdugani A. Abdurazakov

National Institute of Arts and Design Named After K. Bekhzod,
St. Academic Rajabiy, 77, 700031 Tashkent, Uzbekistan

1. Introduction

Up to now, considerable quantities of artifacts and analytical materials on glasswork of Central Asia in ancient and medieval periods[1–5] have been accumulated. Glassware has been uncovered at more than 400 archeological sites, dating from the second century BC to the Middle Ages (15th–17th century AD). A great deal of these finds (more than 600 samples) were subjected to quantitative chemical and spectrographic analyses. On the basis of generalization of the analyzed results, more than 20 chemical composition types of ancient glassy wares spreading on the territory of the Central Asian region have been established. Most of them are alkali(soda and potassium)-calcium-magnesium-alumina-silicate glasses. But at one of the archeological monuments — in ruins of the former capital of Parphian State (the ancient town Old Nisa in southern Turkmen) — glassy wares containing lead oxide in their composition were found. The lead-containing glasses were not discovered anywhere in Central Asia. But this does not mean a lack of demand for lead-containing glasses. Gradually, from the medieval period, lead oxide was applied to produce flushing domestic ceramics — in the condition of the component part of alkaline glazes. Especially in architectural monuments, lead-containing glazes dating from the 11th century were discovered at the same time as alkaline glazes.

This paper is devoted to chemical research on lead-containing glassy wares found at Old Nisa (2nd century BC up to AD) and architectural glazes found at all monuments dating from the 11th to the 17th century AD.

For the study of local and chronological peculiarities of ancient and Middle Ages glasses, their technological parameters in archeology and analytical methods are also widely used in the research. We used quantitative spectrographic and chemical analyses in this study. Thanks to the high precision of both methods, especially the latter, this has increased our knowledge of the glass-making and communication between many countries of the world. This research has been enriched with new radioisotope methods for analyses of lead, oxygen and other elements.

On the basis of analytical researches, the chemical classification of glasses can be established according to the contents presented in a sample in quantity 3% and more of glass-forming components. In this paper, the results of chemical analyses for 8 samples of ancient glasses and 28 samples of Middle Ages glazes from 14 architectural monuments of Central Asia are presented.

2. Experimental

The samples weighing more than 5 g were chosen for analysis. Various types of glass ornaments, beads, eye beads, preformed insets for inlaying, vessel fragments, rhombic tiles, and production waste were analyzed. The samples were pulverized. Some fragments were taken from them, and broken into pieces and pounded into a powdery state with the weight more than 5 g. Silicate analysis was carried out by the traditional method. All these glass finds date back to the 2nd century BC. In the opinion of the specialists, these discovered multicolored rhombic artifacts served as insets for an ornament of wood furniture — the imperial throne.

A sample of glazes was prepared as follows. The samples were removed with a saw from ceramic tiles, with the thickness of the ceramics about 5 mm. Then this glaze layer was mounted on a flat glass plate with an adhesive. After the adhesive reached callosity, the ceramic layer was removed by grinding with an abrasive wheel till the glaze layer appeared. During the processing in hot water, the glaze layer separated from the glass plate, and was then dried and pulverized. The experimental results of chemical analyses are shown in Tables 7.1 and 7.2. According to the contents of basic glass-forming chemical components which were presented in glasses and glazes at the level of 3% and more, the classification of chemical types was performed (Table 7.3).

3. Results and Discussion

According to the contents of basic glass-forming compositions from the analyzed data, the ancient glass from Old Nisa can be divided into five chemical types (Table 7.1). Among them are three compositions (Table 7.1 — 1, 6, 7) with minimum contents of lead oxide (1.28–1.63%); middle — 2.37–2.80% (Table 7.1 — 3, 8); and containing a high percentage of PbO 3.83–4.37% (Table 7.1 — 2, 4, 5). Besides that three samples (Table 7.1 — 2, 5, 8) contained a quantity of the coloring agent assisted by copper oxide (Table 7.1 — 1), also tin oxide (0.04–1.18%) served as the emulsion agent in glass, and in a white sample (Table 7.1 — 3), tin oxide was also used as the colored agent (SnO_2 = 1.40%). Of eight samples, three contained antimony oxide (Sb_2O_3 = 0.50–2.52%). The reason for containing Sb_2O_3 in the glass composition has not been found out yet.

Table 7.2 indicates that Middle Ages lead-containing glazes discovered at architectural monuments of Central Asia can be divided into 10 chemical types, according to the basic glass-forming components. Most of the presented compositions were in the field of glass-making regions in ancient and Middle Ages periods. This

Table 7.1. Chemical analyses of the glasses from Old Nisa (2nd c. BC) (in weight percentage).

No.	Names and color samples	SiO_2	Al_2O_3	Fe_2O_3	CaO	MgO	TiO_2	SO_3	Mn_2O_3	K_2O	Na_2O	FL*	CoO	P_6O	SnO_2	CuO	Sb_2O_3	Total
1	Rhombic tile, red	50,64	10,22	13,46 12,48	5,94	2,98	0,10	0,20	0,68	1,80	10,98	0,92	—	1,47	0,72	Cu_2O = 2.73	—	100,03
2	Rhombic tile, dark blue	52,52	7,64	2,40	7,96	4,40	0,10	0,39	1,28	3,26	10,68	1,30	0,26	3,83	1,18	2,62	0,50	100,13
3	Rhombic tile, white with blue tint	52,28	8,94	3,84	7,69	3,37	0,06	0,36	0,86	3,60	11,64	1,08	—	2,80	1,40	1,68	0,55	100,15
4	Rhombic tile, green	51,96	6,50	2,64	8,96	3,84	0,04	0,28	0,85	2,80	10,20	1,08	—	4,37	1,25	2,41	2,52	99,70
5	Tetrahedral line, dark blue	52,96	8,43	3,38	7,81	3,83	0,08	0,24	0,05	3,02	13,66	1,06	0,25	4,03	0,26	0,52	—	99,88
6	Glass ingot (slag), green	51,28	12,23	4,77	7,21	4,01	0,06	0,32	0,07	2,96	12,78	1,22	—	1,63	0,14	0,92	—	99,60
7	Glass ingot (slag), green	55,73	13,63	2,80	5,86	2,93	0,10	0,40	0,10	3,48	10,86	1,38	—	1,28	0,04	1,02	—	99,63
8	Glass ingot (slag), dark blue	52,04	12,28	3,70	7,61	2,40	0,10	0,44	0,02	3,68	13,08	0,92	0,09	2,37	0,20	0,80	—	99,72

* FL — Fring Loss.

Table 7.2. Chemical analyses of glazes from Central Asia of the Middle Ages (in weight percentage).

No.	Monument name	Glaze color	Date	SiO_2	Al_2O_3	Fe_2O_3	CaO	MgO	Na_2O	K_2O	FL	MnO	SnO_2	PbO	TiO_2	CuO	SO_3	CoO	ZnO	Total
1	Ancient town Afrasiab	Dark blue	11th c.	46,75	3,47	0,89	4,00	1,42	3,47	1,48	1,68	0,03	6,78	27,35	0,06	0,78	0,18	1,26	—	99,60
2	Mausoleum Aysha-bibi	ibd	12th c.	50,31	12,22	2,21	5,96	1,21	4,37	1,89	1,41	0,04	1,50	16,81	0,04	0,03	1,08	0,74	0,03	99,82
3	Same	ibd	12th c.	50,36	10,82	1,88	6,16	2,83	3,33	1,45	1,60	0,02	1,34	18,80	0,08	—	1,20	0,32	0,02	99,91
4	Mausoleum Usta-Ali	ibd	14th c.	49,23	12,87	1,45	5,05	2,83	5,00	2,00	1,68	0,02	2,04	16,85	0,05	—	1,02	0,26	0,02	100,37
5	Mausoleum Amira Burunduk	ibd	14th c.	51,86	8,39	2,13	5,41	2,67	2,02	1,29	1,85	0,05	4,66	18,26	0,04	—	0,64	0,62	0,03	99,86
6	Samarkand	ibd	15th c.	69,40	2,29	1,30	3,43	1,30	10,55	1,40	1,20	0,08	2,85	5,43	—	0,03	—	0,35	Unknown	99,61
7	Medrese Nadir Divan-Begi	ibd	17th c.	60,18	6,74	3,12	4,58	2,18	11,19	2,68	1,30	0,03	7,09	0,36	—	0,03	—	0,29	—	99,77
8	Ancient town Afrasiab	Blue	11th c.	43,82	4,98	2,81	5,32	0,81	4,87	2,49	1,41	0,01	5,80	26,01	—	0,98	0,01	0,59	—	99,91
9	Same	Blue from glazing vessel	11th–12th c.	53,40	1,10	0,74	4,06	2,82	10,57	1,11	Unknown	Few	10,00	12,62	0,07	2,95	0,40	— P_2O_5= 0.11	Unknown	99,99
10	Ancient town New Nisa	ibd	12th c.	45,16	13,20	0,60	5,40	7,20	15,18	—	1,60	—	6,80	2,30	—	1,50	—	—	—	99,94
11	Mausoleum Khodja Akhmad	Blue	14th c.	50,96	3,48	2,72	6,30	3,53	4,53	1,61	1,83	0,07	5,62	16,72	0,03	0,84	1,84	—	0,03	99,75

(Continued)

Table 7.2. *(Continued)*

No.	Monument name	Glaze color	Date	SiO_2	Al_2O_3	Fe_2O_3	CaO	MgO	Na_2O	K_2O	FL	MnO	SnO_2	PbO	TiO_2	CuO	SO_3	CoO	ZnO	Total
12	Mausoleum Usta-Ali	ibd	14th c.	52,81	4,65	1,79	6,85	3,36	2,71	1,79	1,22	0,08	4,85	19,08	0,14	0,40	0,08	—	0,02	99,83
13	Mosque Bibi-Khanim	ibd	14th–15th c.	41,46	8,10	2,74	5,05	2,15	4,08	2,81	1,70	0,05	4,81	24,11	0,06	0,96	1,62	—	0,02	99,82
14	Same	Blue	14th–15th c.	53,95	7,11	2,60	9,16	1,18	6,06	2,50	—	0,10	5,30	7,28	0,32	3,25	1,21	—	—	99,42
15	Medrese Nadir Divan-Begi	ibd	17th c.	56,02	5,39	2,00	4,21	2,60	7,90	2,01	1,90	0,02	4,18	12,38	—	1,26	—	—	—	99,87
16	Complex Shakhi-Zinda	White	14th c.	63,80	2,56	1,72	3,42	2,01	10,32	1,68	1,77	0,02	2,39	9,82	—	—	—	—	—	99,48
17	Mausoleum Burunduk	Brown	14th c.	51,16	2,49	1,21	6,31	3,75	4,58	1,91	1,46	4,00	0,08	21,94	0,09	—	0,80	—	0,02	99,80
18	Mausoleum (No 1 name)	Blue	14th c.	50,15	9,70	3,47	6,43	0,45	5,66	2,69	14,70	0,02	0,12	5,40	0,08	0,94	0,03	—	—	99,82
19	Mausoleum Anvar-Bibi	ibd	14th c.	51,30	8,13	3,00	11,48	—	5,28	2,35	4,52	0,07	5,21	5,01	0,31	0,53	3,14	— P_2O_5= 0,09	—	100,51

Table 7.3. Chemical types of experimented on glasses and glazes of Central Asia.

No.	Chemical type	Place of found samples and number of analyses conformable to tables	Chronological limits existing on chemical types	Quantity of samples, belonging to given types
1	$Na_2O–CaO–MgO–Al_2O_3–SiO_2$	Old Nisa (I,1), New Nisa (II,10)	2nd c. BC–12th c. AD	2
2	$Na_2O–CaO–MgO–Al_2O_3–PbO–SiO_2$	Old Nisa (I,4), Khodja Akhmad (II,11)	2nd c. BC–14th c. AD	2
3	$Na_2O–K_2O–CaO–MgO–Al_2O_3–SiO_2$	Old Nisa (I,3,6,7)	2nd c. BC	3
4	$Na_2O–K_2O–CaO–MgO–Al_2O_3–PbO–SiO_2$	Old Nisa (I,2,5)	2nd c. BC	2
5	$Na_2O–K_2O–CaO–Al_2O_3–SiO_2$	Old Nisa (I,8)	2nd c. BC	1
6	$Na_2O–K_2O–CaO–Al_2O_3–PbO–SiO_2$	Tuman-aka (II,26)	15th c. AD	1
7	$Na_2O–CaO–Al_2O_3–SiO_2$	Nadir Divan Begi (II,7)	17th c. AD	1
8	$Na_2O–CaO–Al_2O_3–PbO–SiO_2$	Afrasiab (II,1,8), Aysha-bibi (II,2,3), Usta-Ali (II,4), Bibi-hanum (II,13,14), Nadir- Divan-Begi (II,15), unknown (II,18), Anvar-bibi (II,19), Tuman-aka (II,25)	11th–17th c. AD	11
9	$[Na_2O]–CaO–Al_2O_3–PbO–SiO_2$	Amir Burunduk (II,5), Usta-Ali (II,12)	14th c. AD	2
10	$Na_2O–CaO–PbO–SiO_2$	Samarkand (II,6), Afrasiab (II,9), Shahi-zinda (II,16)	11th–15th c. AD	3
11	$Na_2O–CaO–MgO–PbO–SiO_2$	Amir Burunduk (II,17)	14th c. AD	1
12	$Na_2O–Al_2O_3–SiO_2$	Sher-Dor (II,24)	17th c. AD	1
13	$PbO–SiO_2$	Nadir-Divan-Begi (II,20,21), Sher-Dor (II,22,23)	17th c. AD	4
	Total: 13 types	Monuments: 15 Samples: 34	2nd c. BC–17th c. AD	34

means that successive traditions developed, stemming from communication with outside regions. Local master glass-makers improved the composition of their glazes by adding a quantity of lead oxide. Such additions lowered the temperature required to melt the glazes and improved their physico-chemical properties. The samples given in Table 7.2 — 1, 2, 5, 6 and 7, 8 indicate exactly that intercommunication. The chemical compositions of samples 9–11 are also near to being glass: the basis of glass-forming is analogous to glasses, but in the recipe of glazes lead oxide was introduced. A relatively high content of lead oxide was observed in all analyses. But only in three cases (Table 7.2 — 7, 10, 24) is the lead oxide percentage comparatively low (within 0.36–2.36%).

Another two types of glazes, 12 and 13, are peculiar; they were exploited in the 17th century, and apparently were experimental products of ceramics-makers. All the analyses of glazes given in Table 7.2 show that they were soda-calcium-alumina-silicate on the whole. Only in samples 20–23 was the composition of sodium oxide insignificant and found within 0.32–0.60%. These glazes also differ from small percentage calcium oxide (1.23–2.66%) and aluminium oxide (1.88–2.40%). According to the contents of alkaline, these glazes are soda and potash mixed. On the whole, the plant ashes served as their raw material. In only two cases (Table 7.2 — 10, 25) was natural soda possibly used. The main colored agent for dark blue glazes was served by cobalt oxide, sometimes in combination with copper oxide. The blue color of glazes was achieved with the use of copper oxide; for white tin oxide (2.59–6.00%); brown oxide manganese (4.00%); green copper oxide (0.96–3.75%); and yellow antimony oxide (3.00–4.50%). In all cases tin oxide was used for stifling of glazes.

4. Conclusions

Based on the above analysis results, it is possible to make the following conclusions:

(1) Lead-containing glasses had limited dissemination in Central Asia, though there is considerable evidence of advanced levels

of glass-making there. Glasses of this kind were found mainly at the ancient archeological monuments.

(2) Lead-containing glasses were used only as decoration; they are of five chemical types.

(3) Over a long period of time — from the 11th to the 17th century — in Central Asia, glazes of 10 chemical types were produced. Two of them are more ancient. Three compositions of glazes relate only to monuments of the 17th century.

(4) The percentage of lead oxide varies in glazes. The highest level is 40%.

(5) The batches for glazes consisted of three components: plant ashes, silica and lead oxide.

(6) Cobalt oxide, copper oxide, arsenic oxide, antimony oxide and tin oxide were used as coloring agents for glasses and glazes. Tin was also used as a stifle of glass color.

References

1. M. A. Bezborodov, *Chemistry and Technology of Ancient and Middle Ages Glasses*. [Nauka i Technika ("Science and Technology"), Minsk, 1969], pp. 188–247.
2. A. A. Abdurazakov, Appearance and the main levels of glass development in Central Asia, in XVth *International Congress on Glass* (Leningrad, 1989), "Archaeometry," pp. 26–31.
3. A. A. Abdurazzakov, History of glassmaking in Central Asia in antiquity and the Middle Ages (main stages). Essay from dissertation paper (Tashkent, 1993), pp. 17–25.
4. A. A. Abdurazzakov, Chemical composition of ceramics and glazes of architectural monuments of Uzbekistan, *History of Financial Culture of Uzbekistan*, issue 21, Tashkent, pub. h. "FAN" Uz ("Science"), 1987, pp. 176–189.
5. M. Kuchkarova *et al.*, Work-out and Instillation of Compositions and Technology Production of Glazes for Reconstruction of Monuments (Tashkent, 1986), p. 20.

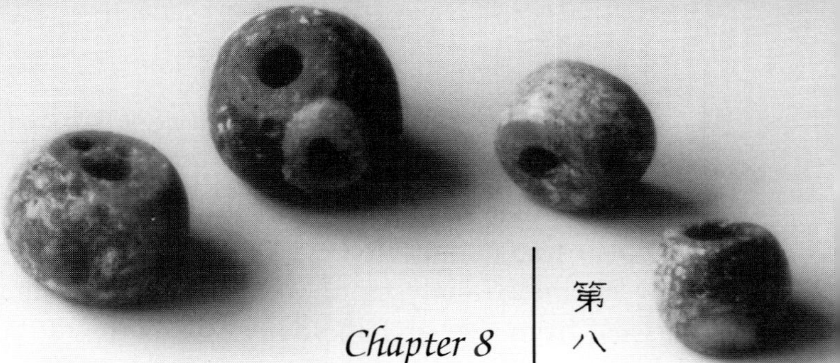

Chapter 8 第八章

Central Asian Glassmaking During the Ancient and Medieval Periods

Abdugani A. Abdurazakov

National Institute of Arts and Design,
Academic Rajabiy, 77, 700031 Tashkent, Uzbekistan

1. Introduction

Chemical analyses of Central Asian glasses have been carried out to study various technological parameters of the manufacture of ancient glasses (chemical composition, raw materials, colorants, methods of melting, and the chronological development of manufacturing methods). Most of these analyses have been done through the initiative of TC-17 and published in its works. The data obtained so far allow assignment of the glasses to separate centers of manufacture during the ancient and medieval periods. On the basis of the chemical classification, 21 types have been established. Among them one type applies to the history of all of Central Asia's glassmaking, but others to only certain periods.

After gaining independence, the Central Asian republics expanded archeological excavations. As a result, valuable new data were obtained, opening the pages of the history of glassmaking in that region. Some such excavations were at archaeological sites in Uzbekistan connected with preparations for the anniversaries of the most ancient cities — Bukhara, Khiva, Termez, Shahrisabz, etc.

Finds of both simple and highly artistic types of glass were made, and also finds connected with glass production in some places. New chemical analyses of 79 samples of ancient and medieval glasses from 24 archaeological sites have been performed. Uzbekistan is located in a region of three river valleys — Surkhandaryo, Kashkadaryo and ancient Khorezm. Comparative research into the development of glassmaking in this region and the link to separate historical periods has been conducted. The investigated samples cover the period from the Late Bronze Age (2nd millennium BC) to the Middle Ages (8th–14th centuries AD). Such a wide chronological approach allows one to investigate the roots of different chemical compositions of glasses and their territorial distributions [1, 2].

2. Experimental

Samples of glass found through archaeological excavations were carefully cleaned. After drying they were crushed. With the prepared powders complete silicate analyses were done using 5 g of the sample, while for spectral analysis it was 0.5 g. Silicate analyses were made by the traditional technique. The results are shown in Tables 8.1–8.3. Placement of the analyses in the tables is in chronological order, from the earliest to later times.

The glasses are grouped according to the levels of their basic glass-forming oxides, those present at 3% or greater. The results are given in Table 8.4. Near the chemical types are indicated the numbers of samples related to each chemical type in the three regions of Uzbekistan specified in Tables 8.1–8.3, along with their sites and dates.

3. Results and Discussion

From the results shown in Table 8.4, it is apparent that within the territory of the three regions of Uzbekistan during that long period (from the 2nd millennium BC to the 14th century AD) there are 16 chemical types. Six of these (5, 9, 10, 11, 12, 14) are individual

Table 8.1. Chemical analysis of ancient and medieval glasses of the Surhandarja valley (in wt%).

No.	Name of monument; name and color of sample	Date	SiO_2	Al_2O_3	Fe_2O_3	CaO	MgO	TiO_2	SO_3	Mn_2O_3	K_2O	Na_2O	L-C	Total
1	Burial ground Sapalli; vase, yellowish-earthen color	18th–16th c. BC	80.00	2.46	2.71	6.21	2.57	N.d	N.d	N.d	1.36	3.33	1.02	99.66
2	Burial ground Dzharkutan; beads, grayish	16th–10th c. BC	64.58	7.17	1.06	5.07	0.85	0.05	—	(0.01)	1.24	12.18	7.47	99.66
3	Setting Talashkan-tepa; bead ring-shaped, greenish	6th–5th c. BC	57.00	10.35	3.37	5.04	1.88	N.d	0.76	(0.02)	2.35	17.52	7.32	99.59
4	Site of ancient town Old Termez; wall of plate, bluish	3rd–2nd c. BC	57.43	10.23	1.72	6.02	2.15	—	—	0.01	4.45	16.69	1.03	99.73
5	Leg of wine glass, bluish	4th c. AD	62.44	10.23	0.97	6.82	1.82	0.06	—	0.03	3.04	13.10	1.46	99.97
6	Leg of wine glass, greenish	4th c. AD	60.48	10.87	0.92	6.78	1.61	0.05	—	0.06	2.88	14.24	1.59	99.48
7	Leg of wine glass, greenish	4th c. AD	62.70	9.62	0.90	6.05	2.50	0.05	—	0.03	3.69	12.79	1.50	99.83
8	Bottom of jar, green	8th–9th c. AD	61.58	11.04	0.25	7.19	2.56	0.05	—	(0.72)	2.53	12.79	1.50	100.20
9	Fannel, colorless	9th–13th c. AD	61.98	10.27	0.27	7.41	2.38	0.05	—	(0.69)	2.40	13.90	1.26	100.55
10	Bottom of wine glass, greenish	11th–13th c. AD	59.41	10.58	0.25	7.11	1.27	0.04	—	(0.03)	4.13	15.24	1.30	99.36
11	Site of ancient town Dalverzin-tepa; bottle, colorless	1st c. BC–1st c. AD	65.50	10.29	2.31	5.40	1.93	0.05	$SnO_2 = 0.13$	0.07	1.99	10.46	1.13	99.56

(Continued)

Table 8.1. *(Continued)*

No.	Name of monument; name and color of sample	Date	SiO_2	Al_2O_3	Fe_2O_3	CaO	MgO	TiO_2	SO_3	Mn_2O_3	K_2O	Na_2O	L-C	Total
12	Piece of plate glass, colorless	1st c. BC–1st c. AD	63.92	4.38	1.14	14.20	0.76	Trace	0.22	0.3	0.93	13.09	1.23	99.95
13	Halo of plate, brown	1st c. BC–1st c. AD	62.28	5.92	1.09	8.18	1.46	Trace	0.23	0.04	3.53	16.50	1.14	100.31
14	Mouth of jug, colorless	1st c. BC–1st c. AD	65.38	4.22	1.56	9.58	0.97	Trace	0.24	0.03	2.27	14.38	1.95	99.88
15	Site of ancient town Balalik-tepa; tube, brown	4th–5th c. AD	57.91	5.76	2.09	8.82	1.26	Trace	0.25	0.02	4.87	17.07	1.92	99.97
16	Castle Zar-tepa; plane glass, colorless	4th–5th c. AD	67.84	8.83	—	8.02	1.14	0.03	—	—	—	12.60	1.38	99.84
17	Site of ancient town Khosijat-tepa; bottom of wine glass, green	4th–5th c. AD	57.07	2.41	4.20	7.25	3.57	—	—	(0.03)	4.73	18.34	1.70	99.50
18	Wall of plate, green	4th–5th c. AD	60.48	3.12	3.05	7.15	3.70	—	—	(0.03)	5.11	15.33	1.53	99.50
19	Piece of alambik, light blue	4th–5th c. AD	61.00	5.38	1.74	5.87	3.85	—	—	(0.02)	3.72	16.31	1.59	99.46
20	Wall of plate with spot ornament, colorless	4th–5th c. AD	64.01	11.03	0.41	5.86	1.57	0.03	—	(0.03)	2.87	12.36	1.27	99.38
21	Mouth of jug, colorless	4th–5th c. AD	64.30	10.84	0.44	5.42	4.15	0.04	—	(0.03)	2.73	10.92	1.37	100.24
22	Halo of cup, colorless	4th–5th c. AD	59.64	9.97	0.17	6.29	2.96	0.04	—	Trace	3.36	15.66	1.19	99.28

(Continued)

Table 8.1. *(Continued)*

No.	Name of monument; name and color of sample	Date	SiO_2	Al_2O_3	Fe_2O_3	CaO	MgO	TiO_2	SO_3	Mn_2O_3	K_2O	Na_2O	L-C	Total
23	Site of ancient town Kara-tepa: piece of bottle, green	10th–12th c. AD	59.58	2.26	1.81	9.27	4.94	0.05	Trace	(0.04)	4.97	14.65	2.08	99.65
24	Halo of pan, green	10th–12th c. AD	59.92	2.15	1.77	9.30	4.90	0.05	Trace	(0.04)	4.22	15.23	2.14	99.72
25	Piece of cup, colorless	10th–12th c. AD	58.54	4.38	1.78	9.73	4.70	0.05	Trace	(0.02)	3.96	14.30	2.24	99.70
26	Wall of plate, colorless	10th–12th c. AD	58.61	4.30	1.88	9.80	4.95	0.08	Trace	(0.03)	3.98	13.96	2.00	99.62
27	Neck of bottle, colorless	10th–12th c. AD	59.12	4.47	2.06	9.85	4.31	0.07	Trace	(0.03)	4.06	13.68	2.16	99.81
28	Wall of plate with spot ornament, colorless	10th–12th c. AD	59.61	2.16	0.92	9.86	4.13	—	Trace	(0.03)	4.51	15.98	2.22	99.71
29	Wall of plate with slanting ornament, colorless	10th–12th c. AD	58.63	2.58	1.47	10.91	4.60	0.06	Trace	(0.02)	4.62	14.63	2.16	99.72
30	Site of ancient town Gormali-tepa; fragment of jug, greenish	10th–12th c. AD	61.47	6.13	6.35	3.18	2.47	0.02	0.26	(2.24)	3.06	12.26	2.51	99.98

(Continued)

Table 8.1. *(Continued)*

No.	Name of monument; name and color of sample	Date	SiO_2	Al_2O_3	Fe_2O_3	CaO	MgO	TiO_2	SO_3	Mn_2O_3	K_2O	Na_2O	L-C	Total
31	Bottom of plate, greenish	10th–12th c. AD	60.62	2.42	5.43	3.67	2.87	0.06	0.38	(1.21)	5.96	16.05	1.47	99.94
32	Bottom of plate, greenish	10th–12th c. AD	60.47	6.79	1.86	4.36	2.11	0.02	0.20	(2.02)	3.86	15.20	3.16	100.05
33	Piece of bottle, colorless	10th–12th c. AD	62.65	3.38	3.70	2.36	1.20	0.02	0.22	(5.62)	3.20	15.73	2.24	100.01
34	Piece of cup, greenish	10th–12th c. AD	62.17	4.67	6.27	3.87	2.23	0.05	0.22	(2.20)	4.49	12.18	1.79	100.03
35	Piece of jug with handle, blue	10th–12th c. AD	64.04	4.99	2.41	2.15	1.58	0.04	0.28	(3.88)	3.84	13.84	2.66	99.96
36	Tube with narrowing end, bluish	10th–12th c. AD	61.57	6.69	3.81	4.86	2.50	0.04	0.20	(1.26)	4.34	13.62	1.09	100.02
37	Mouth with stick strand, colorless	10th–12th c. AD	60.65	5.99	4.65	2.01	1.72	0.06	0.46	(4.88)	3.27	12.82	2.53	99.58
38	Bottom of plate, yellowish	10th–12th c. AD	62.70	4.87	5.13	3.20	2.83	0.04	0.24	(1.04)	4.77	13.63	1.40	99.85
39	Leg of wine glass, greenish	10th–12th c. AD	61.24	4.72	5.31	2.79	1.60	0.02	0.30	(2.08)	4.12	15.85	1.86	99.82

Table 8.2. Chemical analysis of ancient and medieval glasses of the Kashkadarja valley (in wt%).

No.	Name of monument; name and color of sample	Date	SiO_2	Al_2O_3	Fe_2O_3	CaO	MgO	TiO_2	SO_3	Mn_2O_3 (MnO)	K_2O	Na_2O	L-C	Total
1	Site of ancient town Erkurgan; piece of plate, colorless	1st c. BC–1st c. AD	59.40	9.53	1.90	8.00	2.73	N.d	Trace	(0.01)	2.38	14.28	1.44	99.73
2	Slag of glass, black	1st c. BC–1st c. AD	63.32	9.00	6.13	10.75	2.41	0.40	N.d	(0.02)	3.94	3.28	0.32	99.57
3	Piece of bottle, colorless	4th–5th c. AD	64.80	2.38	0.95	4.52	3.40	0.15	0.74	0.03	4.00	15.55	3.56	100.36
4	Bottle with elongated mouth, colorless	4th–5th c. AD	63.20	13.27	3.23	5.35	0.53	0.08	SnO_2 = 0.19	(0.10)	2.03	10.18	1.35	99.51
5	Site of ancient town Aul-tepa; piece of nimbus, colorless	5th–6th c. AD	62.82	3.86	1.79	7.25	3.05	0.26	0.11	0.11	2.36	16.58	1.77	100.49
6	Site of ancient town of Kosh-tepa II; handle of jug, greenish	5th–6th c. AD	62.78	6.05	2.20	4.85	1.36	0.24	0.72	0.03	2.88	17.54	1.14	99.79
7	Bottom of plate, colorless	5th–6th c. AD	68.37	4.72	1.55	8.12	1.42	0.03	0.43	0.49	1.24	12.26	1.21	99.84

(Continued)

Table 8.2. *(Continued)*

No.	Name of monument; name and color of sample	Date	SiO_2	Al_2O_3	Fe_2O_3	CaO	MgO	TiO_2	SO_3	Mn_2O_3 (MnO)	K_2O	Na_2O	L-C	Total
8	Стенка флакона, colorless	5th–6th c. AD	64.20	2.27	1.05	7.09	5.40	0.06	0.39	0.03	1.02	16.17	2.08	99.73
9	Nurkaj-tepa in Kitab; bottom of plate, greenish	5th–6th c. AD	62.78	3.20	3.05	6.20	3.23	0.01	0.39	(0.03)	4.05	16.17	1.17	99.89
10	Tosh-tepa near Kitab; bottom of plate, colorless	8th–9th c. AD	60.58	5.07	2.12	6.21	3.04	0.01	—	(0.03)	3.92	17.32	1.25	99.55
11	Fragment of wall, colorless	8th–9th c. AD	62.36	3.80	1.29	6.30	2.98	0.01	—	(0.03)	3.73	18.37	1.03	99.90
12	Site of ancient town Dogaj-tepa; tin with cell ornament, bluish	9th–10th c. AD	61.69	3.84	0.36	7.00	4.10	0.05	0.10	(0.07)	5.00	15.50	2.35	100.06
13	Bottom of vessel, greenish	9th–10th c. AD	61.61	4.20	0.62	7.00	3.74	0.08	0.10	(1.00)	4.83	15.00	1.80	99.98
14	Setting of Altin-tepa; ornamental body of vessel, violet	10th–11th c. AD	65.82	4.96	0.76	9.62	3.74	0.01	0.25	(1.03)	2.07	10.85	0.82	99.93

Table 8.3. Chemical analysis of ancient and medieval glasses of Khorezm (in wt%).

No.	Name of monument; name and color of sample	Date	SiO_2	Al_2O_3	Fe_2O_3	CaO	MgO	TiO_2	SO_3	Mn_2O_3 (MnO)	K_2O	Na_2O	L-C	Total
1	Palace of Toprak-kala; plane plate, square, colorless	3rd–4th c. AD	67.34	2.29	1.20	8.17	2.01	Trace	—	(0.03)	1.19	15.79	1.06	99.05
2	Palace of Toprak-kala; plane plate, square, colorless	3rd–4th c. AD	65.01	2.91	0.99	6.28	1.83	Trace	—	(0.02)	1.24	19.30	1.93	99.50
3	Palace of Toprak-kala; piece of plate, square, colorless	3rd–4th c. AD	66.32	2.86	1.08	7.72	2.32	Trace	—	(0.04)	0.76	16.44	1.86	99.40
4	Setting of Kurgancha; bottom of plate, green	7th–8th c. AD	53.04	6.83	2.20	0.12	5.30	0.30	0.19	(0.06)	4.03	19.00	1.30	100.37
5	Site of ancient town Khaivan-kala; bottom of plate, colorless	8th–10th c. AD	62.38	3.55	2.18	6.09	3.10	0.07	0.29	(0.06)	5.15	15.42	1.64	99.96
6	Site of ancient town of Khaivan-kala; bottom of plate, greeenish	8th–10th c. AD	61.54	4.62	1.97	9.09	5.15	0.06	0.34	(0.57)	3.88	11.67	0.94	99.98
7	Site of ancient town Khaivan-kala; bottom of plate, green	8th–10th c. AD	58.17	6.15	3.01	8.98	3.81	0.05	0.23	(0.08)	3.42	14.25	1.06	99.95
8	Ingot of glass, green	8th–10th c. AD	51.24	5.58	3.41	9.80	4.31	0.04	0.58	(0.77)	4.46	19.26	0.54	100.02
9	Wall of plate, greenish	8th–10th c. AD	62.84	2.27	1.85	9.26	3.14	0.07	0.22	(0.64)	4.81	13.88	1.62	100.26
10	Ingot of glass, green	8th–10th c. AD	55.44	7.06	3.35	8.02	2.54	0.06	0.20	(0.19)	3.96	18.02	1.00	99.98

(*Continued*)

Table 8.3. *(Continued)*

No.	Name of monument; name and color of sample	Date	SiO_2	Al_2O_3	Fe_2O_3	CaO	MgO	TiO_2	SO_3	Mn_2O_3 (MnO)	K_2O	Na_2O	L-C	Total
11	Wall of plate, greenish	8th–10th c. AD	60.54	2.41	5.90	10.66	4.28	0.07	0.20	(0.69)	2.73	11.12	1.31	99.96
12	Wall of plate, blue	8th–10th c. AD	61.92	5.62	1.28	8.87	3.46	0.08	0.31	(0.77)	3.80	13.63	0.44	100.12
13	Wall of plate, blue	8th–10th c. AD	49.75	4.45	3.60	10.81	4.10	0.05	0.18	(0.68)	5.42	19.77	0.72	100.03
14	Bottom of plate, greenish	8th–10th c. AD	53.16	11.89	3.01	8.20	3.75	0.05	0.14	(0.51)	3.14	14.84	1.12	99.96
15	Ingot of glass, greenish	8th–10th c. AD	58.35	4.22	3.90	10.11	3.52	0.06	0.24	(0.68)	5.24	13.12	0.56	100.01
16	Leg of wine glass, green	8th–10th c. AD	59.84	5.24	0.80	8.12	4.20	0.20	0.10	(0.04)	4.40	14.66	1.20	100.44
17	Site of ancient town Chilpik; bottom of vessel, green	9th–11th c. AD	58.20	5.84	1.90	9.57	2.65	0.14	0.04	0.06	5.19	15.26	0.90	99.75
18	Wall of vessel, green	9th–11th c. AD	57.26	5.64	2.21	10.72	4.07	0.07	0.08	0.07	3.81	14.33	1.30	99.57
19	Site of ancient town Bograkhan; wall of plate, dark blue	9th–14th c. AD	55.62	4.40	2.47	8.70	4.22	0.25	0.23	0.10	1.60	17.55	1.88	100.16
20	Site of ancient town Bograkhan; wall of plate, green	9th–14th c. AD	63.56	5.84	2.45	7.08	3.62	0.29	0.24	0.03	3.81	12.50	0.72	100.13

(Continued)

Table 8.3. *(Continued)*

No.	Name of monument; name and color of sample	Date	SiO_2	Al_2O_3	Fe_2O_3	CaO	MgO	TiO_2	SO_3	Mn_2O_3 (MnO)	K_2O	Na_2O	L-C	Total
21	Bottom of plate, green	9th–14th c. AD	61.76	5.68	2.10	7.88	3.59	0.19	0.25	0.11	4.45	13.35	0.74	100.10
22	Site of ancient town Shah-Senem; bottom of plate, colorless	11th–13th c. AD	68.28	1.40	1.02	7.00	4.35	N.d	0.18	0.07	4.03	13.81	N.d	100.14
23	Cave of Kuksai; bottom of vessel, light green	10th–13th c. AD	68.37	2.18	0.87	6.90	5.02	N.d	0.22	0.10	3.84	12.70	N.d	100.20
24	Hillock of Take-Sengir; bottom of plate, light green	10th–13th c. AD	68.62	2.02	1.17	7.00	4.85	N.d	0.21	0.08	3.86	12.55	N.d	100.36
25	Hillock of Take-Sengir; bottom of plate, light green	10th–13th c. AD	68.41	2.08	1.82	6.58	4.72	N.d	0.06	0.09	3.84	12.57	N.d	100.17
26	Hillock of Take-Sengir; bottom of plate, light green	10th–13th c. AD	68.51	2.06	1.00	6.30	5.27	N.d	0.16	0.08	4.14	12.62	N.d	100.14

Table 8.4. Chemical types of investigated ancient and medieval glasses of Uzbekistan.

	Chemical type of glass	Monuments of the Surhandarja valley (Table 8.1)	Monuments of the Kashkadarja valley (Table 8.2)	Monuments of Khorezm (Table 8.3)	Chronological limits of бытования of glass types	Number of samples concerning given type
1	Na_2O–CaO–SiO_2	1 Sapalli (2nd c. BC) 2 Djarkutan (2nd–1st c. BC)		1,2,3 Toprak-kala (3rd–5th c. AD)	2nd c. BC–3rd–5th c. AD	5
2	Na_2O–CaO–Al_2O_3–SiO_2	6,8,9,10 Old Termez (4th–13th c. AD) 11,12,14 Dalverzin-tepa (1st c. BC–1st c. AD) 16 Zar-tepa (4th–5th c. AD)	1 Erkurgan (1st c. BC–1st c. AD) 6,7 Kosh-tepa II (5th–6th c. AD)	—	1st c. BC–1st c. AD–5th–6th c. AD	11
3	Na_2O–CaO–MgO–SiO_2	3 Talashkan-tepa (6th–5th c. BC)	4 Erkurgan (4th–5th c. AD)	—	5th–6th c. BC–4th–5th c. AD	2
4	Na_2O–CaO–MgO–SiO_2	20 Khosijat-tepa (7th–8th c. AD)	8 Kosh-tepa (5th–6th c. AD)	—	5th–8th c. AD	2
5	Na_2O–CaO–MgO–Fe_2O_3–SiO_2			11 Khaivan-kala (8th–10th c. AD)		1
6	Na_2O–CaO–MgO–Al_2O_3–SiO_2	21 Khosijat-tepa (7th–8th c. AD)	Aul-tepa (5th–6th c. AD) Altin-tepa (10th–11th c. AD)		5th–11th c. AD	3

(*Continued*)

Table 8.4. (Continued)

	Chemical type of glass	Monuments of the Surhandarja valley (Table 8.1)	Monuments of the Kashkadarja valley (Table 8.2)	Monuments of Khorezm (Table 8.3)	Chronological limits of бытования of glass types	Number of samples concerning given type
7	Na_2O–K_2O–CaO–Al_2O_3–SiO_2	4,5,7 OldTermez (3rd–2nd c. BC–4th c. AD) 13 Dalverzin-tepa (1st c. BC–1st c. AD) 15 Balalik-tepa (4th–5th c. AD) 32 Gormali-tepa (10th–12th c. AD)	—	17 Chilpik (10th–11th c. AD)	3rd–2nd c. BC–12th c. AD	7
8	Na_2O–K_2O–CaO–Al_2O_3–Fe_2O_3–SiO_2	30,34,36,38 Gormali-tepa (10th–12th c. AD)	2 Erkurgan (1st c. BC–1st c. AD)	10 Khaivan-kala (8th–10th c. AD)	1st c. BC–1st c. AD–12th c. AD	6
9	Na_2O–K_2O–CaO–Fe_2O_3–SiO_2	31 Gormali-tepa (10th–12th c. AD)	—	—	10th–12th c. AD	1
10	Na_2O–K_2O–Al_2O_3–Fe_2O_3–SiO_2	39 Gormali-tepa (10th–12th c. AD)	—	—	10th–12th c. AD	1
11	Na_2O–K_2O–Al_2O_3–Fe_2O_3–Mn_2O_3–SiO_2	33,37 Gormali-tepa (10th–12th c. AD)	—	—	10th–12th c. AD	2
12	Na_2O–K_2O–Al_2O_3–Mn_2O_3–SiO_2	35 Gormali-tepa (10th–12th c. AD)	—	—	10th–12th c. AD	1

(Continued)

Table 8.4. *(Continued)*

	Chemical type of glass	Monuments of the Surhandarja valley (Table 8.1)	Monuments of the Kashkadarja valley (Table 8.2)	Monuments of Khorezm (Table 8.3)	Chronological limits of бытования of glass types	Number of samples concerning given type
13	**Na_2O–K_2O–CaO–MgO–SiO_2**	23,24,28,29 Kara-tepa (10th–12th c. AD)	3 Erkurgan (4th–5th c. AD)	9 Khaivan-kala (8th–10th c. AD) 22 Shah-Senem (11th–13th c. AD) 23 Kujusai (12th–13th c. AD) 24,25,16 Take-Sengir (12th–13th c. AD)	4th–13th c. AD	11
14	**Na_2O–K_2O–CaO–MgO–Fe_2O_3–SiO_2**	17,18 Khosijat-tepa (7th–8th c. AD)	—	—	7th–8th c. AD	2

(Continued)

Table 8.4. *(Continued)*

	Chemical type of glass	Monuments of the Surhandarja valley (Table 8.1)	Monuments of the Kashkadarja valley (Table 8.2)	Monuments of Khorezm (Table 8.3)	Chronological limits of бытования of glass types	Number of samples concerning given type
15	**Na_2O–K_2O–CaO–MgO–Al_2O_3–SiO_2**	22 Khosijat-tepa (7th–8th c. AD) 25,26,27 Kara-tepa (10th–12th c. AD)	10,11 Tosh-tepa (8th–9th c. AD) 12,13 Dogai-tepa (9th–10th c. AD)	4 Kurgancha (7th–8th c. AD) 5,6,12,16 Khaivan-kala (8th–10th c. AD) 18 Chilipik (9th–11th c. AD) 19,20,21 Bograhan (9th–14th c. AD)	7th–14th c. AD	17
16	**Na_2O–K_2O–CaO–MgO–Al_2O_3–Fe_2O_3–SiO_2**	19 Khosijat-tepa (7th–8th c. AD)	9 Nurkai-tepa (8th–9th c. AD)	7,8,13,14,15 Khaivan-kala (8th–10th c. AD)	5th–10th c. AD	7
	Total: 16 types	**Samples: 39** **Monuments: 9**	**14** **7**	**26** **8**	**2nd c. of BC–14th c. AD**	**79** **24**

types and are found only in certain monuments (Khayvan-kala, Gormaly-tepa and Khosiyat-tepa). Their occurrence, as seen, is connected with experimental searches of the glassmakers and attempts at the introduction of new raw materials or sources. It is possible to assume that at the specified locations the future will reveal the existence of local glassmaking workshops. (However, direct traces of manufacturing, such as remains of furnaces and stages of semifinished items, have not yet been found — with the exception of Khayvan-kala, where a significant quantity of "slag" was uncovered.)

Of the remaining ten chemical types of glass, five groups (1, 2, 3, 4, 6) are listed as soda glasses and five others (7, 8, 13, 15, 16) as mixed alkalis. As can be seen, there are ancient glasses of simple composition and glasses in the process of development, as glass-making gradually became more complicated, and multicomponent. It is not difficult to see that alongside the basic chemical composition of the glass are variants with increased contents of iron oxide and manganese oxide, sometimes with the two components together. Apparently, these compositions have an experimental character.

The most ancient composition in Table 8.4, for the regions studied, has three components: sodium, potassium and magnesium. It appeared for the first time in the Late Bronze Age in southern areas of the republic and seems to have had connections with the ancient Western civilizations, particularly with West Asia by the school glassmaking. The first sample of those is made from silica and ashes of plants. At the same time glasses of the given compositions are found in the territory of ancient Khorezm (Toprak-kala) and resemble the Roman glasses of the same period. They differ from other Central Asian glasses in their low contents of Al, Mg and K oxides and higher Na_2O:MgO and Na_2O:K_2O ratios.

Another chemical type of glass (Na_2O–CaO–Al_2O_3–SiO_2) is found mainly in the southern areas of Uzbekistan and not among the remains at ancient Khorezm. It also is one of the ancient compositions of glass, now found at five locations. In a number of places (Old Termez and Erkurgan), ingots have been found along with the glass, testifying to the presence of manufacturing at those places.

The fourth type of glass (Na_2O–CaO–MgO–SiO_2), one of the ancient compositions used in ancient Egypt and Mesopotamia, is found in only small numbers and only at two locations in the south of Uzbekistan. This type dates back to the earlier part of the Middle Ages (5th–8th centuries). The glass products of the sixth chemical type (Na_2O–CaO–MgO–Al_2O_3–SiO_2) are also found in small quantities and their occurrence is probably connected with development by the glassmakers of local compositions of glass.

Wide circulation is notable in the ancient period of glass products of the seventh chemical type (Na_2O–K_2O–CaO–Al_2O_3–SiO_2). They are mainly concentrated in southern areas of Uzbekistan. They are high in their alumina contents. [Editor's note: The remainder of this paragraph is not clear, but appears to refer to the second and seventh types of compositions. It has not been edited.] The roots of the origin of these kind of glass in the territory of northern Bactria are connected with the rise of productive forces which occurred during the Kushan period in southern areas of Central Asia. The cultural–economic connections of the countries on the Great Silk Road played a positive role in that. The high-content alumina characteristic circumstances of ancient Indian glasses again show the influence of the experience of glassmakers of India in Central Asia.

Glass products of the 15th chemical type (15), and the related version with high iron oxide (16), represent compositions of local Central Asian production. The first composition is found at eight sites, and the second at three sites. Earlier similar to 15 samples met in others again 40 objects (more than 200 samples), and 16 at 5 points (about 15 samples) Central Asiatic of the territory. Six components at the chemical composition with 15 arose during the ancient period in southern areas of Uzbekistan, and since the early Middle Ages they came to be among the main compositions of glass in Central Asia. At many archaeological *sites* alongside finds of glass artifacts were the remains of glass manufacturing, proving its existence where the finds were made. Except for finds concerning the ancient period at 21 sites of a valley the ornaments and utensils, and at 4 sites remains of glass industry were found. All these finds are dated from the ancient period to the late Middle Ages (13th–17th centuries). The same finds are marked in the valley of Kashkadarya — in 12, and in Khorezm — in 29 — utensils and in 3 remains of manufacturing.

4. Conclusions

(1) The glasses analyzed can be divided into three groups: soda lime glass, mixed-alkali lime glass and glass containing high iron and/or manganese.

(2) In those regions, glassmaking originally continued the use of the ancient compositions from Egypt, Mesopotamia and Rome. At some time Bactria glasses of the Na_2O–CaO–Al_2O_3–SiO_2 and Na_2O–K_2O–CaO–Al_2O_3–SiO_2 types began. These glasses were presented in numerous sites in Uzbekistan. The glass compositions might indicate an Indian influence.

(3) From the early Middle Ages, new compositions were in use there (as well as in another region of Central Asia): Na_2O–K_2O–CaO–MgO–SiO_2 and Na_2O–K_2O–CaO–MgO–Al_2O_3–SiO_2. Glass artifacts marked by these compositions were widely circulated, not only throughout Central Asia but also far beyond its limits, to Siberia, India, China, Japan, etc.

References

1. A. A. Abdurazakov, The origin and main stages of development of glassmaking in Central Asia, in: *Proceedings of the XVth International Congress on Glass* (Leningrad, 1989; Archaeometry Section), pp. 26–31.

2. A. A. Abdurazakov, Indian and Central Asian connections: a study based on chemical analyses of glasses, archaeometry of glass, in *Proceedings of the Archaeometry Session of the XIVth International Congress on Glass* (New Delhi, 1986; Calcutta, 1987), pp. 37–43.

Chapter 9 第九章

Scientific Study of the Glass Objects Found in Japan from the Third Century BC to the Third Century AD

Takayasu Koezuka
Nara National Research Institute for Cultural Properties,
Nara, 630-8577, Japan

Kazuo Yamasaki
Professor emeritus, Nagoya University, Nagoya, 464-860, Japan

1. Introduction

We have so far analyzed some 1500 ancient glass objects, which were excavated in Japan.[1] Some were analyzed with X-ray fluorescence or ICP, but these methods are not suitable for most of the excavated relics, because there is a possibility of failing to grasp important information. Thus there is a need to develop an identification method suitable for the majority of relics. Toward that end, we began in 2003 to use an imaging plate that is sensitive to very weak radiation, and succeeded in distinguishing potash glasses from soda-lime glasses. This method, which for convenience is called autoradiography (AR), is used in combination with the established techniques of computed radiography (CR), enabling one to identify the material of many glass beads at one time. After examination by these methods, more important objects are further subjected to detailed X-ray fluorescence and ICP analyses.

2. Experimental

To investigate glass relics we have used the newly developed imaging plate method along with X-ray fluorescence and ICP analyses. The imaging plate method combines the CR and AR techniques.

The CR technique replaces the X-ray film base with an imaging plate prepared by Fuji Photo Film Co. The imaging plate is supersensitive and has an expanded dynamic range, and its support for digital imaging offers strong prospects. In recent years, for most glass beads, the CR method has been used to distinguish alkali silica glasses from lead silica glasses, by simply observing differences in shading of black-and-white images. Distinguishing between potash glasses and soda-lime glasses, however, is difficult with the CR method. The distinction is made successfully by using the imaging plate, as it has enough sensitivity to measure very weak radiation from the natural radioactive potassium isotope. After placing the glass sample on an imaging plate for several days, the plate is scanned to a laser beam (He–Ne, 632.8 nm). The radiation accumulated from each specimen is imaged by subjecting it to photo-stimulated luminescence (PSL), in order to identify the constituent materials.

The equipment used for the experiments and measurements includes a microfocus X-ray enlargement system (μFX-1000), an imaging analyzer (BAS-5000), an exposure box made of lead and copper, and the analytical software.

The reference material of the AR method is potassium carbonate (K_2CO_3), compressed pieces of Japanese standard rocks JG-1a and JB-1a (10 mm diameter; 3 mm thick). The samples are placed on the imaging plate, and the elapsed time (accumulated radiation) and PSL values are measured. The data obtained are shown in Fig. 9.1. A good linear relationship is observed between the PSL values and the exposure time.

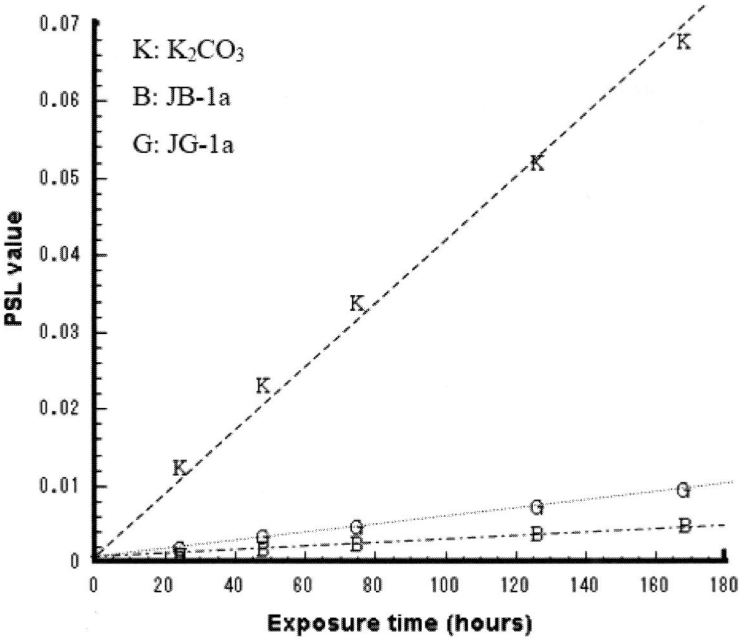

Fig. 9.1. Progression of PSL values (pixels2) for reference materials exposed to the imaging plate.

Table 9.1. Distinction of glass materials using the CR method.

Glass systems	Soda glass	Potash glass	Lead glass	Lead glass
PSL value	2.16	2.02	0.15	0.19

For excavated relics, the CR technique was used to distinguish lead silica glasses from alkali silica glasses. When a relic of known material and about the same thickness is irradiated by X-rays and the amount of X-ray transmission is measured as the PSL value, lead silica glasses absorb a high amount of radiation, while alkali silica glasses absorb one tenth of or less than that amount, making it very easy to distinguish between the two (Table 9.1).

It is also possible to identify alkali silica glasses by using AR, as potash silica glasses show more blackness in black-and-white images, and present higher PSL values than soda-lime glasses. In fact, strong radiation is detected for the potash glasses because they contain about 15% potassium oxide (K_2O), which in turn contains the radioactive isotope ^{40}K. When we expose glass relics of known material and the reference material to the imaging plate, soda-lime glasses show higher PSL values than basalt (Japanese standard rock JB-1a), and similar or slightly lower values than granite (Japanese standard rock JG-1a). For potash glasses, the PSL values were about twice as high as those for granite (Fig. 9.2.1). These AR technique shows that a large number of glass relics can simply be identified by exposing them and observing the difference in shading of the images (Fig. 9.2.2).

3. Results

In Fig. 9.3 the CR/AR images of more than 200 light blue and dark blue beads of about 5 mm diameter are shown which were excavated from the Tamura site in Kochi prefecture. It is not possible to detect lead silica glasses in CR images. In the investigation with AR, all of the beads turned out to be potash glasses. When the beads taken from the sites dating from the 3rd century BC to the 3rd century AD are examined, one can see that nearly all the glasses from that period are potash glasses aside from lead silica glasses.

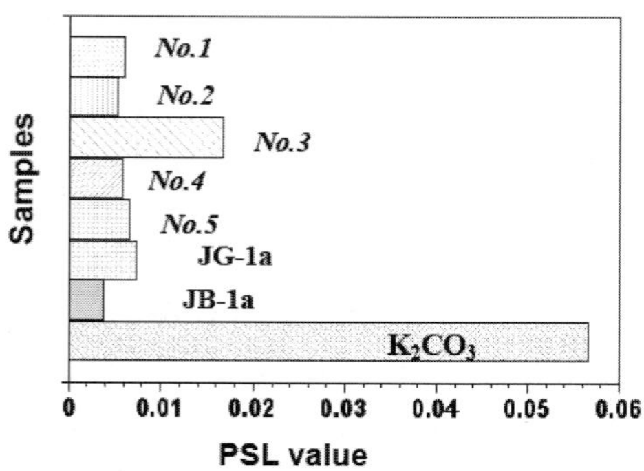

Fig. 9.2.1. PSL values obtained by AR.

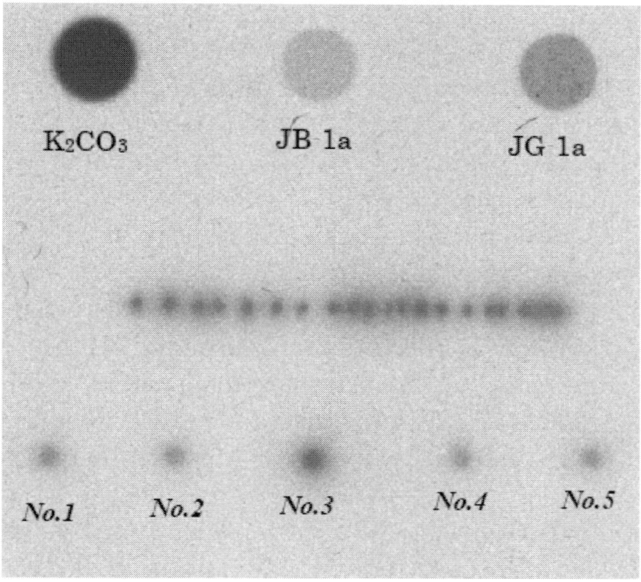

Fig. 9.2.2. Autoradiographic images of glass beads.

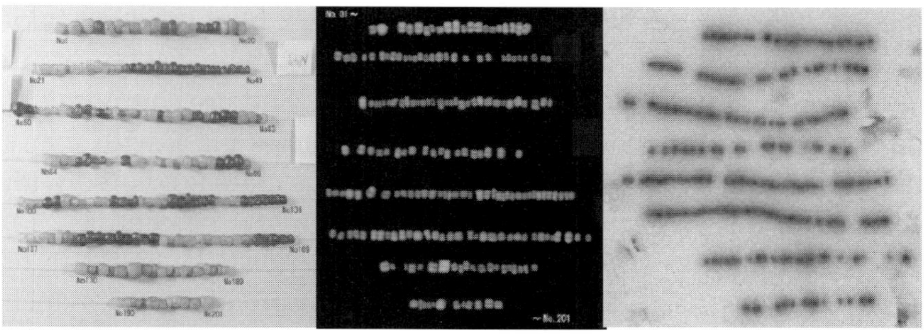

Fig. 9.3. Glass beads excavated from the Tamura site (2nd–3rd century AD). Left: Relics. Center: CR image. Right: AR image.

Potash glass beads are distributed from Okinawa to southern Hokkaido, until the 3rd century AD. Most potash glasses excavated in Japan are either light blue or dark blue in color. The color of the light blue glass beads is due to copper and that of the dark blue beads to cobalt. The distribution of these two colors is not uniform.

The ratio of the dark blue to the light blue beads is about 1:1 for Kyushu and about 1:4 for the Kinki district. Among the beads excavated from the site of around the 4th to the 5th century, light blue potash glasses are not found. Light blue beads are all soda-lime glass, while dark blue beads are produced from both potash and soda-lime glasses. Hence dark blue and light blue potash glasses traded in Japan seem to have been produced in different localities.

Most potash glasses found in Japan consist of small beads, with the exception of a bracelet (10 cm in diameter) excavated from the Ohfuro Minami tombs in Kyoto prefecture (2nd century AD). This bracelet is light blue in color, extremely transparent, and is colored with iron. A relic found from Bandon Ta Pet in Thailand has a similar shape to this bracelet, but it is not known if the piece is made of potash glass.

In Japan, soda-lime glass appears only from the late Yayoi period (the 1st to the 2nd century AD), and such glasses are very scarce. The pieces are mostly in the form of opaque red beads (called *mutisalah*) around 3 mm in diameter.[2] Also, some blue and yellow soda-lime glass beads have been found. An unusual case is the discovery of a few dozen segmented dark blue beads of two-layered construction. These segmented soda-lime glass beads are about 4.5 mm in outer diameter, 2 mm thick, and are composed of dark blue glass as the outer layer and colorless glass with many bubbles as the inner layer. The outer dark blue glass is colored with cobalt and the manganese oxide content is a few percent (1–2%). This fact indicates that the outer dark blue glass layer made of soda-lime glass is colored by the same type of cobalt ore, which was used to make the dark blue potash glasses.

As the cobalt ore containing fairly low manganese oxide was generally used to make the blue soda-lime glass excavated in Japan, these segmented soda-lime glasses are unusual. In all, 276 *mutisalah* beads were excavated from the Hirabaru site in Fukuoka prefecture (of the latter half of the 3rd century AD), but at other sites only about a dozen *mutisalah* beads were found.

Mutisalah beads are distributed mainly in the northern part of Kyushu, in the latter part of the Yayoi period (the 1st to the

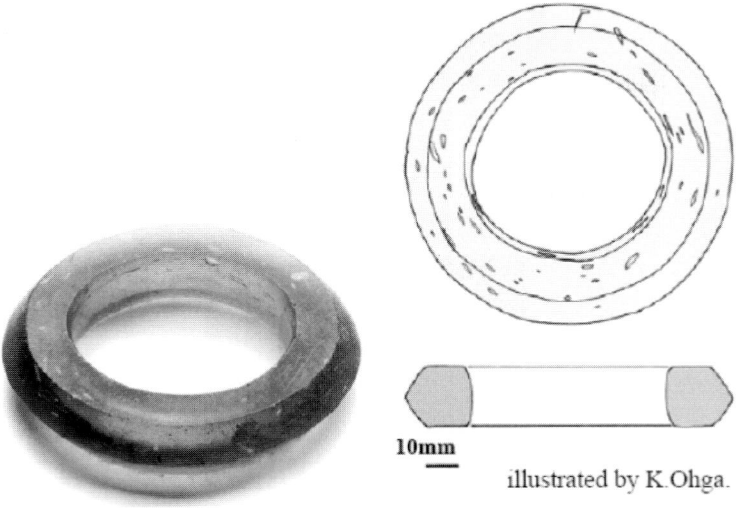

Fig. 9.4. Large bracelet made of potash glass (site of the Ohfuro Minami tombs 2nd–3rd century AD).

3rd century AD). Also, small glass tubes that are supposed to be unfinished *mutisalah* products have been found. It has become clear that the circulation of *mutisalah* beads ceased around the 4th century AD and started again in the 5th century AD. According to our studies, all the *mutisalah* beads found in Japan are made of soda-lime glass.

To clarify the route through which *mutisalah* beads were brought to Japan, we investigated *mutisalah* beads excavated from the Nangnang (Lolang) earthen ramparts in the northern part of the Korean peninsula (about the 3rd century BC to the 3rd century AD). Using the CR/AR technique, only one out of 48 beads was found to be potash glass (Fig. 9.5). It is difficult to distinguish potash glass *mutisalah* beads with the naked eye. X-ray fluorescence analysis revealed that this potash glass bead contains rather high magnesium oxide (Table 9.2). For that period, not a single *mutisalah* bead made of potash glass has been found in Japan.

As potash glass objects are easily distinguished by using the AR technique, more examples are expected to be found in the future.

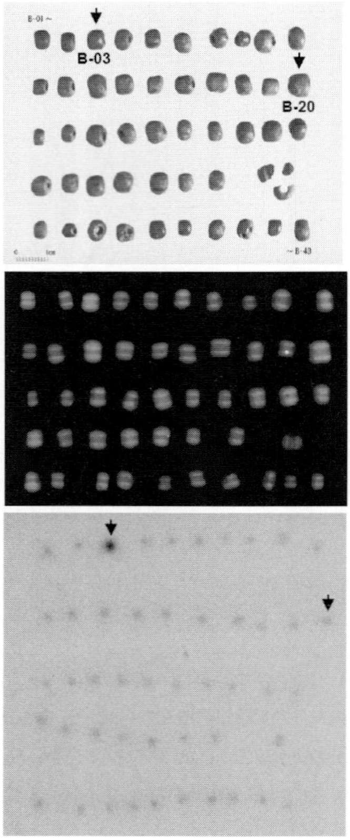

Fig. 9.5. CR/AR image of *mutisalah* beads from the Nangnang earthen ramparts.

Table 9.2. X-ray fluorescence analyses (*mutisalah* sample: B-03, B-20) **(%)**.

Sample	Na_2O	MgO	Al_2O_3	SiO_2	K_2O	CaO	Fe_2O_3	CuO
B-03	1.0	3.8	3.0	72.5	13.9	1.0	1.2	1.9
B-20	15.0	2.1	5.3	70.3	2.4	2.3	1.0	1.0

4. Conclusion

In order to learn where the glass beads were produced and distributed, it is important to find their materials. The areas of production of potash glasses and soda-lime glasses have not been identified yet, particularly in Asia. In the present study, new nondestructive techniques using CR and AR have been developed and remarkable results in distinguishing potash glasses from soda-lime glasses have been obtained.

References

1. T. Koezuka and K. Yamasaki. Scientific studies on the glass beads found in the Yayoi period of Japan, in *Scientific Research in the Field of Asian Art* (Freer Gallely of Art, 2003).
2. The first *mutisalah* bead reported in Japan is the one kept in the Oku-cho Archeological Museum, Okayama prefecture. In July 1954 Dorothy Blair and K. Yamasaki visited this museum and several pieces were donated. One of these was sent to the Corning Museum of Glass and examined qualitatively by spectrochemical analysis; K. Yamasaki, Chemical studies on the ancient glass beads found in Tsushima and at the Toro site; in *Scientific Studies on Japanese Antiques and Art and Craft*, No. 8 (1954), pp. 13–16 (in Japanese). Later, in 1980, K. Yamasaki analyzed this bead by the atomic absorption method and got the following data: Na_2O 16.3%, K_2O 3.1%, Fe_2O_3 0.91%, Cu_2O 1.70%.

Chapter 10 | 第十章

Chemical Analysis of the Glass Vessel in Toshodaiji Temple Designated a National Treasure Through a Portable X-Ray Fluorescence Spectrometer — Where Did the Glass Vessel Come From?

Akiko Hokura, Takashi Sawada and Izumi Nakai

Department of Applied Chemistry, Tokyo University of Science,
1-3 Kagurazaka, Shinjuku, Tokyo 162-8601, Japan

Yoko Shindo

Section of Islamic Archaeology and Culture,
The Middle Eastern Culture Center in Japan,
5-1-4 Nishiogi-kita, Suginami, Tokyo 167-0042, Japan

Takashi Taniichi

Okayama Orient Museum,
9-31 Tenjin-cho, Okayama, Okayama 700-0814, Japan

1. Introduction

The glass receptacle for Buddha's ashes which is designated a national treasure is stored in Toshodaiji Temple, in Nara, Japan.[1] This temple was built in 759 AD by a famous monk from China named Ganjin (Jianzhen in Chinese). He was invited to Japan by the emperor to train monks and improve Japanese Buddhism. It was reported that Buddha's ashes of 3000 grains (*Śarīra*) had been brought by Ganjin from China during the Tang Dynasty. The vessel is precious from a historical viewpoint, and it has not been shown to the public except for a few exhibitions. There has only been one time during a long history when the glass vessel was subjected to analysis by a scientist.[2] From the measurement of the coefficient of β-ray backscattering, it was reported that the glass did not contain PbO. There was no information about other elements. Recently, we have had a chance to analyze the glass vessel using a portable X-ray fluorescence (XRF) spectrometer in order to reveal the chemical composition for reproducing the artifact. Another purpose of the analysis is to track the country of origin for the glass vessel based on its chemical composition. Where did the glass vessel come from? We try to find out the answer by instrumental analysis and typological study.

A chemical composition would potentially enable us to characterize ancient glass artifacts and to determine the raw materials used, and the fabrication process and the place where they were made. For these purposes, precise quantitative analysis is necessary. Many researchers have utilized analytical instruments in the laboratory through destructive sample preparation for a limited number of samples. However, in order to study precious artifacts, such as a national treasure, nondestructive analysis has been desired.

Nakai *et al.* developed a portable XRF spectrometer in 2001[3] and have analyzed a lot of archaeological artifacts on the excavated site, e.g. the Abusir area,[4] Sinai Peninsula and Fustāt[5–7] in Egypt, and Kaman-Kalehöyük in Turkey.[8] The instrument has an excellent optics system equipped with monochromator and capillary. Thus, various elements can be detected high-sensitively by a silicon drift detector (SDD). The instrument is portable, yet it can produce high-resolution spectra (FWHM = 134.7 eV at Mn Kα). Sawada *et al.* have been carrying out nondestructive analysis of glasses using this instrument at excavation sites of southern Sinai Peninsula, Egypt, in 2001 and 2002.[5–7] The archaeological sites Rāya and Wādī al-Ṭūr have been excavated by the team of the Middle Eastern Cultural

Center in Japan (director: Dr M. Kawatoko), and it is estimated that these sites were in use from the 6th to the 12th century.[9] Rāya was a port city, so the excavated glasses exhibit diverse types of decoration. These excavated Islamic glasses have been typologically classified by Y. Shindo.[10] XRF analysis using the portable spectrometer has been found to be effective for qualitative and semiquantitative analyses of major and trace elements in Islamic glass artifacts. The number of samples analyzed was over 400 and they were characterized based on their chemical compositions, such as natron glass or plant-ash glass.[5–7] It is believed that Rāya thrived as a trading city, because many kinds of items have been excavated, and the East–West trade was carried out via this port. Many Chinese pottery vessels were excavated at Rāya, indicating that the trading area at that time was fairly wide.

In this article, we elucidate the chemical composition of the glass receptacle for Buddha's ashes in Toshodaiji Temple and consider the association of it with the glass artifacts excavated at Rāya and other Islamic glasses.

2. Experimental

2.1. *Sample artifact*

The glass vessel for Buddha's ashes, which is designated a national treasure, is housed in a metallic tower and stored in Toshodaiji Temple (Nara, Japan). It contains Buddha's ashes with a metallic cap. A photograph of the vessel is shown in Fig. 10.1. The glass is transparent and the color is light greenish amber. The diameter of the globular body is about 11 cm. Its base is slightly concave and is covered with Japanese paper. The shape of the rim cannot be seen, because it is covered by the metallic lid.

2.2. *XRF analysis*

The portable XRF spectrometer OURSTEX 100FA, developed by Nakai *et al.*, was used for the analysis. The details on it were given

Fig. 10.1. The glass vessel stored in Toshodaiji Temple.

Table 10.1. Measurement conditions.

Equipment	OURSTEX 100FA
X-ray tube	Air cooling, Pd target
Exciting voltage	40 kV
Tube current	1.0 mA for Pd-Kα X-ray 0.25 mA for white X-ray
Measuring time	300 s
Detector	Silicon drift detector (SDD) with digital signal processor (DSP)
Atmosphere	Air

in our previous papers.[3,4] This instrument was set up in a room of Toshodaiji Temple on 10 February 2004. The analytical conditions are given in Table 10.1. The analysis was carried out in air. The measurement points were on the shoulder of the vessel. Since the X-ray tube current was very low (below 1.0 mA), the analyzed sample was not damaged.

The intensity of the XRF signal — in units of cps (counts per second) — was obtained by subtracting the background from the peak of each element in the spectrum. The obtained intensity was normalized by that of the scattered X-ray of the palladium Kα line (Compton scattering), which was used as an excitation X-ray source. Elemental quantification from the XRF intensity to the oxide concentration was carried out using a calibration curve method utilizing the normalized intensity. Details of data analysis were reported in the previous paper.[5] A calibration curve of each element was obtained by using the standard reference material of glasses by NIST (SRM621 Soda-Lime, Container, SRM1830 Soda-Lime, Float, SRM1831 Soda-Lime, Sheet), and the calibration glasses synthesized in the following way.

2.3. *Synthesis of glass*

To determine the chemical composition, the calibration glasses were synthesized for Na, Mg, Al, K, Ca, Ti, Mn, Fe, Ni, Cu, and Sr. Raw materials were weighed, and homogenized in an alumina mortar, and the mixture was placed in a crucible (alumina, SSA-H), and then heated up to 1400°C in an electric furnace.

As a preliminary test to duplicate the glass vessel, weighed amounts of raw materials calculated based on the analytical results were also homogenized, melted, and quenched. In addition, test-piece glass obtained was subjected to the XRF analysis.

3. Results

The glass vessel was subjected to the XRF analysis, and the spectrum obtained by Pd Kα is shown in Fig. 10.2. To determine the light elements, the spectrum was also measured by white X-ray. As is seen in the figure, the glass vessel contains a certain amount of transition metal elements. The chemical composition obtained from XRF intensity using the calibration curve is summarized in Table 10.2. Because precise determinations of Na and Al were difficult using this instrument, the concentrations of these elements were estimated through comparison

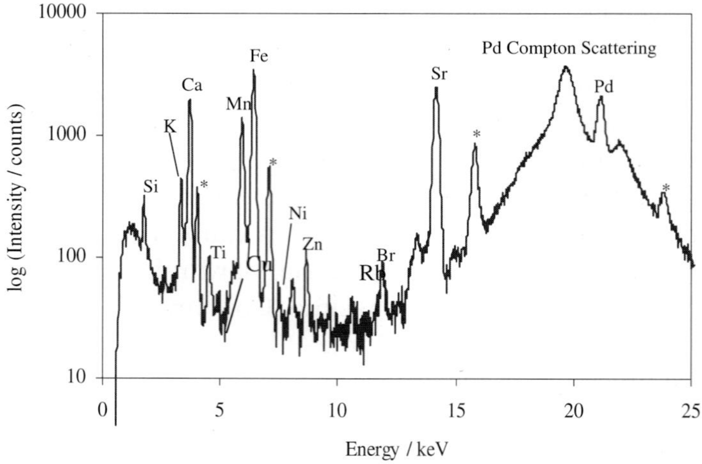

Fig. 10.2. The spectrum of the glass vessel. Excitation source: Pd Kα. *:Kβ.

Table 10.2. Analytical results for the glass receptacle for Buddha's ashes, *Hakururi Shari Ko*.

Element	Concentration (%)
Na_2O	15[a]
MgO	5.9[b]
Al_2O_3	2[a]
SiO_2	60[c]
Cl	< 10[a]
K_2O	2.2[b]
CaO	8.4[b]
TiO_2	0.16[b]
MnO_2	0.9[b]
Fe_2O_3	1.4[b]
NiO	0.006[b]
CuO	< 0.01[a]
Br	trace[a]
SrO	0.07[b]

[a] Estimated by peak shape. [b] Obtained by each calibration curve. [c] Obtained by subtracting other elemental compositions from 100%.

of the shape of the spectra for each element with those of the calibration glasses. It was found that the glass vessel does not contain PbO.[2] In the present study, the peak of Pb was not detected. The minimum detection limit (MDL) for PbO was 0.0067%,[4] and consequently it was estimated that the content of PbO was less than that.

On the other hand, the test-piece glass was prepared based on the analytical results for the glass vessel. The raw materials used for the synthesis were as follows: Na_2CO_3 3.847 g, MgO 0.9 g, Al_2O_3 0.3 g, K_2CO_3 0.572 g, $CaCO_3$ 2.463 g, TiO_2 (anataze) 0.03 g, MnO_2 0.12 g, α-Fe_2O_3 0.18 g, $SrCO_3$ 0.008 g, SiO_2 9.459 g — total about 15 g. Its color was quite similar to that of the original glass vessel. This test-piece glass was also subjected to the XRF analysis, and the chemical composition was determined. The data agreed with those on the glass vessel.

4. Discussion

4.1. *Chemical composition of the glass vessel*

From the analytical results shown in Table 10.2, we tried to classify the glass vessel as one of the well-defined chemical types — the natron glass or plant-ash glass.[11] Because the concentration of K_2O and of MgO were over 2%, it was judged to be plant-ash glass. As described above, this glass vessel does not contain PbO. This agrees with the previous report.[2] In Table 10.2, we can see that the glass vessel contains a certain amount of colorant elements. The color of the vessel, greenish amber, could have been caused by the observed levels and the chemical state of iron and manganese ions. Further study of the coloration mechanism of glasses will be carried out by X-ray absorption spectrometry to elucidate the chemical state of colorant elements such as iron and manganese ions. At this moment, it has been confirmed that the portable XRF instrument is the most powerful tool for nondestructive analysis of major-to-trace elements in glass samples.

4.2. *Typology of the glass vessel*

It was reported that a similar type of glass vessel was excavated in Fustāt, Egypt by an American excavation team.[12] It has a transparent green tinge, with the rim folded outward. The diameter of the body is 14.5 cm and this glass vessel is therefore larger than that in Toshodaiji Temple. Another similar vessel has been reported from Syria (Jebel Sais), but it has a pain rim.[13] On the other hand, there is a glass vessel excavated from the ruins of reliquary tower in China.[14] According to the literature, this was estimated by typology to be produced somewhere in the Iranian Plateau.[14] It is of the flask type but its shape and thickness are somewhat different than that in Toshodaiji Temple. However, it contains a certain amount of potassium and is thought to have been used for Buddha's ashes. Therefore, a precise quantitative analysis of that piece would be desirable to reveal any possible relationship with the glass vessel in Toshodaiji Temple.

4.3. *Comparison of the glass vessel with the glass objects excavated in Rāya, Egypt*

To compare the glass vessel in Toshodaiji Temple with the glass artifacts excavated at Rāya, the relationship between [SrO] and [CaO] and that between [SrO] and [TiO_2] are shown in the compositional diagrams in Figs. 10.3 and 10.4, respectively. It was reported for glasses in Rāya that two characteristic groups were presumed to be derived from the difference in the raw material used as the calcium source and the sand impurities.[5–7] In the [SrO] vs. [CaO] plot, the glasses in Rāya are clearly classified into two groups (A and B): the glasses in Group A correspond to natron glass, and those in Group B correspond to the plant-ash glass according to the SrO content.[5–7] Furthermore, Group A (natron glasses), with a lower SrO content, has a higher TiO_2 content; in contrast, Group B (plant-ash glasses), with higher SrO, has lower TiO_2. In Figs. 10.3 and 10.4, the glass vessel in Toshodaiji Temple seems to have a compositional characteristic similar to those of

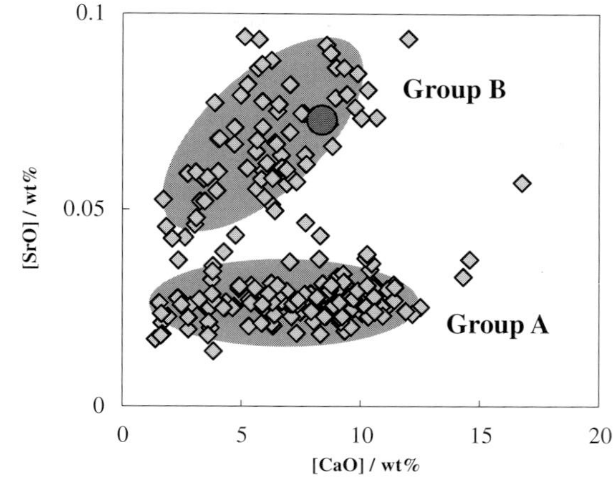

Fig. 10.3. Relationship between [SrO] and [CaO]. ●: the glass vessel, ◆: the glasses excavated at Rāya.[7]

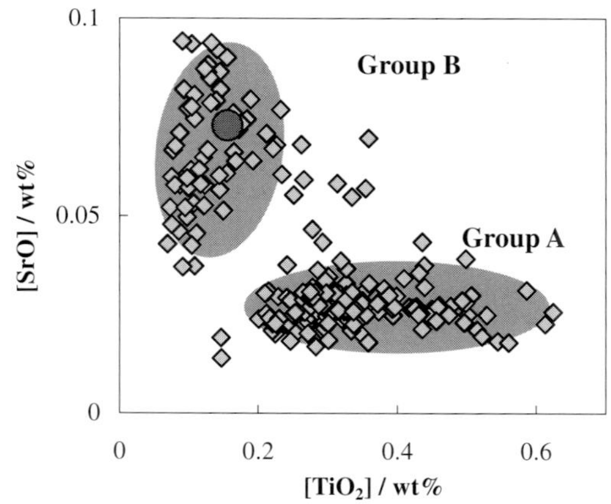

Fig. 10.4. Relationship between [SrO] and [TiO_2]. ●: the glass vessel, ◆: the glasses excavated at Rāya.[7]

Group B. This could increase the possibility that the glass vessel was produced in an Islamic country using plant ashes as an alkaline source.

It was considered that the natron glasses produced in the Mediterranean region, such as Caesarea of Israel and Fustāt of

Egypt, dated from the eighth century or earlier; while the plant-ash glasses mostly dated from the ninth century or later.[11] On the other hand, the glass of Nishapūr, in eastern Iran, contains potassium and magnesium at high concentrations and is believed to be made of soda derived from plant ashes.[11] It was also reported that the glasses made in ancient China contain a high amount of lead.

Consequently, it was concluded that the glass vessel had come to Japan all the way from somewhere in the world of Islam via China. Further details of the production site may be revealed to utilize the information on varied trace elements as a fingerprint of glass.

Acknowledgments

We express our sincere thanks to Toshodaiji Temple for giving us the opportunity to analyze the precious artifact. We also thank the Nara National Museum for helping us with the analysis. We are grateful to Takeomi Sakoda (Kurashiki University of Science and the Arts) for useful discussion regarding preparation of the test-piece glasses. This study is financially supported by the Tokyo Broadcasting System, as part of TBS Toshodaiji Temple Project 2010. A. Hokura has received financial support from Corning Research Grant 2002.

References

1. T. Yoshimizu (ed.), *World Glass Arts*, Vol. 5 (Kyuryudo, Tokyo, 1992).
2. T. Asahina, F. Yamazaki, I. Otsuka, T. Hamada, K. Saito and S. Oda, *Scientific Papers on Japanese Antiques and Art Crafts* **6**, 14–18 (1953).
3. I. Nakai, S. Yamada, A. Hokura, Y. Terada, Y. Shindo and T. Utaka, *X-Ray Spectrometry*, in press.
4. T. Sanada, A. Hokura, I. Nakai, S. Maeo, S. Nomura, K. Taniguchi, T. Utaka and S. Yoshimura, *Advances in X-Ray Chemical Analysis* **34**, 289–306 (2003).
5. T. Sawada, A. Hokura, S. Yamada, I. Nakai and Y. Shindo, *Bunseki Kagaku* **53**, 153–160 (2004).

6. T. Sawada, A. Hokura, I. Nakai and Y. Shindo, *Annales du 16e Congrès de l'Association Internationale pour l'Histoire du Verre*, in press.
7. T. Sawada, A. Hokura, I. Nakai and Y. Shindo, *Archaeological Survey of the Rāya/al-Tur Area on the Sinai Peninsula, Egypt, 2003,* ed. M. Kawatoko (The Middle Eastern Culture Center in Japan, Tokyo), in press.
8. M. Masubuchi and I. Nakai, *Anatolian Archaeological Studies*, in press.
9. M. Kawatoko, *Archeological Survey of the Rāya/al-Tur Area on the Sinai Peninsula, Egypt, 2002*, ed. M. Kawatoko (The Middle Eastern Culture Center in Japan, Tokyo, 2003).
10. Y. Shindo, *Annales de l'Association, Internationale pour l'Histoire du Verre* **15**, 180–184 (2001).
11. R. H. Brill, in *Glass of the Sultans*, eds. S. Carboni and D. Whitehouse (Metropolitan Museum of Art, New York, 2001).
12. G. T. Scanlon and R. Pinder-Wilson, *Fustāt Glass of the Early Islamic Period, Finds Excavated by the American Research Center in Egypt, 1964–1980*, No. 11b (Fox Communications and Publications, London, 2001).
13. B. von Klaus, Das omayyadische Schloß in Usais (II), in *Mitteilungen des Deutschen Archäologischen Instituts Abteilung Kairo*, Band 20, p. 173, Fig. 40, No. 358 (Abb40, 1965).
14. T. Yoshimizu (ed.), *World Glass Arts*, Vol. 4 (Kyuryudo, Tokyo, 1992).

Chapter 11 第十一章

On the Glass Origins in Ancient China from the Relationship Between Glassmaking and Metallurgy

Qian Wei

Institute of Historical Metallurgy and Materials,
University of Science and Technology Beijing, 100083, China

1. Introduction

Glass is an amorphous solid structure with the character of liquid material, formed by the process of melting and then cooling down. As an artificial material, it has played an important role in the history of human civilization. Archeological evidence shows that glass technology was invented the earliest around 2500 BC in Egypt or Mesopotamia. In the West, glass was used as decorations at first, and then as glassware and construction material. It played a positive role in the process of promoting Western civilization, which is comparable with the brilliant Chinese civilization cultivated by Chinese porcelain and bronze.

As the major carrier of China–West cultural and technological exchanges, glass has attracted universal attention. With more and more discoveries of ancient glass products, the concern over the origin of the glass technology of China is increasing, which has also been puzzling scholars at home and abroad. Ancient Chinese glass in archeological discoveries shows that the date of glass technology in China is later than that in the West. From scientific analysis, the earliest date for the invention of artificial glass in China is some where from the Western Zhou Dynasty (1046–771 BC) to the early Spring and Autumn Period (770–476 BC). Current studies have shown that material from ancient Chinese glass components is somewhat similar to that from Western glass, but there is still a big difference between them. Was ancient Chinese glass imported or produced locally? Does it have an independent origin of ancient Chinese technology of glass? All these issues are worth further study. In particular, China during the Shang (17th–11th c. BC) and the Zhou Dynasty (1046–256 BC) led the world in metallurgical technology and created a splendid bronze civilization. However, with regard to the same physical and chemical properties similar to the relationship between glass and metallurgical slag, there were no systematic studies. This article discusses briefly the origin of Chinese glass technology from the relationship between ancient glassmaking and metallurgy.

2. The Metallurgical Origin of Glass Technology

The problem of the origin of glass technology is a hard and hot focus in the international research and academic field. It is now widely recognized that there are at least five possibilities for the origin of glassmaking:

(1) *Natural glass*. The glassy obsidian formed through rapid cooling after a volcanic eruption and the meteoric stones produced by a sky meteor can stimulate the earliest understanding of glass.
(2) *Pottery*. After mastering the fire technique, perhaps in the pottery process, when the material components matched the raw material composition of the glass, then through a high-temperature process, the original glass was obtained.
(3) *Glazing*. After mastering the pottery technology in high-temperature ceramics, the glaze material with the quality of glass was inadvertently formed. The emergence of glazed pottery

and porcelain gave evidence that there were the technical conditions for the origin of the glass.

(4) *Metallurgy*. The slag produced during the process of metal smelting, after cooling down, could form colorful substances with the quality of glass, which may be one motivation for the origin of glass technology.

(5) *Alchemy*. There was a good grasp of the chemical reaction, and the raw materials used were close to those needed to produce the glass. In particular, the unique high-leaded glass production in China's Tang Dynasty may have a close connection with the origin of the glass technology.

Metallurgy, like glassmaking, is a very important technological invention in the history of human beings. Both belong to the pyrotechnics in high-temperature physical and chemical processes, so there is a certain technological correlation. According to the existing archeological data, metallurgy originated in West Asia in 5000 BC, much earlier than glassmaking. The influence of metallurgical technology on glass technology has become a major topic in the international academic field.

The recent research on the origin of glass technology in Western countries is also focused on the links between the origin of glass and metallurgical technology: some Bronze Age glass of Germany's Lower Saxony and Hesse was tested. It was found to contain high antimony which the tester thinks is closely related to the nearby copper-containing antimony.[1] Professor Rehren of University College London studied the glassmaking techniques in the Middle East region, including the ancient Egyptian ones, and noticed the relationship between metallurgical slag and the origin of glass.[2,3] The recent discoveries of ancient Egyptian glass art of ancient glass which were made by American scholars show that copper slag, corrosion products or blue glass containers are a potential source of the coloring agent, and the use of lead–antimony mineral glass is likely a potential source of the yellow and green glass coloring agent.[4]

Many Chinese scholars have also supported the belief in the metallurgical origins of the glass, including: Boda Yang, who argues that the origin of the glass in Chinese ancient history is inseparable from the development of the smelting operation in the Shang and Zhou Dynasties of China, which is likely to have been accidental during the production process of refined copper and slag smelting, but there is no direct scientific evidence[5]; Kuanghua Zhao, who speculates through the ancient Chinese literature that the ancient traditional leaded glass was developed with the development of the galena-smelting process, but there is also no scientific evidence[6]; Shuanglin Zhou, who analyzed the glassy slag unearthed from the ruins of Yangcheng iron-making residue, at Dengfeng, Henan province, and concluded that the metal fabrication industry promoted the development of China's glass, but this only provides evidence of the development of the glass later than the Han Dynasty and does not solve the problem of the origin of the glass.[7]

3. Analysis of the Glass Unearthed at the Kiziltur Cemetery in Baicheng

The Kiziltur Cemetery in Baicheng county, Xinjiang Region, northwest China, is among the important remains of the Bronze to Early Iron Age in the south of the Tianshan Mountain, dating from 1100 to 600 BC (see Fig. 11.1). From 1990 to 1992, the Xinjiang Institute of Cultural Relics and Archeology carried out excavation for reservoir construction and discovered a burial site with more than 160 tombs, including painted pottery, bronze, iron, stone, bone, glass and many other artifacts.[8] The author analyzed the unearthed beads and made some relative research.[9] Of five selected samples from three different tombs, one is a more complete bead and the remaining ones are bead debris. The analysis of the samples and the results are shown in Tables 11.1–11.3.

The microstructure analysis indicates that the long rectangular crystal ingredients of samples G1, G2 and G5 are similar and seem to be the same zircon minerals. The analysis of the crystal core of samples G1 and G3 shows that they are the same stable magnesium silicates. Both sample G2 and sample G4 have rich copper which is a noneutectic substance of copper oxide. Rich lead–antimony

Fig. 11.1. Location of the Kiziltur Cemetery and other sites mentioned in this article.

Table 11.1. Glass bead samples unearthed at the Kiziltur Cemetery.

Sample number	G1	G2	G3	G4	G5
Tomb number	91BKKM21:4	91BKKM21:5	91BKKM3:9	91BKKM27:B	91BKKM27:B
Appearance statement	fragment	fragment	fragment	complete	fragment
Outlook color	black	bright blue	yellow	blue	bright green
Colorant element	Fe	Cu	Cu, Sb	Cu	Cu, Fe

inclusions were found in sample G3; they are the main colorant particles for yellow glass. Rich metallurgy was found in sample G4, and it has some sort of link with copper smelting. Rich antimony inclusions were found in sample G5, and they are related to the use of the raw materials.

Table 11.2. Electronic probe analysis of glass samples from the Kizilur Cemetery (wt%).

	SiO_2	Al_2O_3	Fe_2O_3	TiO_2	Na_2O	K_2O	CaO	MgO	Sb_2O_5	CuO	CoO	PbO	BaO	P_2O_5	SO_3	Cl
G1	62.9	1.43	1.05	0.21	17.7	4.53	6.14	4.28	n.d.	n.d.	n.d.	n.d.	n.d.	0.27	0.72	1.11
G2	64.1	1.22	0.59	0.28	17.7	1.76	5.49	4.51	1.55	n.d.	n.d.	n.d.	n.d.	0.18	0.39	0.72
G3	61.2	1.85	0.83	0.23	16.8	2.78	4.94	3.40	1.12	n.d.	n.d.	4.38	0.04	0.22	0.56	0.41
G4	62.2	2.06	1.05	0.20	19.7	2.32	6.01	2.87	n.d.	1.24	n.d.	n.d.	n.d.	0.20	0.48	0.63
G5	60.0	1.11	0.58	0.16	18.4	2.18	7.32	5.06	1.65	0.82	n.d.	n.d.	n.d.	0.22	0.44	0.73

n.d. means "not detected". These samples were analyzed with a JEOL JXA 8600 Super Probe electron probe microanalyzer (WD-EPMA) in the archeological science laboratory at UCL.

Table 11.3. Component microanalysis of the glass samples unearthed at the Kizilur Cemetery.

No.	Analyzed parts	Element compositions (wt%)											
		Si	Ca	K	Na	Mg	Al	Pb	Ba	Zr	Fe	Cu	Sb
G1	Long crystal	19.1	0.18	n.d.	n.d.	n.d.	n.d.	1.45	n.d.	79.0	n.d.	n.d.	0.26
	Crystal core	53.0	4.52	0.47	n.d.	38.1	1.16	0.12	0.24	n.d.	1.91	0.42	n.d.
	Quartz crystal	96.0	0.76	0.60	n.d.	0.80	n.d.	n.d.	0.04	1.58	0.25	n.d.	n.d.
	Boundary	87.6	2.04	1.64	0.52	2.00	0.70	n.d.	1.36	1.41	2.62	0.12	n.d.
G2	Long crystal	18.5	0.71	0.26	0.30	0.11	n.d.	1.26	0.01	78.4	0.44	n.d.	n.d.
	Inclusion 1	68.6	10.1	3.38	1.51	7.57	0.36	0.54	0.98	n.d.	1.55	3.50	1.76
	Inclusion 2	70.6	10.9	3.49	0.54	5.29	n.d.	n.d.	0.84	0.50	1.55	3.98	2.28
	Cu-rich particles	3.15	0.30	0.28	n.d.	2.35	0.82	n.d.	0.15	0.18	0.41	92.4	n.d.
G3	Rich Pb–Sb phase	9.35	3.86	0.53	2.30	1.64	n.d.	43.9	n.d.	n.d.	0.45	n.d.	38.0
	Crystal core	44.9	14.4	0.71	n.d.	33.0	1.31	n.d.	0.83	0.37	1.52	0.13	2.84
	Inclusion	53.4	14.8	1.66	1.33	7.01	n.d.	9.04	0.60	n.d.	0.97	0.26	10.9
	Glassy phase	51.4	12.6	5.23	3.50	3.56	n.d.	6.34	1.59	n.d.	1.52	n.d.	14.2
G4	Rich Cu phase	2.11	0.36	0.21	n.d.	1.33	n.d.	n.d.	n.d.	n.d.	1.22	93.0	n.d.
	Rich Cu–Fe phase	11.3	1.68	1.03	n.d.	3.08	n.d.	n.d.	0.31	0.33	16.9	65.0	0.41
	Glassy phase	60.5	10.9	6.37	5.15	5.91	n.d.	n.d.	0.07	0.07	0.78	1.19	9.08
G5	Long crystal	19.1	1.12	n.d.	n.d.	n.d	n.d.	0.34	2.11	76.2	n.d.	n.d.	0.33
	Rich Sb phase	39.1	4.25	n.d.	n.d.	2.18	n.d.	0.06	n.d.	0.06	1.08	1.92	51.4

n.d. means "not detected." These samples were analyzed through Cambridge LINK–AN10000 scanning electron microscopy equipped with an energy-dispersive analyzer of X-rays (SEM-EDAX) at the University of Science and Technology in Beijing.

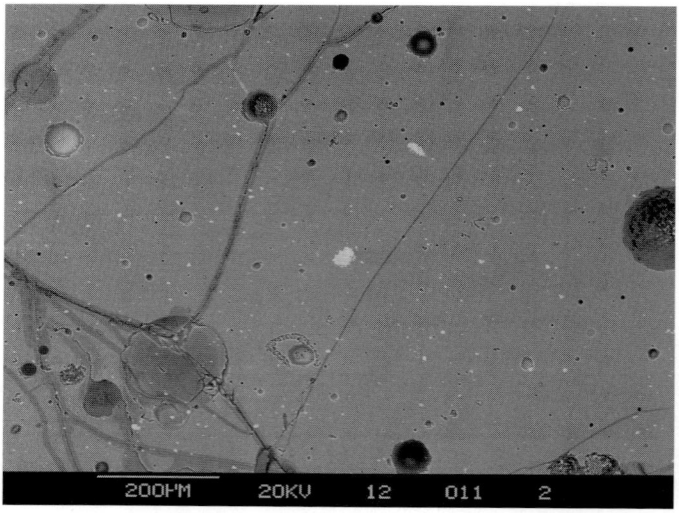

Fig. 11.2. SEM microstructure of sample G3. The bright white particles are Pb–Sb-rich inclusions.

Figure 11.2 is a scanning backscattered electronic reflection of sample G3 taken by scanning electron microscopy (SEM). It shows that the glass is comparably uniform, sometimes mixed with distribution of small particles. The bright particles are rich in lead and antimony, which play a dominant role in yellow coloring. The bubbles and cracks in this sample observed through the electron microscope are more obvious.

Figure 11.3 is a scanning backscattered electronic reflection of sample G4 taken by electron microscopy. The big crystal particles in the intermediate matrix are colorant particles, the brightest part is the copper-rich phase, the second-brightest part is a transition layer of the rich metallurgy phase, and the matrix is the glass phase with the dark bubbles and cracks. The white color gradually turns into a dark color from the colorant particles to the matrix, so we can clearly see the effect of coloring.

Microstructure analysis shows that these samples, to a greater or lesser extent, contain the copper-, iron- and antimony-rich phases; all these might have something to do with the copper

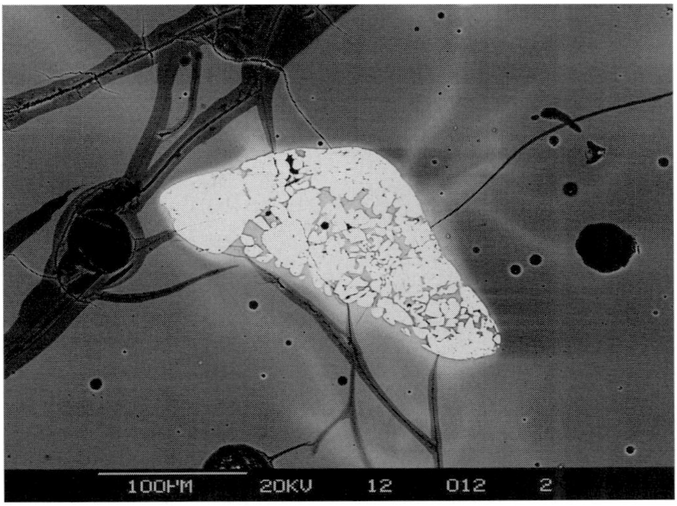

Fig. 11.3. SEM microstructure of sample G4. The white part is the copper-rich phase, and the gray parts among the white ones are the Cu–Fe-rich phase.

smelting. As we know, the raw material of copper smelting often includes iron sulfide and antimony sulfide as well as copper sulfide (the former is called chalcopyrite, the latter tetrahedrite), which are common in nature and are associated with the copper mine. The most common element of the slag produced in copper smelting is iron. In addition, there are a few copper and antimony elements in the slag. Generally, if the copper slag contains a higher level of the iron oxide phase, it is easy to form iron olivine of crystalline silica with a low melting point, and the major components of the glassy phase are calcium and magnesium oxide. The high calcium content in these glass samples also shows that it is much more likely to be formation of the glass by rapidly cooling down the copper smelting slag.

Among the glass samples examined, one is a lead glass containing some antimony (sample G3); the lead is up to more than 4%. Obviously, this is not caused by unconsciously mixing a small amount of lead with the glassmaking raw materials. It might be similar to the process of metal alloying, which is adding certain minerals consciously to change the properties of materials so as to achieve some useful results. The rich resource of lead mines around the cemetery provides a wealth of lead raw materials. A spoon unearthed from the same cemetery was tested and the result showed that it was leaded tin bronze with an above-10% lead weight.[10] A bronze button unearthed from the 4th Chawuhu Cemetery in Hejing county (see Fig. 11.1), with a similar age and geography, was tested and found to be lead bronze; this shows that people there were aware of the role of lead mines earlier on and used them in smelting galena; so the use of galena in glass-melting should not be too difficult to understand.

4. A Study of the Slag Unearthed from Smelting Ruins Near the Kiziltur Cemetery

An analysis of the furnace slag that was unearthed from the ruins of about 200 BC located at the Keriya River site in the south of the hinterland of the Tarim Basin (see Fig. 11.1), about 250 km from the Kiziltur Cemetery, has shown the existence of glassy slag.[11] The observation through an orthogonal polarizing microscope showed slag in the formation of glass, in which there are a large number of bubbles. The transparent gray material is a mixture of SiO_2, Al_2O_3 and CaO with red Cu_2O distributed at the center, containing a few metal copper particles, with a few blue CuS particles and brown ferrous oxides. The analysis result of SEM-EDAX shows that the average slag proportion is: Cu 12.7%, Mg 4.62%, Fe 11.9%, S 6.20%, Si 44.4%, Ca 13.0%, Al 5.50%, P 1.60%.

An investigation of a series of colorful glassy slag from Kangcun village, Kuche county, Xinjiang Region (see Fig. 11.1), has been carried out by Jianjun Mei and Thilo Rehren.[12] Although the carbon-14 dating of the slag at Kangcun was back to the 18th century AD, the examination of the slag also showed an interesting link with the glassmaking. The observation of the slag proved that it is more like glassy material. The results of that analysis showed a certain element of soda and potash, which might have been the result of

Table 11.4. Semiquantitative bulk XRF analysis of the Kangcun samples (wt%).[12]

	SiO_2	CaO	Al_2O_3	FeO	K_2O	Na_2O	MgO	SrO	CuO	SO_3
KC-01	84.0	3.5	3.5	5.9	1.0	1.4	0.4	0.1	0.1	0.0
KC-04	58.5	24.5	8.5	1.5	2.0	2.0	1.2	0.2	0.9	0.0
KC-05	66.5	17.0	9.0	0.9	1.0	2.6	1.3	0.1	0.3	0.0
KC-06	66.5	17.5	8.5	1.8	1.6	2.2	1.4	0.0	0.9	0.0
KC-07	65.0	20.5	7.0	1.2	1.8	2.0	0.7	1.7	0.5	0.1
KC-10	66.0	13.5	8.0	1.2	1.9	2.6	0.9	2.1	3.8	0.0
KC-14	71.5	13.0	7.5	1.1	1.4	2.7	0.7	1.7	0.7	0.0

Note: These samples were analyzed through X-ray fluorescence spectrometry (XRF) in the archeological science laboratory at UCL.

using fuel ash in the smelting process. It seems to be some sort of link with the glass (Table 11.4).

5. The Relationship Between Glassmaking and Metallurgy

Metallurgical waste is often used in the production of glass in modern industry, which has demonstrated the relationship between glassmaking and metallurgy.[13, 14] As to the relationship between ancient glass technology and metallurgical technology, we can study it at least from the following aspects:

(1) *Metallurgical raw materials and glassmaking raw materials*. The main raw materials used in metallurgy are all kinds of natural metal mineral and flux (limestone). Many minerals in metallurgical raw materials can be transformed into various oxides in the process of metallurgy, some are mixed with metal inclusions to be a mixture, and some mixed with slag to be the composition of slag. The raw materials of glass also require the use of various natural ores, which need to be refined in order to become useful minerals. Many sources of minerals have a direct relationship with metal ores, so knowledge of metallic minerals may promote the production of glass.

(2) *Metallurgical slag and glass.* Glass and metallurgical slag with glassy matrix have similar physical and chemical properties and quite similar composition. For instance, the soda-lime glass system generally contains a lot of calcium oxide, which coincides with calcium features of the metallurgical slag. Many of the major components of glass can be found in metallurgical slag. There are a large number of links between leaded glass and smelting lead. Even the relatively normal-composition of glass (soda and potash) previously thought to be less in metallurgical slag, can now also be detected in it.

(3) *Metallurgical crucible and glass crucible.* Although very few sites for the production of glass have been uncovered, glass crucibles already found were largely influenced by the metallurgical crucible. It is even right that the metallurgical crucible was used in the production of glassmaking. Smelting foundries or crucibles usually have a strong resistance to high-temperature performance, and also play a good corrosion resistance role in acidic oxide (primarily silica). This is essential for the process of glassmaking, because the main ingredients of glass are various silicates that generally have lower alkalinity.

(4) *Metallurgical trace elements and glass coloring agent.* The glass coloring agent is an important aspect of glass technology. Many studies have shown that the colorant itself has many links with metallurgical trace elements. Various elements in metallurgical minerals are likely to have been found and used in the metallurgical process by early glassmakers. Slag left in the smelting crucible could later be used to make glass, so some of the slag material might be easily mixed with glass, forming coloring particles in the high-temperature process, and then gradually spreading to the whole glass; this is particularly notable in the early glass with poor transparency.

6. New Researches on the Origin of Chinese Glass Technology

China has a long history, with an ancient civilization lasting thousands of years. Unearthed ancient glass, especially its origin, has naturally aroused the keen interest of foreign scholars and attracted widespread attention. In the 1930s, Beck and Seligman analyzed the collected ancient Chinese glass, and PbO and BaO were found in it; the glass is different from ancient Egyptian and Mesopotamian glass.[15, 16] Thus they concluded that Chinese glass has the possibility of an independent origin. Scholars from Britain, Russia, the United States, Japan and other countries analyzed the composition of ancient Chinese glass as well and basically agreed that the content of PbO and BaO was a fundamental characteristic of the glass. They have also studied the possible track of the spread of ancient Chinese glass. The Corning Museum of Glass in America has also conducted many meaningful researches on the glass. Using the method of the lead isotope ratio, Brill studied ancient Chinese glass so as to identify its origin.[17] He found that ancient Chinese glass from the Warring States Period (476–256 BC) to the Han Dynasty (206 BC–220 AD) has higher lead isotope ratios than that uncovered abroad. Hall found that the glass of the Sarmatia culture of Russia and the Kazakhstan border is similar to ancient Chinese glass.[18]

Chinese scholars began to pay attention to the problem of the origin of glass technology from the 1950s. The Warring States glass unearthed in Changsha and Luoyang was analyzed and found to be lead barium or lead silicate glass by Hanqing Yuan. In the 1980s, due to the constant richer archeological data and the involvement of technical experts on glass, the research on the origin of Chinese glass underwent rapid development. In 1983 and 1984, two academic conferences, the Symposium on Ancient Chinese Glass and the International Symposium on Glass (archaeological glass), were held respectively in Changsha and Beijing. The publications after the conferences are still of great importance.[19] Since the 1990s, the archeological community and the scientific community have placed much importance on the origin of the Chinese glassmaking technique. In particular, the K_2O–SiO_2 series of the Han Dynasty unearthed in Guangdong and Guangxi, and the Na_2O–CaO–SiO_2 series unearthed in Xinjiang, are very important for the research on the influence of the maritime Silk Road and the northwestern Silk Road on the origin and spread of the glassmaking technique.

The representative Chinese studies include the following: Fuxi Gan, on the basis of chemical analysis, holds that ancient Chinese glass is lead–barium glass, different from the soda calcium glass of the West, and thus concludes that Chinese glass has an independent origin.[20] The summary of his recent research on ancient Chinese glassware reiterates the importance of the studies on the origin of ancient Chinese glass technology.[21] Meiguang Shi has systematically analyzed the ancient leaded glass unearthed in China and holds that people in the Western Zhou Dynasty (1046–771 BC) had mastered the use of lead compounds, and the ceramic sintering temperature could reach 1200°C, the glass melting temperature.[22] Zhuhai Cheng proposed considering the Western Zhou Dynasty colored glaze unearthed at Baoji, Fengxi, Shaanxi province, and at Luoyang, Shanxian, Henan province, as the early embryo of ancient Chinese glass.[23] Through scientific analysis and testing, Fukang Zhang argued that the colored glaze beads unearthed at the Western Zhou cemetery are multiquartz beads with some appearance of glass.[24] He found that they are faience, and not ancient Chinese glass. He rejects the glaze origin of ancient Chinese glass. Jiayao An holds that the earliest Chinese embedded glass beads date back to the late Spring and Autumn Period and the early Warring States Period. The beads show similar technique, emblazonry and chemical composition to those from West Asia, and were probably spread to the central plain region of China by nomads.[25] Using the lead isotope ratio method, Xiaocen Li believes that the lead barium glass originated in Yunnan, southwest China.[26] Using isotope ratios on ancient Chinese lead (barium) glass, Zhonghong Jiang determined the locality of the origin of ancient Chinese glass.[27] Qinghui Li summed up the various analyzing and testing techniques for archeological glass, and conducted a more systematic study of some recently unearthed ancient Chinese glass using PIXE.[28, 29]

These research results rely on archeological excavations, and most dates are accurate. Some researchers started with glass systems, and some conducted studies through gradual improvement of the analyzing and testing tools and began to make use of lead isotope ratios. The results generally show that lead barium glass is a unique system of ancient Chinese glassware, different from that of Western lime-soda-potash glass. Ancient Chinese glass may have a complicated origin. There exists the possibility of import and local production, but further scientific analyses are needed.

New archeological discoveries at the Kiziltur Cemetery may help re-examine the data and finds. Currently, in the rotten woods of the burial site at the Kiziltur Cemetery, nine series of carbon-14 dating data have been obtained. In the western part of the cemetery, M9 is aged 2494 ± 61 years, M11 2578 ± 92 years, M14 2569 ± 65 years, M15 2522 ± 62 years, M22 2569 ± 72 years, and M27 2787 ± 62 years. In the eastern part of the cemetery, M20 is aged 2893 ± 66 years, M13 2873 ± 63 years, and M8 3327 ± 83 years. This list gives the absolute date data. Through tree-ring calibration, they are all around 1110–600 BC. Combining the data on burial pottery, it is considered that the Kiziltur Cemetery is closely related to the Chawuhu Cemetery (990–625 BC), the Luntai Cemetery (950–62 BC), etc. The archeological studies showed that the Kiziltur Cemetery itself represents one type of Chawuhu culture in central Xinjiang in the first millennium BC. Therefore, archeological dating can also show that the Kiziltur Cemetery dates from the Western Zhou Dynasty (1046–771 BC) or the early Spring and Autumn Period (770–476 BC). These glass beads and debris should be the earliest-ever glass unearthed in China, based on scientific study.

Worldwide, it is generally believed that the earliest glass unearthed dates back to around 2500 BC in Egypt or Mesopotamia. Glass then gradually spread to the entire Near East and Middle East region. For example, in Nuzi, Iraq, a number of glass beads date back to 1700–1400 BC. The appearances of these glass beads are similar to those unearthed at the Kiziltur Cemetery. The matrix analysis of the glass in Mesopotamia has shown that it is all soda-lime glass, which is also similar to that at the Kiziltur Cemetery. In addition, Indian glass is given much attention. According to *Hanshu — Xiyu Zhuan* (*The History of the Han Dynasty — The Memoir of the Western Regions*), Kapisa (a region up to the Indian River) produced a great number of jade glasses in the first several centuries after the Christian era. The early Xinjiang glass mentioned in this article contains low aluminum and magnesium, sometimes similar to early Indian glass. This perhaps is a hint that there is some link between Xinjiang glass and ancient India. Western glass has a history 1500 years earlier than that of central China. Given the fact that the cultural and scientific exchange on the Silk Road around 1000 BC was already busy, it is understandable that glass, as an important artifact in the West, spread to China. Meanwhile, the earliest glass produced in Xinjiang by a number of local craftsmen using local minerals is earlier than the glass produced in central China. The Kiziltur glass beads are immature glass, at the stage of exploration and development. They were directly influenced by the Western technology and may even have a link with the early glass in the South Asia subcontinent. We hope that the new archeological discoveries and research in Xinjiang will confirm the route through which glass spread from the West to central China through Xinjiang Region.

7. Methodologies of the Research on the Origin of Chinese Glass Technology

It is beneficial for the systematic study of the origin of the Chinese glassmaking technique to combine the latest domestic and foreign researches, to test the composition and analyze the microstructure of the unearthed glass in China and survey the relevant mining sites, to emphasize the influence of the ancient metallurgical technology and others on glass production, and to conduct simulation of glass production. That should be carried out in the following steps:

(1) Sorting of glass archeological data: to sum up the information on unearthed glass before the Han Dynasty (206 BC–220 AD) and classify and count the glass by type of object, color, state, region and time.
(2) Analysis of unearthed glass relics: to examine the composition, structure, microstructure and optical properties of early unearthed glass. Due to the complexity of ancient glass production, it is necessary to study the microstructure and composition of glass and to emphasize the study of the coloring mechanism of melting impurities and trace elements.
(3) Surveying of and research on glass production remains: we look forward to finding ancient crucibles or hearths of glass production, of which we can conduct comprehensive and detailed studies and do sample analyses.
(4) Influence of metallurgical technology on the origin of glass: conduct systematic surveys and studies of related remains, pay attention to the crucibles used in the melting of metal, analyze in particular the composition and structure of metallurgical slag, and conduct comparative studies of related glass artifacts.
(5) Influence of ceramic technology on the origin of glass: combining with the existing research findings on ancient ceramic technology, conduct analyses of and researches on the materials of glass and the pottery production temperature, and research the relation between the ceramic technology and the glass technique.
(6) Research on the provenance of glass: using the lead isotope ratios method, focus on the early Chinese lead (barium) glass, and, combining with the geological and mineral research findings and lead isotope ratios of bronzes, explore the origin of early glass.
(7) Comprehensive comparative study: based on the results of the above studies, combining with domestic literature data and simulation, explore the origin of glass technology.

The scientific study of ancient Chinese glass has just started and needs full cooperation from all departments, the combination of various experimental methods and the establishment of databases. The study of glass after the Han Dynasty can be focused on the composition analysis while the study of glass before the Han Dynasty should be focused on the study of microstructure. With the progress of study, it has been found that the classification of ancient Chinese glass of the south and the north can hardly manifest the true features; thus, it is recommended that we treat Chinese glass as a whole and create reasonable regional division. Metallurgy archeological and ceramic archeological scholars should actively join in the research and explore common things and seek links, which can contribute to the in-depth study of the origin of Chinese glass. We hope the combination of new archeological discoveries and scientific analyses will solve the age-old mystery of the origin of ancient Chinese glass.

8. Conclusions

Metallurgy is one of the five possible origins of glassmaking. The scholars in China and abroad have done some research on it. The glass beads unearthed from the Kiziltur Cemetery in Baicheng county, Xinjiang Region, northwest China, are the earliest glass found in China through scientific analysis. Examination of those samples showed that the metallurgical slag or some alloying agent was used in the process of glassmaking from the 11th to the 6th century BC in central Xinjiang Region. The trace elements in the glass are similar to those in bronzes in the same cemetery, which can be found among the metallurgical remains nearby. Unfortunately, glassmaking crucibles were not found at the cemetery or among related remains at that time. The composition of the glass at the Kiziltur Cemetery is similar to that of the glass in Mesopotamia, and gives some hints of the connection with Indian glass. It is obvious that the Silk Road has been the main route of the technological transmission of glassmaking from West Asia to central China since the beginning of the first millennium.

Further studies of the origins of the Chinese glass should be carried out, after detailed planning.

Acknowledgments

I would like to thank Mr Ping Zhang of the Institute of Culture Relics and Archeology of Xinjiang, for his excavation at the Kiziltur Cemetery and his kind invitation to analyze the early glass bead unearthed from that cemetery. I would also like to thank Ms Yixian Lin of the Institute of Historical Metallurgy and Materials, University of Science and Technology in Beijing, for useful discussion and some experimental work on this article. Finally, I thank Prof Fuxi Gan and Dr Qinghui Li of the Shanghai Institute of Optics and Fine Mechanics, Chinese Academy of Sciences, for inviting me to present the main opinions of this article at two symposiums, in Urumqi and Shanghai in 2004 and 2005, repectively.

References

1. G. Hartmann, Chemistry and technology of prehistoric glass from Lower Saxony and Hesse, *J. Archaeol. Sci.* **24,** 547–559 (1997).
2. T. Rehren, Rationales in Old World base glass compositions, *J. Archaeol. Sci.* **27**, 1225–1234 (2000).
3. T. Rehren, New aspects of ancient Egyptian glassmaking, *J. Glass Studies* **42**, 13–24 (2000).
4. J. L. Mass, M. T. Wypyski and R. E. Stone Malkata, and Lisht glassmaking technologies: towards a specific link between second millennium BC metallurgists and glassmakers, *Archaeometry* **44**(1): 67–82 (2002).
5. B. D. Yang, About several problems on the research on Chinese ancient glass, *Culture Relics* (in Chinese) **5**, 76–78 (1979).
6. K. H. Zhao, The study on the origin of Chinese traditional glass and the contribution of the alchemy, *Studies in History of Natural Sciences* (in Chinese) **10**(2): 145–146 (1991).
7. S. L. Zhou *et al.*, Analysis on the glass samples unearthed from iron smelting furnaces at Yangcheng site in East Zhou Dynasty in Henan province. *Archeology* (in Chinese) **7**, 76–79 (1999).

8. Xinjiang Institute of Culture Relics and Archaeology, The report of the first excavation at Kizilur Cemetery at Baicheng County in Xinjiang Region, *Archeology* (in Chinese) **6**, 14–28 (2002).
9. W. Qian, P. Zhang and Q. M. Li, Analysis and research on the glass beads unearthed from Kizilur Cemetery (1100 BC–600 BC), in *Proceedings of the Fifth Symposium on the History of Science and Technology of Chinese Minor Nationalities* (Guangxi Nationalities Press, 2002) (in Chinese), pp. 138–145.
10. P. Zhang, I. Abduresul and W. Qian, Metallurgical study on the bronzes unearthed from the Kizilur Cemetery at Baicheng, Xinjiang, in *Proceedings of the Fifth Symposium on the History of Science and Technology of Chinese Minor Nationalities*, (Guangxi Nationalities Press, 2002) in Chinese, pp. 130–137.
11. W. Qian *et al.*, Metallurgical studies on the metal relics unearthed in the Keriya Valley in Xinjiang Region, *Studies in Western Region* (in Chinese) **4**, 1–11 (2000).
12. J. J. Mei and R. Thilo, Copper smelting from Xinjiang, NW China. Part 1: Kangcun village, Kuche county, c. 18th century AD, *Historical Metallurgy* **39**(2), 96–105 (2005).
13. Z. Y. Liang and R. Lian, Production of construction glass using the slag from metallurgical industry, *Construction Materials in Shandong Province* (in Chinese) **2**, 25–26 (1997).
14. L. D. Teng *et al.*, Study on the theory of the micro-crystal glass corelization of furnace slag, *Transactions of Silicate Studies* (in Chinese) **1**, 14–18 (1995).
15. H. C. Beck and C. G. Seligman, Barium in ancient China, *Nature* **133**(6), 982 (1934).
16. C. G. Seligman *et al.*, Early Chinese glass from pre-Han to Tang's time, *Nature* **138**, 721 (1936).
17. R. H. Brill, *Chemical Analysis of Early Glass*, Corning Museum of Glass Press, New York, (1999).
18. M. E. Hall, Chemical analyses of Sarmatian glass beads from Pokrovka, Russia, *J. Archaeol. Sci.* **25**, 1239–1245 (1998).
19. F. X. Gan (ed.), *Studies on the Ancient Chinese Glass — Proceedings of the International Symposium on the Glass Research in Beijing in 1984*, (Chinese Construction Press, Beijing, 1986) in Chinese.

20. F. X. Gan, Z. F. Huang and X. R. Xiao, On the origins of the ancient Chinese glasses, *J. Chin. Cer. Soc.*, (in Chinese) **12(6)**, 99–103 (1978).
21. F. X. Gan, Several aspects on the research on the ancient Chinese glass, *J. Chin. Cer. Soc.*, (in Chinese) **32**(2), 182–188 (2004).
22. M. G. Shi *et al.*, Study on a series of leaded glass in ancient China. In: F. X. Gan (ed.), *Studies on the Ancient Chinese Glass — Proceedings of the International Symposium on the Glass Research in Beijing in 1984*, (Chinese Construction Press, Beijing, 1986) in Chinese.
23. Z. H. Cheng, Primary study on the development of Chinese glass, *J. Chin. Cer. Soc.*, (in Chinese) **9**(1), 79–84 (1981).
24. F. K. Zhang *et al.*, Study on ancient Chinese glasses, *J. Chin. Cer. Soc.*, (in Chinese) **11**(1), 67–75 (1983, 1983).
25. J. Y. An, Three items of glass archeology, *Cultural Relics* (in Chinese) **1**, 89–96 (2000).
26. X. C. Li, On the provenance of the origins of lead barium glass in China, *Studies in History of Natural Sciences* (in Chinese) **15**(2), 144–150 (1996).
27. Z. H. Jiang and Q. Y. Zhang, Research on the ancient Chinese lead (barium) glass by using the isotope ratios method, *J. Chin. Cer. Soc.*, (in Chinese) **26**(1), 109–113 (1998).
28. Q. H. Li *et al.*, Research on the early glasses unearthed at Baicheng and Tacheng in Xinjiang, *J. Chin. Cer. Soc.*, (in Chinese) **31**(7), 663–668 (2003).
29. Q. H. Li *et al.*, PIXE technology applied in the study on the composition analysis of ancient Chinese glass, *J. Chin. Cer. Soc.*, (in Chinese) **31**(10), 950–954 (2003).

Chapter 12

第十二章

The Inspiration of the Silk Road for Chinese Glass Art

Lu Chi

Institute of Visual Art, Fudan University, Shanghai 200433, China

Archeological study has proven that the Chinese were able to produce glass in the Western Zhou period, which was more than 3000 years ago. From the end of the Spring and Autumn Period, foreign glass products and technology came to China and impacted Chinese glass development. At that time, two developed cultures existed at opposite ends of the old world: China and Greece, both during the period of the Warring States. Meanwhile, the Achaemenidae Empire unified Syria and Iran. A dangerous trade route named the "Scythian Path" tied the East to the West. This path across the grasslands was occupied by nomadic tribes, and made the cultural exchange possible. It brought the first stone of other mountains to Chinese antique glass art. From then on, Chinese antique glass art had a long, jadelike period. This was after the eye bead time.

1. The Silk Road Opened the Glass Road

The Silk Road connected the economy, culture and technology of the continents of Asia and Europe, and became the most important land route for Sino-Western cultural exchange. This was during the

Han Dynasty. Between the 2nd century BC and the 2nd century AD, there were four empires along this Asia–Europe inland main route (Fig. 12.1). From east to west, they were:

(1) Han Dynasty (206 BC – 220 AD), in East Asia
(2) Kushan (45–226 AD), in Middle Asia
(3) Parphia (3rd century BC – 226 AD), in West Asia
(4) Roman Empire (30 BC – 284 AD), in Europe

During the Christian period, the four empires entered a power-extending phase. The visit of the Han Dynasty's ambassadors Zhang Qian and Gan Yin changed the indirect contact among the Chinese, Indian, West Asian and Greek–Roman cultures. There was direct exchange and dialogue. From then on, the developments of those four cultures were not isolated.

The area of West Asia called Mesopotamia, which included Syria, Iran, Anatolia and Elam, was the cradle of glassmaking. This area started to make glass around 3000 BC. Greece and Rome occupied the central station in the glass development history. This culture was steeped in glass much more than any other culture, and gave glass art a golden age. From the 3rd century BC to the

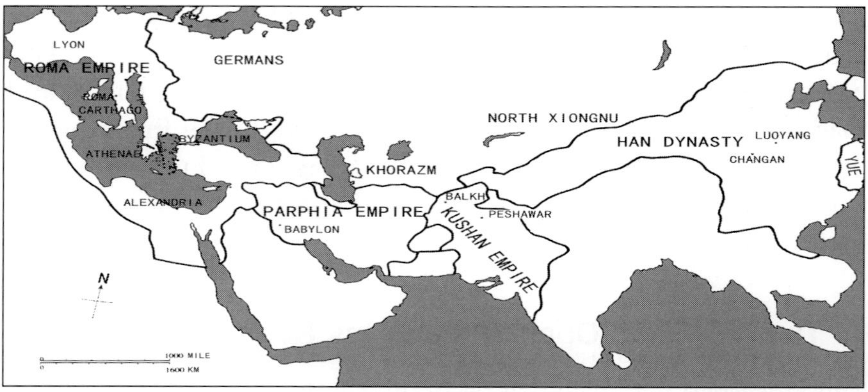

Fig. 12.1. The Han Dynasty, Kushan, Parphia and the Roman Empire around 100 AD.

19th century AD, Rome kept a leading position in many areas of glassmaking. So, from a developmental angle, exchange with those areas could bring many benefits to Chinese glassmaking at that time.

2. The Impact of the Silk Road on Chinese Antique Glass

Since the Chinese culture had a strong affinity for jade, and the rooted traditional ceramic culture was booming, Chinese antique glass always grew under a full-bodied shadow of jade and ceramics. Imitation of jade became the main use of glass material, according to shape, color, ornament and function. From glass *bi* disks (Fig. 12.2), *huan* loops, sword suits and plugs in the early age to glass band suits, jerry and ornaments in the later times, glass was used as a substitute material for jade. Some glass imitated jade exactly in color, texture and quality. The imitation of jade gave the Chinese antique glass a big disadvantage with regard to material language and article expression. It also hampered the growth of Chinese antique glass art. For the incunabulum, glass always took the role of a man-made substitute for natural jade. It found a special material language relatively quickly in the cultures of West Asia, Egypt, Greece and Rome. To control a material's language, you need to understand the material's character and discover the technology as a base. For example, the Romans created the gold leaf sandwich glass technology in the 2nd century BC (Fig. 12.3). This technology sandwiches the gold leaf between two pieces of flat glass and combines it with the glass in the slumping stage. During the same period, the rod-casting technology was also very popular in the Western world. Glass rods were cut, arranged, and fused as a flat with pattern, then slumped in a mold. Those attempts made glassmaking technology develop from the coring form and casting form to slumping, fusing, blowing, and lamp work. Thus glass began to own its article language, station and value. The value of Chinese antique glass seems to have been locked into its substitution for jade. Glass could not get rid of jade's opaque color. After casting,

Fig. 12.2. Glass *bi* disk, Warring States.

Fig. 12.3. Gold sandwich glass bowl, 2nd century BC, Rome.

glass requires a lot of cold work, like jade. Undoubtedly, the Silk Road imported fresh air into the Chinese style of glass art. Although the West did not impact China in some areas of thought like today, the colorful, bright, fine glassware from the West was something that the Chinese could not refuse. As the social needs increased, the technology came into China easily.

The main impact of the Silk Road on Chinese antique glass can be summarized as follows.

2.1. *Transformation of the glass chemical composition*

From the analysis of the material's chemical composition, Chinese antique glass can be divided into four stages:

(1) *From the Western Zhou in the 5th century BC to the Spring and Autumn Period (770–481 BC)*. For this period glass composition belongs to high silicon oxide and has a lot of agglomerate of quartz crystal, which could be considered as the embryo of Chinese antique glass.
(2) *From the Warring States (481–221 BC) to the Sui Dynasty in the 6th century BC*. At this stage, Chinese glass began to have its own system. Early in this period, $PbO–BaO–SiO_2$ glass was the main system. Later, $PbO–SiO_2$ system glass took its place. Some $K_2O–SiO_2$ glass also appeared in this period.
(3) *From the Tang Dynasty in the 7th century to the Yuan Dynasty in the 13th century*. The main composition still was the $PbO–SiO_2$ system; $Na_2O–CaO–PbO–SiO_2$, $K_2O–CaO–SiO_2$ and $Na_2O–CaO–SiO_2$ also were found in this period.
(4) *From the Ming Dynasty in the 14th century to the Qing Dynasty in the 19th century*. The main composition included $K_2O–PbO–SiO_2$, $Na_2O–CaO–PbO–SiO_2$, $K_2O–CaO–SiO_2$ and $Na_2O–CaO–SiO_2$.

From the evolutionary changes of Chinese antique glass' chemical composition, the following conclusion can be made:

The Chinese antique glass system was the $PbO–BaO–SiO_2$ and $PbO–SiO_2$ system. This kind of glass is brittle and cannot withstand the strong transformation of temperature. This type of glass should not be treated as a material for daily use. Because of its brittleness, Chinese antique glass rarely entered the normal people's lives. It did not have a booming development, unlike Chinese ceramics. From the Tang Dynasty on, the composition of the Na_2O–CaO system glass imported into China made glass stronger, and more suitable for daily use.

2.2. *The importation of the blowing technique*

Early in the Han Dynasty, Chinese people had glass for daily use, as ear-cups, bowls, plates and baskets. The casting technique and the complex, time-consuming cold work made glass available only for the nobility and the higher classes. During the Northern Wei (386–534) period, the blowing technique imported along the Silk Road became the main form of technology in China. As recorded in the history book *Wei Shu — Darouzhi*, West Asian businessmen brought the blowing technique to China. During Emperor Shizu's reign, they set up the first big glass-blowing mill at Pingcheng (present-day Datong), the capital of Northern Wei. The blue glass bottle that was found at Dingxian (Hebei province) in 1964 is daily-use glassware made by the blowing technique (Fig. 12.4). The application of the blowing technique enhanced the glass productivity. It also greatly shortened the production cycle of glass vessels, and supplied technological support for the development of daily glassware.

Fig. 12.4. Blowing glass bottle, Northern Wei (386–534).

2.3. *The diversity of the art style*

As a substitute for jade, Chinese antique glass kept the style of the jade art of the same period. The Silk Road brought different foreign styles to China and greatly impacted the style of Chinese glass art. Till the Wei, Jin, and Southern and Northern Dynasties, the style of Rome and Sassanidae began to impact Chinese glass' form and color. Completely different from the traditional massive and dark style, this glassware became light and transparent. New colors also came into use. During the Tang and Sui Dynasties, Chinese glass art followed a Persian style. The standard colors — blue, green, light green and yellow–green — were supplemented by new colors such as milky white, yellow, brown–yellow and brown. Based on carving and stick sculpture decoration, craftsmen developed a fine ornamental technique. This can be seen in the net ornamental glass bottle of the Tang Dynasty, which was found at Lintong (Shanxi province) in 1985. The net ornament was made by the double-stick technique. Those facts from the West greatly enriched Chinese antique glass' art style, and provided inspiration and enlightenment for Chinese glass composition.

3. The Inspiration of the Silk Road for Contemporary Chinese Glass Art

Glass became an independent art material in the 1960's. Contemporary Chinese glass art not only has a successful relationship with traditional Chinese culture, but is also part of the worldwide contemporary glass art movement. From the 1960's on, with the development of the glass studio movement, colleges in the Western world have set up glass art courses. In China, Tsinghua University and Shanghai University began to set up glass art studios in 2000 in order to provide glass art courses. From the idea to the technology, modern glass art comes from contemporary Western art. A most important lesson is to maintain the communication between the Western world and

Fig. 12.5. Detail of Colin Reid's glass work.

China. From the Silk Road's impact on Chinese antique glass, we can see that communication not only can change the history, but also can create the future. When Western glass art ideas come to China, Western and Chinese ideas will impact each other. In British glass artist Colin Reid's work, Chinese characters are elegantly displayed (Fig. 12.5). This is a lively, suitable, refined expression of the glass art language. Undoubtedly, the Silk Road has also inspired modern Chinese glass art. We Chinese should build a new Silk Road to exchange glass art ideas. This will let the world understand us, and also allow us to understand the world.

References

1. A. Toynbee, *A Study of History*, transl. by B. C. Liu and X. L. Guo (Shanghai People Press, 2000) in Chinese, p. 444.
2. K. Cummings, *Techniques of Kiln-Formed Glass* (A & C Black, London, Philadelphia; University of Pennsylvania Press, 1998), p. 24.

3. A. Macfarlane and G. Martin, *The Glass Bathyscaphe*, transl. by K. N. Guan (Commercial Press, Beijing, 2003), pp. 10, 15, 117.
4. J. Y. An, *The Art of Glass Along the Silk Road — China Dawn of a Golden Age, 200–750 AD* (The Metropolitan Museum of Art, New York; Yale University Press, New Haven, London, 2004), p. 57.
5. J. H. Zhou, Culture and Chemistry (Orient Press, Beijing, 2000) in Chinese, p. 41.
6. B. D. Yang, Collection of Chinese Arts: Section of Art and Craft, (Cultural Relic Press, Beijing, 1987) in Chinese, p. 16.

Chapter 13

第十三章

Faience Beads of the Western Zhou Dynasty Excavated in Gansu Province, China: A Technical Study

Zhang Zhiguo and Ma Qinglin

China National Institute of Cultural Property, Beijing 100029, China

1. Introduction

One of the oldest artificial substances, faience was first made (probably in Egypt) 5500 years ago, a millennium before glass was invented. While similar to glass in some ways, it differs in others. Faience is a glazed, nonclay ceramic material.[1,2] It is composed mainly of crushed quartz or sand, with small amounts of lime and either natron or plant ash. The coated glaze on the body is generally a bright blue–green, due to the presence of copper.

Faience was made in many places, such as Egypt, China, Iran and Mesopotamia. In China, plenty of faience artifacts were made, with different shapes, such as beads, tubes and sticks, which were mostly excavated in the provinces of Shaanxi, Gansu, Henan and Shanxi. Most of them have been identified to be in the period from the Western Zhou Dynasty to the Warring States. In general, ancient Chinese faiences have high potassium and low sodium, because they are not composed of natron but plant ash. On the contrary, faiences of West Asia and Egypt commonly have high sodium and low potassium.

At Yu Jia-wan in Chongxin county, Gansu there is a cemetery of the Western Zhou Dynasty (1046–771 BC) that was excavated in 1982, 1984 and 1986 by archeologists. A lot of cultural relics were unearthed, such as bronzes, pottery, jade and faiences.[3] In the present work, fragments of three faience beads belonging to the middle period of the Western Zhou Dynasty are analyzed.[4]

2. Experimental

2.1. *Objects investigated*

The faience bead specimens are shown in Figs. 13.1 and 13.2. They are all beads from a necklace and have a greenish appearance. Cross-section samples through the glaze and into the body were prepared and analyzed directly by Raman microscopy, and then coated with carbon for measurements by scanning electron microscopy and energy-dispersive X-ray spectrometry (SEM-EDX).

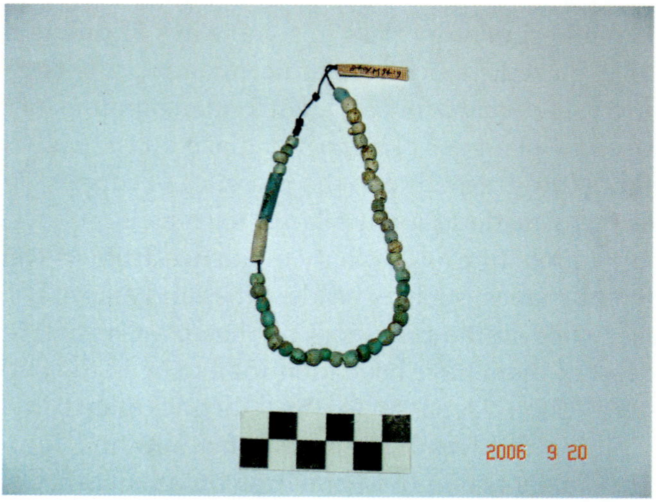

Fig. 13.1. Faience necklace excavated at Yu Jia-wan.

Fig. 13.2. Cross-sections of faience beads (black scale corresponds to 1 cm).

2.2. *SEM-EDX*

SEM-EDX measurements were made by using Hitachi S-3600N (scanning electron microscopy) and EDAX Genesis 2000XMS (energy-dispersive X-ray spectrometry). Samples were coated with a thin layer of carbon to improve the conductivity. EDX analysis was performed at various points or areas throughout the cross-section.

2.3. *Micro-Raman spectroscopy*

A Renishaw System 1000 Raman microscope was used for the Raman measurements. This system comprises a Leica DMLM microscope equipped with a ×50 objective, a spectrometer with a 1200- or 1800-grooves-per-millimeter grating and an NIR-enhanced, Peltier-cooled CCD camera. An air-cooled argon ion laser with a wavelength of 633 nm served as the excitation source for the hematite in the faience specimen. The laser power for the sample can be modified according to the samples.

3. Results and Discussion

3.1. *Manufacturing technology for faience*

Faience frequently has two distinct body layers: a coarse, often discolored core covered by a brilliant white layer over which the glaze was applied.[5] At temperatures of around 600, the edges of the clay platelets in faience begin to fuse in a process known as sintering, but most of the particles are still angular (Figs. 13.3–13.5).

Raw materials of faience can be classified into alkaline, acidic and neutral, according to their chemical properties. Alkaline raw materials of glaze include sodium oxide (Na_2O), potassium oxide (K_2O), lead oxide (PbO), calcium oxide (CaO), zinc oxide (ZnO), etc., which help to reduce the melting temperature of the silica and combine with it to make the glaze. Acidic raw materials of glaze are mainly silica (SiO_2) in quartz sand, which constitutes the basic network former of glaze. Neutral raw materials of glaze are alumina (Al_2O_3) coming from clay, which can increase the viscidity and density of the glaze. In faience, the proportions of ingredients are

Fig. 13.3. Scanning-electron-microscope photograph of a section through faience GCYF-1. The upper part is the surface layer, and the undersurface is the interior of the faience.

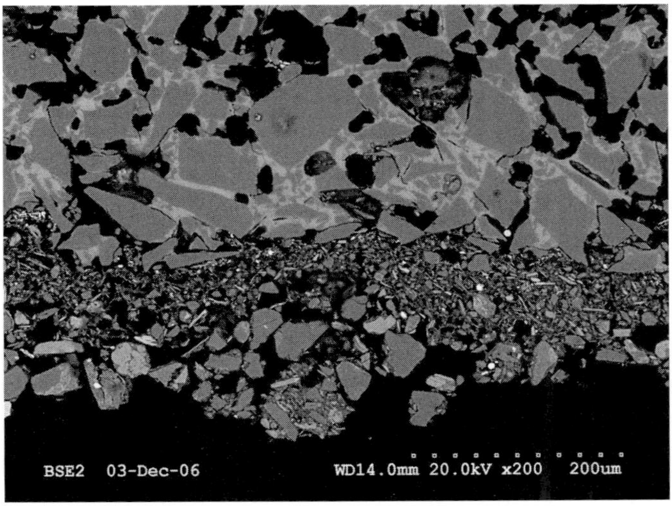

Fig. 13.4. Scanning-electron-microscope photograph of a section through faience GCYF-1. The upper part is the interior of the faience, and the undersurface is the core.

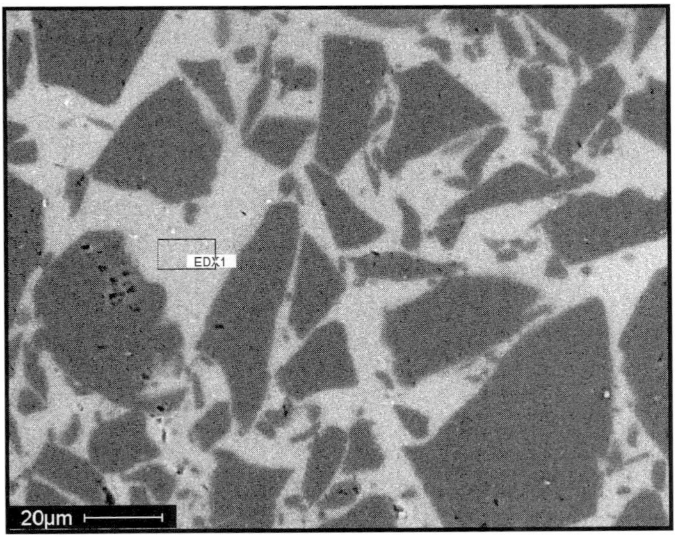

Fig. 13.5. Scanning-electron-microscope photograph of a section through faience GCYF-1. The area labeled EDX1 is the glass phase, and the black area is quartz.

different, and the higher proportion of silica, probably combined with a lower firing temperature and a shorter firing duration, produces a crystalline material. However, the fluxes help to develop some degree of interstitial glass, which fuses the mass together.

Vandiver divided the various faience glazing techniques into three broad categories[6–8]: efflorescence, cementation and application.

Efflorescence is a so-called self-glazing method in which the glazing materials, in the form of soluble salts, are mixed with the raw crushed quartz and alkalis of the body. As the water in the body evaporates, the salts migrate to the surface to form a kind of scum. In firing, this precipitated layer melts and fuses to become a glaze. The glazing methods for GCYF-1, GCYF-2 and GCYF-3 are all efflorescence, mainly because there still remain some interstitial glass interior layers, and there are obvious and narrow, well-defined interfaces between the glaze and the underlying body. For samples GCYF-1 and GCYF-3, the glaze is only distributed on the surface of the faience, because there are clays in their cores before firing. But, for sample GCYF-2, the glaze is distributed on both the external surface and the interior surface of the faience, because there are no clays in its core before firing (Figs. 13.6–13.8).

3.2. *Determination of chemical compositions*

The polished sections of the three faience bead samples were examined by SEM-EDX, in order to determine their microstructures and concentrations; the backscattered electron mode in which the different phases can be distinguished on the basis of their atomic number contrast is used. SEM-EDX data for the surface layer of the three faience bead samples showed that the Cu, Na and K contents are higher than those of the interior (Figs 13.9–13.11, Table 13.1), and formed the SiO_2–K_2O–Na_2O-CuO glaze on the surface of the faience beads.

Bulk analyses of the glaze of GCYF-1 show concentrations in the range of 77.0–79.4% for Si, 6.9–7.6% for Na, 2.3–2.6% for

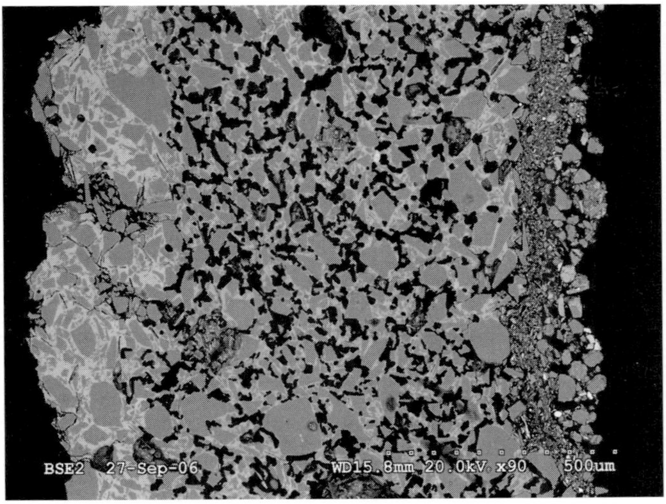

Fig. 13.6. Scanning-electron-microscope photograph of a section through faience GCYF-1, glazed by efflorescence. Glaze and interstitial glass are shown in gray, quartz in dark gray, and voids in black.

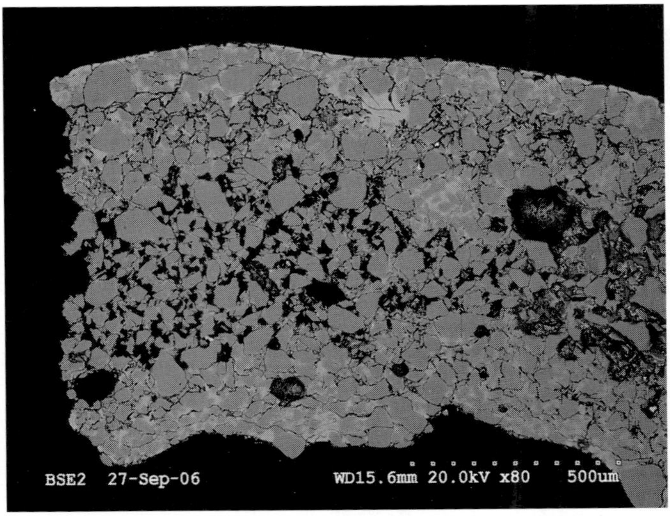

Fig. 13.7. Scanning-electron-microscope photograph of a section through faience GCYF-2, glazed by efflorescence. Glaze and interstitial glass are shown in gray, quartz in dark gray, and voids in black.

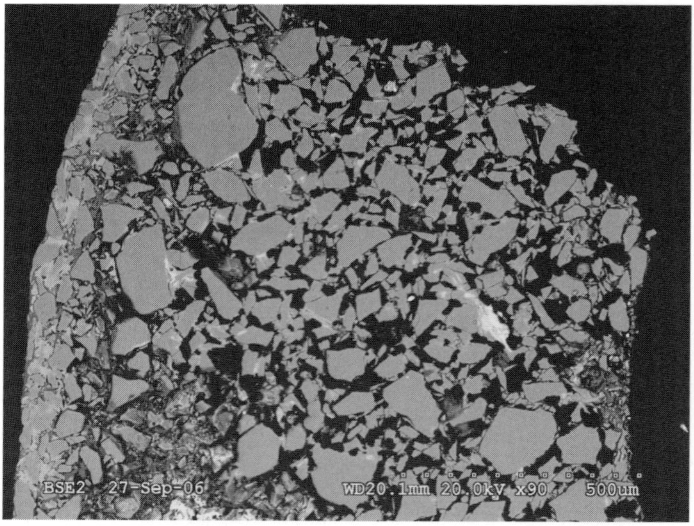

Fig. 13.8. Scanning-electron-microscope photograph of a section through faience GCYF-3, glazed by efflorescence. Glaze and interstitial glass are shown in gray, quartz in dark gray, and voids in black.

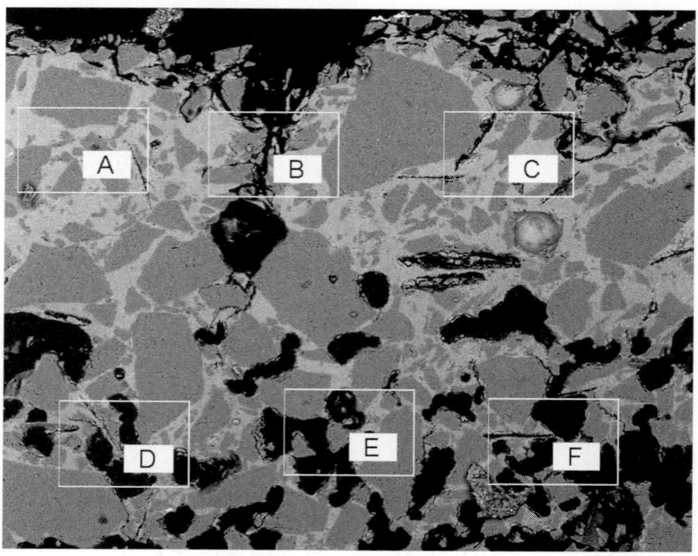

Fig. 13.9. Backscattering electron image from SEM of GCYF-1.

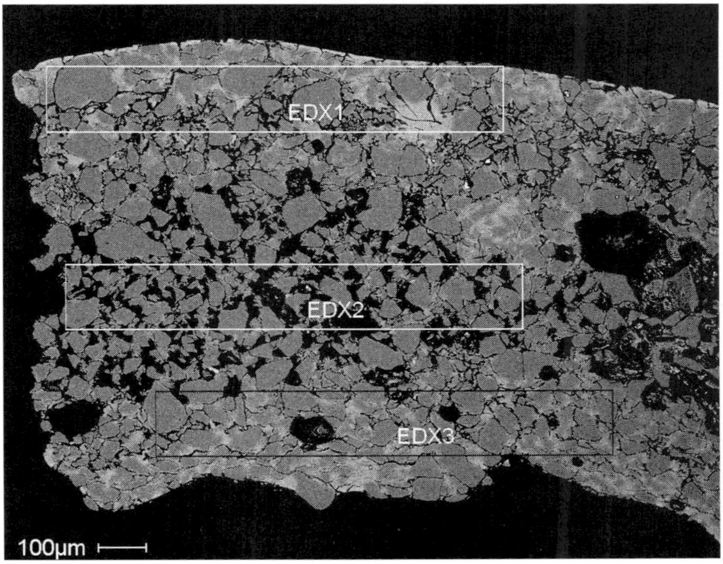

Fig. 13.10. Backscattering electron image from SEM of GCYF-2.

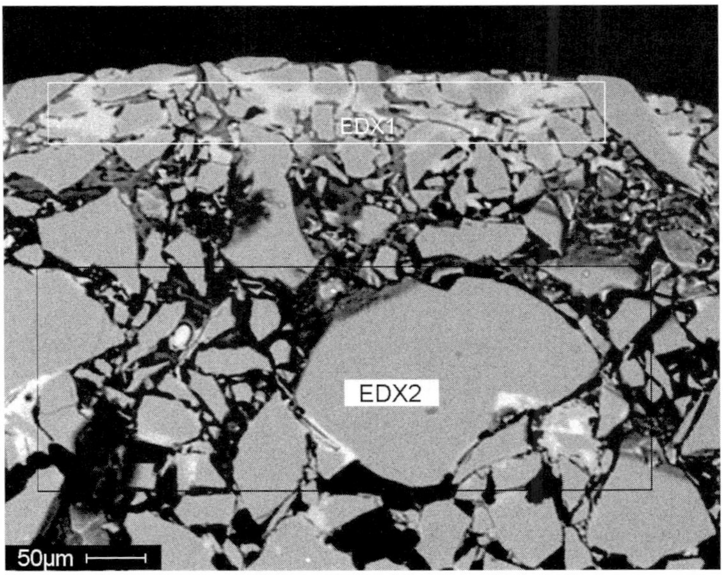

Fig. 13.11. Backscattering electron image from SEM of GCYF-3.

Table 13.1. SEM-EDX data for three faience bead samples (elemental compositions given in wt% and expressed as metal oxides).

Spot	Na	Mg	Al	Si	P	S	Cl	K	Ca	Fe	Cu
9-A	7.6	0.9	2.6	79.4	—	0.3	0.7	2.3	0.4	1.8	3.9
9-B	7.6	1.0	2.5	77.0	—	0.7	0.9	2.4	0.7	2.2	5.0
9-C	6.9	0.7	2.4	78.7	—	0.3	0.8	2.6	0.9	2.3	4.5
9-D	4.8	1.0	3.3	83.2	—	0.4	0.7	2.2	0.3	1.5	2.6
9-E	2.6	1.1	1.6	87.5	—	0.4	0.7	1.2	0.7	1.8	2.4
9-F	4.4	1.2	3.8	82.6	—	0.6	0.7	2.0	0.7	1.7	2.3
10-1	0.2	—	1.2	92.8	—	—	0.9	0.2	0.6	1.2	3.0
10-2	0.4	—	1.8	93.0	—	—	0.6	0.4	0.5	1.9	1.4
10-3	0.2	—	2.0	92.5	—	—	0.6	0.2	0.6	1.2	2.7
11-1	0.5	0.6	1.8	91.4	0.4	0.4	0.6	0.3	1.2	—	2.9
11-2	0.4	0.5	2.0	93.2	0.2	0.5	0.3	1.0	0.7	—	1.2

K and 3.9–5.0% for Cu (Fig. 13.9-A, B and C in Table 13.1), and analyses of the body of GCYF-1 show concentrations in the range of 82.6–87.5% for Si, 2.6–4.8% for Na, 1.2–2.2% for K and 2.4–2.6% for Cu (Fig. 13.9-D, E and F in Table 13.1). The differences of the concentrations from the glaze to the body are all based on silica as the network former; copper oxide as the colorant; lime, alumina and magnesium oxide as the stabilizers that limit solubility and hence weathering; and soda and plant ash as the alkali flux. There is high sodium and low potassium in sample GCYF-1, which is similar to the feature in West Asia and Egypt in element composition but converse to the feature of high potassium and low sodium in China. The research indicates that the manufacturing technology for faiences in this area may have been influenced by West Asia and Egypt to some degree. The distribution of the main elements in faience bead GCYF-1 is shown in Fig. 13.12.

There is no obvious difference in the K and Na contents between the surface layers and the interior layers for GCYF-2 and GCYF-3, but the difference in the Cu content is still obvious (Figs. 13.10 and 13.11, and Table 13.1).

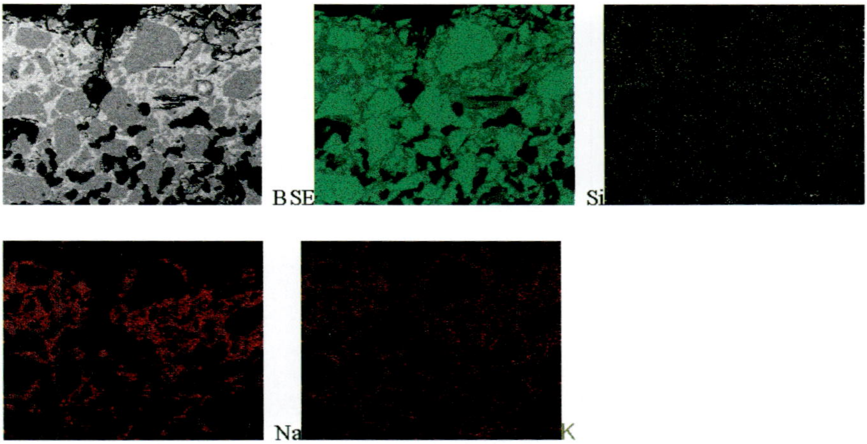

Fig. 13.12. Distribution of the main elements in faience bead GCYF-1 by SEM.

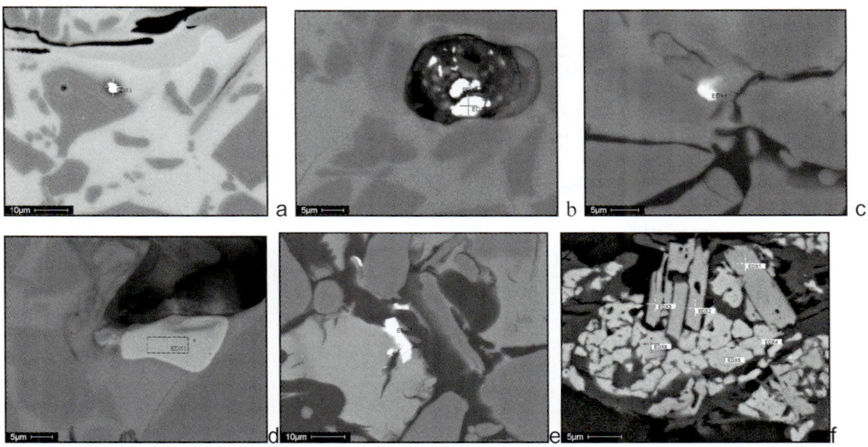

Fig. 13.13. Backscattering electron image from SEM of GCYF-1.

3.3. *Determination of microelements in faience*

The SEM-EDX data for microelements in faience beads obtained in some areas or spots are shown in Fig. 13.13 and Tables 13.2 and 13.3. Based on the EDX analysis, the particles in the position of Fig. 13.13a (EDX1) and of Fig. 13.13b (EDX1 and 2) all have the approximate

Table 13.2. SEM-EDX data for faience bead GCYF-1 (elemental compositions given in wt%, and expressed as metal oxides).

Spot	Na	Mg	Al	Si	P	S	K	Ca	Fe	Cu	Ba	Pt	Ti	Pb	Possible phase
13a-1	8.4	1.4	2.4	40.8	1.2	11.4	0.9	1.2	1.1	1.6	29.7	—	—	—	$BaSO_4$
13b-1	1.2	—	0.7	1.2	—	21.0	—	0.1	—	—	75.8	—	—	—	$BaSO_4$
13b-2	1.0	—	1.4	7.0	—	20.1	—	0.8	—	—	69.8	—	—	—	$BaSO_4$
13c-1	—	—	—	21.5	—	—	—	—	—	—	—	78.5	—	—	Pt
13d-1	—	—	—	—	—	—	—	—	—	—	—	—	100	—	TiO_2
13e-1	—	1.4	1.8	4.8	13.4	—	—	19.8	2.7	2.1	—	—	0.5	53.6	Pb,$Ca_3(PO_4)_2$
13f-1	—	—	0.9	1.0	—	—	—	—	98.1	—	—	—	—	—	Fe_2O_3
13f-2	—	—	1.0	2.0	—	—	—	—	97.0	—	—	—	—	—	Fe_2O_3
13f-3	—	—	1.3	3.0	—	—	—	—	95.7	—	—	—	—	—	Fe_2O_3
13f-4	—	—	2.1	2.2	—	—	—	—	70.1	25.6	—	—	—	—	Fe_2O_3, CuO
13f-5	—	—	2.6	1.0	—	—	—	—	68.4	28.1	—	—	—	—	Fe_2O_3, CuO
13f-6	—	—	0.5	3.9	—	—	—	—	69.0	26.6	—	—	—	—	Fe_2O_3, CuO

Table 13.3. SEM-EDX data for faience bead GCYF-2 (elemental compositions given in wt% and expressed as metal oxides).

Spot	Al	Si	Cl	Ca	Fe	Cu	Possible phase
15a	1.5	51.2	10.6	—	—	36.7	Chloride of copper
15b	21.1	36.1	—	27.2	15.6	—	Hematite, anorthite

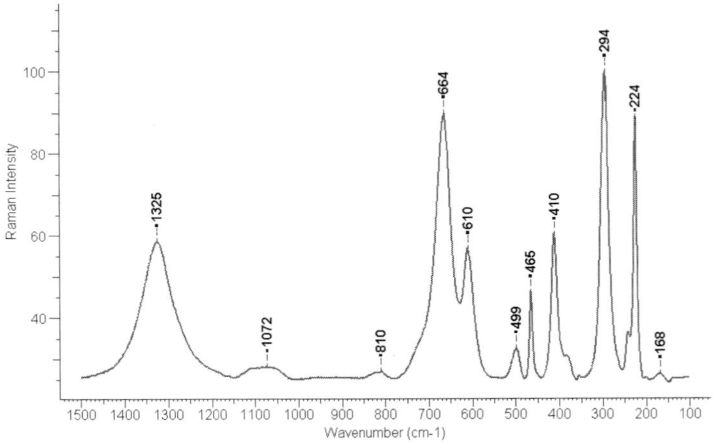

Fig. 13.14. Raman spectroscopy of Fe_2O_3 in GCYF-1 (Fig. 13.13f).

composition of barite ($BaSO_4$). The particle in the position of Fig. 13.13e (EDX1) is mainly lead and $Ca_3(PO_4)_2$. One author has detected trace barium and lead elements in one faience bead excavated from the archeological site in Li county, Gansu, and identified it as Chinese purple ($BaCuSi_2O_6$) by Raman spectroscopy —the earliest Chinese purple (766 BC) that has ever been detected.[9] Barium, copper and silica are the basic elements for synthesizing Chinese purple or Chinese blue ($BaCuSi_4O_{10}$).[10,11] Once the proportions of BaO, CuO and SiO_2 in the objects accord with the stoichiometry of Chinese purple or Chinese blue, there exists the possibility of forming them, and so the trace Ba and Pb elements detected in the glaze of faience GCYF-1 are very interesting.

The bar crystals in gray in the position of Fig. 13.13f (EDX1-3) are identified to be hematite by Raman spectroscopy (Fig. 13.14). Based

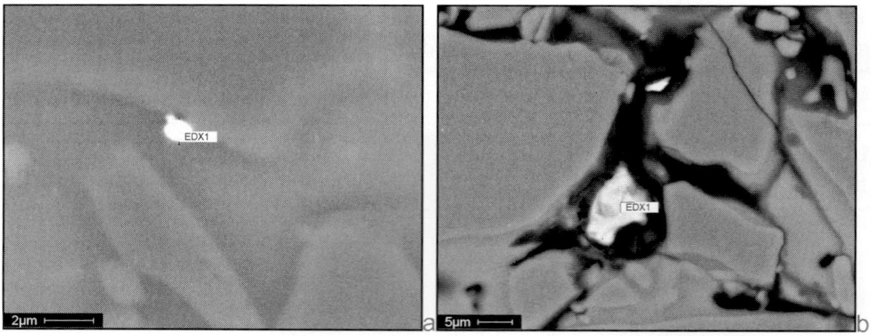

Fig. 13.15. Backscattering electron image from SEM of GCYF-2.

on the EDX analysis, the particles in white in the position of Fig. 13.13f (EDX 4-6) are probably a mixture of hematite and tenorite.

In the microarea analysis of faience bead GCYF-2, some compounds such as Pt, TiO_2, $Ca_3\ (PO_4)_2$ and Fe_2O_3, were detected — similar to GCYF-1. Except for these compounds, the particle in area EDX1 (Fig. 13.15(a), Table 13.3) approximately corresponds to the composition $Cu_2(OH)_3Cl$. Possibly, ancient craftsmen may have used $Cu_2(OH)_3Cl$ as one of the copper colorants, because chloride was detected in many areas of these three faience bead specimens (Table 13.1).

4. Conclusions

The glazing method for faience beads at Yu Jia-wan in the middle period of the Western Zhou Dynasty is efflorescence. One faience bead has high sodium and low potassium, which is similar to the feature in West Asia and Egypt in element composition but converse to the feature of high potassium and low sodium in China. The detection of barium and lead in one faience bead is very interesting; it illuminated that there existed the material possible for fabricating Chinese purple or Chinese blue in the later period. The research indicates that the manufacturing technology for faience beads in this area may have been influenced by West Asia and Egypt to some degree.

Acknowledgments

We would like to thank Director Yang Huifu and researcher Wei Huaiheng of the Cultural Relics and Archeology Institute of Gansu province and Researcher Zhang Long of the Shenzhen Museum for providing the samples. Thanks are also due to our colleagues at the CNICP, for their help in the research.

References

1. F. X. Gan, *J. Chin. Ceram. Soc.* **2,** 182–188 (2004).
2. F. X. Gan, *Development of Chinese Ancient Glass* (Shanghai Science and Technology Publishers, 2005), in Chinese, pp. 80–82.
3. R. Tao, Trace to Chongxin's source, *Pingliang Daily* (in Chinese) 7 May 2005.
4. Cultural Relics team of Gansu province, The briefing on the tomb of the Zhou Dynasty excavated from Yu Jia-wan of Chongxin county in Gansu province, *Archeology and Cultural Relic*s (in Chinese), **1**, 1–7 (1986).
5. P. T. Nicholson, *Materials and Technology. Gifts of the Nile: Ancient Egyptian Faience* (Tames and Hudson, London), pp. 50–63.
6. M. S. Tite, I. C. Freestone and M. Bimson Faience: an investigation of the methods of production, *Archaeometry*, **25**, 17–27 (1983).
7. P. B. Vandiver, Technological changes in Egyptian faience, in *Archaeological Ceramics*, eds. J. S. Olin and A. D. Franklin (Smithsonian Institution Press, 1982), pp. 167–179.
8. M. S. Tite, Pottery production, distribution and consumption — the contribution of the physical sciences, *J. Archaeol. Method Theory* **6**, 181–233 (1999).
9. Q. Ma, A. P. Ferdinand, R. W. P. Wild and H. Berke, Raman and SEM studies of man-made barium copper silicate pigments in ancient Chinese artifacts, *Studies in Conservation* **2**, 1–19 (2006).
10. H. Berke, Chemistry in ancient times: the development of blue and purple pigments, *Angew. Chem. Int. Ed.* **41**, 2483–2483 (2002).
11. E. W. FitzHugh and L. A. Zycherman, A purple barium copper silicate pigment from early China, *Studies in Conservation* **37**, 145–154 (1992).

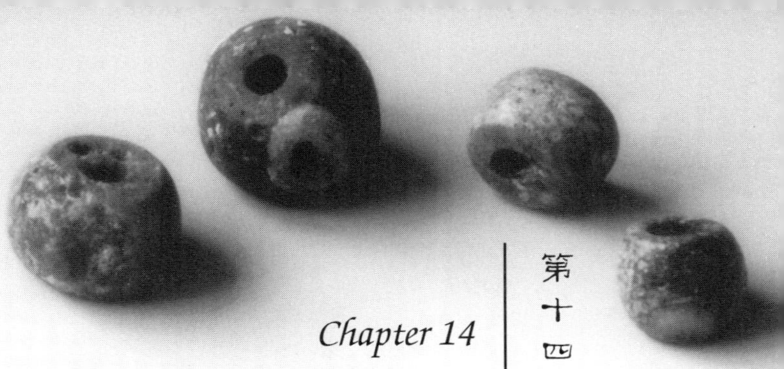

Chapter 14 第十四章

Scientific Research on Glass Fragments of the 6th Century AD in Guyuan County, Ningxia, China

Song Yan and Ma Qinglin

China National Institute of Cultural Property, Beijing 100029, China

1. Introduction

Guyuan county is located in the south of Ningxia, China (Fig. 14.1). It is a historic city and was once an important place along the ancient Silk Road. A series of significant archeological excavations have been carried out here since the 1980s, making Guyuan an attention-focusing region. The cultural relics unearthed in this region have revealed its long history and cultural intercommunications among different cultures, and have attracted the interest of many researchers.

Tian Hong was a very famous general and official of the Beizhou Dynasty (557–581 AD).[1] He died in 575 AD and was buried in Guyuan. His tomb was excavated by a Chinese–Japanese archeological team in 1996.[2] Though the tomb had ever been robbed by ghouls in the past, some rare relics, including hundreds of glass beads (some of them are shown in Fig. 14.2[2]) were unearthed. At the same time, fragments which probably came from some glass vessels were discovered. They provided good samples for scientific research on ancient glasses. In this article, the chemical

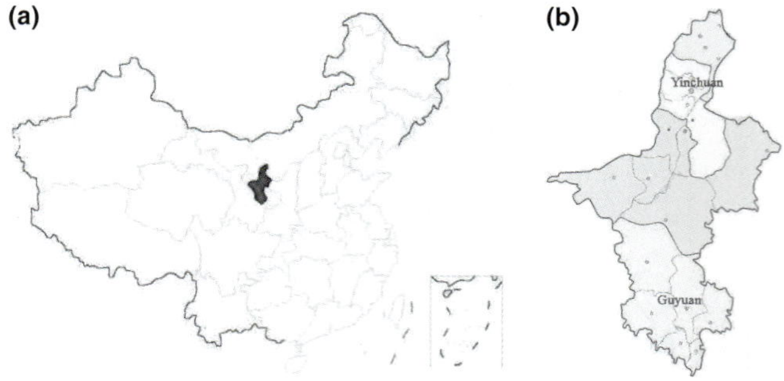

Fig. 14.1. Location of Guyuan county. (a) Location of Ningxia in China (b) Location of Guyuan in Ningxia.

Fig. 14.2. Photos of glass beads unearthed from Tian Hong's tomb.

components, microstructure and weathering production of the glass fragments from Tian Hong's tomb are analyzed by PLM, EDXRF, XRD and SEM/EDS. The weathering phenomena and weathering reasons of the glasses are also discussed.

2. Sample Description

Glass fragments were selected and cleaned with absolute ethanol. Under the optical microscope the glasses present a diverse green, with white or white–yellow weathering resultants on the surfaces. Some fragments are thin and semitransparent, with air bubbles inside [Fig. 14.3(a)]. However, some are thick and opaque, looking like jade [Fig.14.3(b)]. Considering their color, thickness and appearance,

(a) (b)

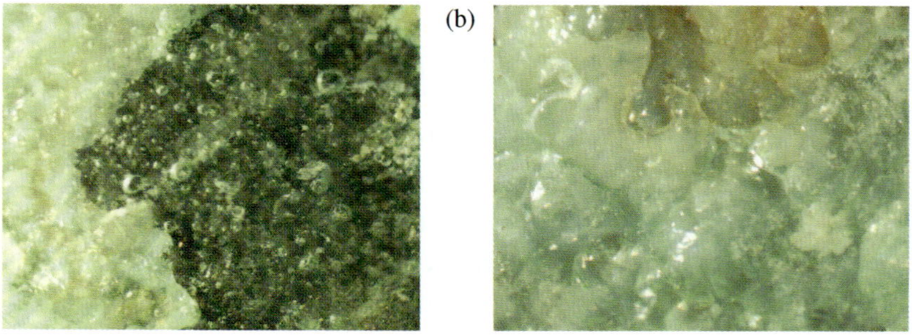

Fig. 14.3. Surface appearance of the glass fragments.

Table 14.1. Chemical components of glass fragments determined by EDXRF (wt %).

No.	PbO	SiO_2	Al_2O_3	CaO	CuO	Fe_2O_3	P_2O_5
1	77.53	18.94	1.49	1.41	0.34	0.29	—
2	86.88	8.99	—	1.66	2.04	0.43	—
3	83.78	7.45	—	1.67	6.1	—	1.00
4	87.79	10.81	—	0.54	0.45	0.41	—
5	79.59	17.94	1.24	0.85	0.38	—	—

these glass fragments probably came from different artifacts or different parts of one artifact.

3. Results and Discussion

3.1. *Chemical components*

Biggish glass fragments were selected and cleaned. Their chemical components were determined by EDXRF. The results (Table 14.1) show that the main components of these fragments are PbO and SiO_2, which indicate that they are PbO–SiO_2 glass. The relative percentage contents of PbO and SiO_2 are 77.53–87.79% and 7.45–18.94% (wt%), respectively. Other trace components include Al_2O_3, CaO, CuO, Fe_2O_3, etc. Compared with the percentage contents of PbO reported in other literatures,[3–6] the PbO contents of these glass fragments are higher, but the contents of SiO_2 are lower. Moreover, the percentage contents of PbO in No. 1 and No. 5 are obviously lower than those of the other three samples (Nos. 2–4), but their SiO_2 contents are higher.

The above samples were analyzed by XRD.[7] The results show that the No. 1 and No. 5 samples are still vitreous, but the No. 2 and No. 4 samples are partially weathered. The weathering resultant is cerusite ($PbCO_3$), confirmed by XRD. The results suggest that the PbO contents of the weathered samples (Nos. 2–4) are higher than the PbO contents of the samples which remain vitreous (Nos. 1 and 5). And weathering of the glasses was probably due to the formation of $PbCO_3$, which changed the components (such as the increase in PbO and the decrease in SiO_2) and structures of the glasses.

To avoid the effect of the weathering resultants on the surface, a green glass fragment which remains well was selected. The sample was embedded in resin and polished. Then, it was covered with a thin film of carbon and observed under SEM. Backscatter electron (BSE) images of the cross-section show that the main body of the sample remains vitreous and there is little weathering except at the edges. Four regions of the cross-section were selected (Fig. 14.4) to get the elemental components by EDS. The results show that the average contents of PbO and SiO_2 are 76.53% (wt%) and 20.92% (wt%), respectively, which approach the values reported in some of the literature.[8] Moreover, the sample contains small quantities of CuO, Fe_2O_3, Al_2O_3 and Na_2O. The results indicate that the glass

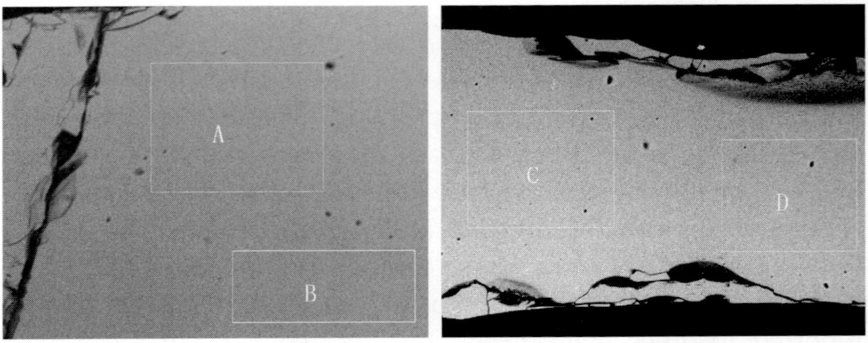

Fig. 14.4. BSE images of one sample.

fragment is typical $PbO–SiO_2$ glass with very high lead oxide, and the colorants of the green glass are CuO and Fe_2O_3.

3.2. *Microstructure and weathering*

Five more glass fragments were embedded in resin and polished, and then observed by PLM. The images of the fragments show: (1) one sample displays a homogeneous dark green under a dark field and there are no obvious holes on the cross-section [see Fig. 14.5(a)], which indicates that the glass body remains quite well; (2) on the cross-sections of two samples there exist nearly parallel strips [Fig. 14.5(b)]; because of the difference in the refractive index, the strips can be observed clearly; (3) there are obvious weathered layers and holes on the cross-section of two samples [Fig. 14.5(c)], revealing that they have effloresced greatly.

After PLM observation, these five glass fragments were covered with a thin film of carbon and analyzed under SEM, by which the inner character of the fragments could be seen clearly. Three typical BSM images are shown in Fig. 14.6. On the cross-section of one

Fig. 14.5. PLM images of glass fragments.

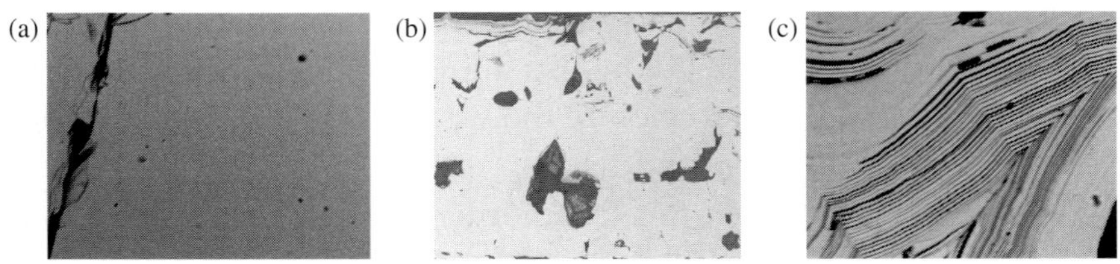

Fig. 14.6. BSM images of glass fragments.

nonefflorescent or slightly efflorescent sample [Fig. 14.6(a)] there are a few small holes, which probably resulted from the breaking of air bubbles. Breakage and weathering appear merely on the edges of the sample. However, two images of other samples [Figs. 14.6(b) and 14.6(c)] show that there are biggish holes and visible weathered striations or layers on the cross-sections, indicating that severe weathering has taken place in these areas. The weathered striations have also been reported by other literature.[9,10] This is a common efflorescent phenomenon that appears on ancient glasses.

Studies of glass weathering have been reported in much literature.[11–16] According to these reports, hydrolyzing reaction is an important reason for the weathering of glasses, especially for glasses which contain oxides of alkali metals (such as Na_2O and K_2O). The contents of alkali metals in the glass fragments from Tian Hong's tomb are quite low. Hence, except for the effect of hydrolyzation, the weathering of these glass fragments was probably due to some other factors:

(1) Water in the tomb could have attacked the [SiO_4] frameworks and wrecked them.
(2) Tian Hong's tomb was ever robbed by ghouls. Hence, the closed space of the tomb was opened up to some extent. The outer surroundings, such as temperature, humidity or corrosive gases, could have eroded the glasses gradually and weathered them. Moreover, the surface of the glass probably broke up under outside forces, resulting in holes and interstices on the surface. Then, the substances clinging to the surface of the glass would have entered the inner part of the glasses easily and reacted with the components of the glasses, accelerating the weathering.
(3) The disfigurements in the glass, such as air bubbles and stripes formed during production, also allowed outside substances to enter the glass and accelerate the weathering, especially under the effect of the surroundings.
(4) One crucial factor in the weathering of the glasses from Tian Hong's tomb is their components, which contain more than 75% PbO. PbO in the glasses can react with water and CO_2 or

other carbonates in the environment and form $PbCO_3$. Since lead carbonate is nonsoluble and quite stable, it can separate out easily, which can further accelerate the reaction and form more $PbCO_3$. The formation of $PbCO_3$ changes the composition and structure of the glasses and results in their weathering. Meanwhile, the tropism growth of lead carbonate is a possible factor in the formation of weathered layers.

4. Conclusion

The glass fragments from Tian Hong's tomb belong to the $PbO–SiO_2$ glass system. The percentage content of PbO is generally higher than 75 wt%, and the content of SiO_2 is about 20 wt%. The colorants of the green glasses are CuO or CuO and Fe_2O_3 compounds.

Weathering appears in both the interior and the exterior of the glasses. In the areas of severe weathering, weathered layers are observed, which changed the chemical components and structure of the glasses. The components, the inner disfigurements (such as air bubbles and stripes existing widely) of the glass fragments and the influence of the outer surroundings are the main reasons for the quite severe weathering of the glasses. The main weathering resultant is $PbCO_3$.

Acknowledgments

We would like to thank Mr Yan Shizhong, Deputy Director of the Guyuan Museum, Ningxia, who supplied the archeological glass fragments. This research was supported by the China National Institute of Cultural Property (CNICP).

References

1. *History of Zhou — Biography of Tianhong.*
2. Yuanzhou united archeological team, *Tian Hong's Tomb of the Beizhou Dynasty — Excavation Report of the Yuanzhou United Archeological Team* [M] (Miancheng, Japan, 2000).

3. R. H. Brill and J. H. Martin, (eds.), *Scientific Research in Early Chinese Glass* (The Corning Museum of Glass, 1991).
4. F. Li, Q. H. Li, F. X. Gan *et al.*, Chemical composition analysis for some ancient glasses by the proton induced X-ray emission technique, *J. Chin. Ceram. Soc.* (in Chinese) **33**(5), 581–586 (2005).
5. M. G. Shi, O. L. He, Z. D. Wu *et al.*, Study on some ancient flint glasses of China, *Bull. Chin. Ceram. Soc.* (in Chinese) **1**, 17–23 (1986).
6. Q. H. Li, J. Z. Huang, F. Li *et al.*, A report on the analysis of the chemical compositions of some glasses artifacts from the Warring States period, *Sciences of Conservation and Archeology* (in Chinese) **18**(2), 8–13 (2004).
7. F. X. Gan, Some considerations about research on Chinese ancient glasses, *J. Chin. Ceram. Soc.* (in Chinese) **32**(2), 182–188 (2004)
8. F. X. Gan, *Development of Technology of Chinese Ancient Glasses* (Shanghai Science and Technology Publishers, 2005) in Chinese, pp. 286, 300.
9. Z. D. Wu, F. Z. Zhou and M. G. Shi, Preliminary study on the microstructure, composition and weathering of some ancient glass, *J. Chin. Electr. Micros. Soc.* (in Chinese) **4**, 65–70 (1986).
10. C. Y. Wang and Y. Tao, The weathering of silicate glasses, *J. Chin. Ceram. Soc.* (in Chinese) **31**(1), 78–85 (2003).
11. R. H. Brill, The record of time in weathered glass, *Archaeology* **14**(1), 18–22 (1961).
12. H. Römich, *Historic Glass and Its Interaction with the Environment* (The Conservation of Glass and Ceramics: Research, Practice and Training, 1999), pp. 5–14.
13. G. A. Cox, O. S. Heavens, R. G. Newton and A. M. Pollard, A study of weathering behavior of medieval glass from York Minister, *J. Glass Studies* **21**, 54–75 (1979).
14. H. Römich, *Laboratory Experiments to Simulate Corrosion on Stained Glass Windows* (The Conservation of Glass and Ceramics: Research, Practice and Training, 1999), pp. 57–65.
15. R. G. Newton and A. B. Seddon, The durability of silicate glass in the presence of a saturated leachant, *Corrosion Sci.* **33**(4), 617–626 (1992).
16. F. X. Gan, Optical Glass. (Scientific Publisher, Beijing, 1964) (in Chinese).

Chapter 15 第十五章

Glass Artifacts Unearthed from the Tombs at the Zhagunluke and Sampula Cemeteries in Xinjiang

Wang Bo and Lu Lipeng

Archeology Team, Xinjiang Uygur Autonomous Region Museum, Urumqi 830000, China

At present, the ancient glass research activity in Xinjiang is still focusing on the collection of archeological materials. Analytical methods of natural sciences have been gradually applied in the study of ancient glass artifacts in order to reveal their structure and composition.[1] Therefore we can discuss their origins and disseminative routes from the development and variation of the ancient glass objects, and understand the vital significance of ancient cultural exchanges along the Silk Road.

A great deal of glass artifacts have been excavated at the Zhagunluke and Sampula cemeteries along the southern edge of the Tarim Basin in Xinjiang.[2] In the past, due to a lack of very clear understanding of glass essence, most of the unearthed glass and glasslike (vitreous materials) beads were classified as *liaozhu* (glazed beads) and so on, apart from some obviously recognized glass artifacts. Perhaps some beads of faience and frit may have been confused with glass beads. Therefore, it is necessary to re-examine and further analyze the glass artifacts unearthed from the Zhagunluke and Sampula cemeteries, and correct some mistakes that occurred in the previous research on the ancient glass artifacts of Xinjiang.

1. Ancient Glass Artifacts Excavated at the Zhagunluke Cemetery

The Zhagunluke cemetery is situated at Zhagunluke village of Tuogelakeleke country in Qiemo county. It consists of five graveyards. Since 1985, the No. 1 and No. 2 graveyards have been unearthed, and in total 169 tombs have been excavated, of which 167 tombs are in the No. 1 graveyard and 2 tombs in the No. 2 graveyard. The cultural remains of the Zhagunluke graveyard can be divided into three cultural phases: only one tomb belongs to the first cultural phase and dates back to the Western Zhou period (about 3000 years ago); 140 tombs belong to the second cultural phase and date from the Spring and Autumn Period to the Western Han Dynasty (8th century BC–1st century AD); 28 tombs belong to the third cultural phase and date from the last period of the Eastern Han Dynasty to the Northern and Southern Dynasties (3rd–6th century AD).

In 1996 and 1998, some ancient glass artifacts were unearthed from the tombs of the second and third cultural phases at the No. 1 and No. 2 Zhagunluke graveyards. An introduction of these glass artifacts is given below, according to the excavation time and the excavated graveyard.

1.1. *Glass beads and a glass cup unearthed at the No. 1 graveyard in 1996*

In 1996, six glass beads and a glass cup were excavated from tombs 96QZIM14 and 96QZIM49 respectively at the No.1 graveyard of Zhagunluke.

1.1.1. *Glass beads*

The glass beads can be divided into two kinds of colored beads: blue regular rounded beads with a longitudinal section of the short truncated convex bicone (similar to the beads of an abacus, so we call them "abacus-bead-shaped" in the following context); and similarly shaped glass eye beads with a blue background.

Three blue abacus-bead-shaped beads were found (96QZIM14: 28-1-3L). They are single-colored (blue), with some air bubbles inside. Their sizes are similar: 0.5–0.7 cm in length and 0.8–0.95 cm in diameter.

There are three other beads of the same shape, with white stratified concentric circle eyes against a blue background (96QZIM14:43-1-3G). Their sizes are 1.1–1.6 cm in length and 1.6–1.7 cm in diameter.

1.1.2. *An aqua glass cup decorated with rows of elliptical facets*

The glass cup was unearthed from tomb 96QZIM49:4. Its rim is slightly broken but can be restored. The cup was made by the blowing method; it is glaucous green in color and is translucent and of good quality. It has a normal round shape, with a cut decoration of three rows of elliptical facets engraved with 13, 13 and 7 circles respectively in each row in the belly, and a deep-cut circle at the bottom. It is 6.8 cm high, the diameter of the mouth is 6.8 cm, and the diameter of the bottom is 1.3 cm (Photo 2.11).

1.2. *Glass beads unearthed at the No. 1 graveyard in 1998*

In 1998, 22 glass beads were excavated from three tombs coded as 98QZIM24, 98QZIM147 and 98QZIM133 at the No. 1 graveyard of Zhagunluke. These beads have three colors: blue, white, and stratified circles of eye patterns against a blue background.

(a) The 20 blue glass beads were discovered on a necklace at tombs 98QZIM124:9 and 98QZIM147:17. While differing slightly in shape, they could be thought of as being of the same type as regular rounded beads and also in the same group of the circular transverse section, and then divided into two sub-groups — round and abacus-bead-shaped.

Four of the beads are blue round beads (98QZIM124: 9-1–4). They appear to be glass in the sense of quality. The shapes of these glass beads are similar, and they are circularly

Photo 15.1. Necklace of glass beads (98QZIM147:17).

Photo 15.2. Stratified glass eye bead with a blue background (98QZIM133:2-2).

rounded and have a plain perforation in the middle. Their sizes are nearly uniform, 0.4–0.6 cm in length and 0.5–0.8 cm in diameter.

Sixteen of them are blue abacus-bead-shaped beads (98QZIM147:17). Their sizes are varied, with the biggest being 0.4 cm in length and 0.5 cm in diameter (Photo 15.1).

(b) There is a white circularly rounded bead (98QZIM124:9-5). It has a plain perforation in the middle and its size is 0.4 cm in length and 0.5 cm in diameter.

(c) There is a stratified circles eye bead with a blue background (98QZIM133:2-2). It is a standard convex bicone bead with a convex bicone longitudinal section (rhombic) and a circular transverse section. The bead is opaque blue, with some obvious bubbles. There are eight oval eyes and five layers of each, with the colors yellow, deep yellow, red, white and black from outside to inside, asymmetrically distributed on the surface, as a complex motif (Photo 15.2).

1.3. *Glass beads unearthed at the No. 2 graveyard in 1996*

In 1996, five glass beads were discovered at tomb 96QZIIM2 in the excavation of the No. 2 graveyard of Zhagunluke. The beads have two colors: blue and green.

(a) There are three regular rounded beads with different shapes. They can be divided into two subgroups by longitudinal section: blue abacus-bead-shaped beads and blue olive-shaped beads.

The two blue abacus-bead-shaped beads (96QZIIM2:27 and 96QZIIM2:85) are transparent blue, with elapse plain perforations. Their sizes are identical, 0.55 cm in length and 0.7 cm in diameter (Photo 15.3).

The blue olive-shaped bead (96QZIIM2:39) is transparent medium blue, with some bubbles and cracks on its surface. It is 0.55 cm in length and 0.7 cm in diameter (Photo 15.4).

Photo 15.3. Blue short truncated convex bicone beads (96QZIIM2:27).

Photo 15.4. Blue long truncated convex bicone bead (96QZIIM2:39).

(b) There are two green glass beads with obvious differences in shape. They can be divided into two types by transverse sections: flattened abacus-bead-shaped and regular hexagonal-faceted.

The abacus-bead-shaped bead (96QZIIM2:89) is well-preserved and transparent olive green, with some bubbles on its surface. It is 0.4 cm in length and 0.75 cm in diameter.

The hexagonal-faceted bead (96QZIIM2:55) is 3.1 cm in length and 1.2 cm in diameter, with two broader sides and four narrower sides. It is transparent aqua green, with some obvious cracks (Photo 15.5).

Photo 15.5. Hexagonal-faceted bead (96QZIIM2:55).

2. Glassware Unearthed at the Sampula Cemetery

The Sampula cemetery is located in the terrace to the south of Sampula village, 14 km southwest of Lop county. This ancient cemetery is divided into two parts from east to west by a modern cemetery in the middle. The eastern part is in the Gobi area and its landscape has been preserved fairly well. In contrast, the western part has been disturbed by modern construction and agricultural activities such as roads and fruit gardens. It is known as "Shayi Bahe" ("a garden on the river bed").

In total, 68 tombs and 2 sacrificial horse pits were excavated at the Sampula cemetery, in 1983, 1984 and 1992.[3] It was divided into two excavation sites, No. 1 and No. 2, during the excavation.

The No. 1 excavation site is situated in the eastern part of the cemetery. Fifty tombs were excavated and glass bead ornaments were unearthed from 13 of them (84LSIM01, 84LSIM02, 84LSIM13, 84LSIM23, 84LSIM25, 84LSIM30, 84LSIM34, 84LSIM35, 84LSIM37, 84LSIM42, 84LSIM44, 84LSIM45, 84LSIM49).

The No. 2 excavation site is situated in the western part of the cemetery. Eighteen tombs were excavated and glass bead ornaments were found in two of them (92LSIIM3, 92LSIIM6). The tombs of the Sampula cemetery can be classified into two cultural phases: the earlier phase dates from the 1st century BC to the 3rd century AD, which is comparable to the period from the late Western Han Dynasty to the end of the Eastern Han Dynasty in the Central Plains of China; the later phase dates from the mid-3rd century AD to the end of the 4th century AD, which is comparable to the period of the Wei and Jin Dynasties in the Central Plains.

Most of glasses unearthed are ornaments, such as necklace beads, bracelets and eardrops. Separate glass beads and ornaments were also found, some of which may be called "ear pendants." The total number of these glass beads is 1405.

2.1. *Glass beads*

These glass beads exhibit many different colors. They can be classified into 13 kinds, such as monochromic beads on opaque orange, brown–purple, olive–amber, blue, amber, white, green or black; eye beads; banded mosaic (cloud-pattern) beads; swirled two-color beads; silver-foiled beads; gilded beads, etc.

(a) There are 357 opaque orange glass beads, which were discovered in tombs 84LSIM01:c91 (a necklace consisting of coral beads and glass beads), 84LSIM01:c108 (opaque orange glass beads), 84LSIM01:c115 (a necklace consisting of opaque orange glass beads), 84LSIM23:2 (a necklace consisting of opaque orange glass beads and Job's-tears seeds), 84LSIM49:114M (a bracelet consisting of opaque orange glass beads), 84LSIM49:115L (a necklace consisting of opaque orange glass beads) and 92LSIIM6:366 (a necklace consisting of opaque orange glass beads). The color and luster of the opaque orange glass beads are bright. The beads have two different forms: one is barrel-shaped, and the other is triangular-cylinder-shaped.

The number of barrel-shaped glass beads is 356; they were discovered in tombs 84LSIM01:c91 (a necklace consisting of coral beads and glass beads), 84LSIM01:c108 (opaque orange glass beads), 84LSIM01:c115 (a necklace consisting of opaque orange glass beads), 84LSIM23:2 (a necklace consisting of opaque orange glass beads and Job's-tears seeds) (Photo 15.6),

Photo 15.6. Necklace consisting of opaque orange glass beads and Job's-tears seeds (84LSIM23:2).

Photo 15.7. Necklace consisting of brown–purple glass in the shape of a square tube (84LSIM45:6-1B).

84LSIM49:114M (a bracelet consisting of opaque orange glass beads), 84LSIM49:115L (a necklace consisting of opaque orange glass beads) and 92LSIIM6:366 (a necklace consisting of opaque orange glass beads). Both walls of these glass beads are straighter and their sizes are similar, 0.27–0.3 cm in length and 0.27–0.54 cm in diameter.

One glass bead is curved triangular-cylinder-shaped (84LSIM01:c108); it is 0.7 cm in length and 0.6 cm in diameter.

(b) There are 221 glass beads in brown–purple. They were discovered in tombs 84LSIM45:6-1B (a necklace consisting of brown–purple glass beads in the shape of a square tube) and 92LS:M3:295 (a necklace consisting of brown–purple glass beads in the shape of a square tube). The brown–purple glass beads are all square tubes in shape and their sizes are similar. They are dull in luster and have low transparency, but show some purplish light. Their sizes are 0.2–0.4 cm in length and 0.15–0.2 cm in width of the wall (Photo 15.7).

(c) There are 420 glass beads in deep brown, and many of them were discovered in tombs 92LSIIM3:298 (a string of glass beads), 84LSIM45:6-2B (a necklace consisting of glass beads), 84LSIM42:3 (eardrops consisting of glass beads) and 84LSIM37:1 (a necklace consisting of glass beads). The luster of the glass beads is dark and gives an impression of glass quality. They are different in shape and can be divided into two types: abacus-bead-shaped and gourd-shaped.

The number of abacus-bead-shaped glass beads is 418. They were discovered in tombs 84LSIM45:6-2B (a necklace consisting of glass beads) (Photo 15.8), 84LSIM42:3 (two strings of ear pendant glass beads) and 92LSIIM3:298 (a string of glass beads) (Photo 15.9). They are all small, 0.1–0.4 cm in length and 0.2–0.4 cm in diameter.

Of the two gourd-shaped glass beads (84LSIM45:6-2B), one is of three-segment connection, and 0.4 cm in length and 0.2 cm in diameter, and the other is of two-segment connection, and 0.2 cm in length and 0.2 cm in diameter.

Photo 15.8. Necklace consisting of reddish olive–amber glass beads (84LSIM45:6-2B).

(d) There are 51 blue glass beads, with some differences in both color and shape. They can be divided into four types: abacus-bead-shaped, barrel-shaped, gourd-shaped and cabinet-shaped.

The abacus-bead-shaped glass beads are 48 in quantity. They were discovered in tombs 84LSIM45:7B (a necklace consisting of glass beads with banded mosaic patterns), 84LSIM37:1 (a necklace consisting of glass beads), 92LS:M6:365 (a necklace consisting of glass beads and carbon beads). Their sizes are similar, 0.2–0.6 cm in length and 0.3–0.8 cm in diameter.

One glass bead is a barrel-shaped bead (84LSIM37:1). It is 0.5 cm in length and 0.3 cm in diameter (Photo 15.10).

Photo 15.9. String of reddish olive–amber glass beads (92LSIIM3:298).

Photo 15.10. Necklace consisting of glass beads in the shape of a flat and square rhombus (84LSIM37:1).

Another one is a gourd-shaped bead (84LSIM37:1). It is 0.5 cm in length and 0.3 cm in diameter.

There is a glass bead in the shape of a "cabinet" (84LSIM01:c105). It is a separate find. Its shape is rectangular, with a flat top and 4 legs. A hole was bored in the mid-waist. It is 1.3 cm in length, 0.8 cm in width and 0.6 cm in thickness (Photo 15.11).

(e) There are 24 glass beads in amber. They were discovered in tombs 84LSIM01:c92 (a string of shell beads and glass beads), 84LSIM45:7B (a necklace consisting of glass bead with a banded mosaic pattern), 84LSIM34:2 (a necklace consisting of black–amber and white glass beads) and 84LSIM37:1 (a necklace consisting of glass beads). Their shapes are rather different and can be divided into three types: abacus-bead-shaped, square-cylinder-shaped and rhombus-shaped.

The abacus-bead-shaped glass beads are 16 in quantity. They were discovered in tombs 84LSIM01:c92 (a string of shell

Photo 15.11. Glass bead in the shape of a "cabinet" (84LSIM01:c105).

Photo 15.12. Necklace consisting of glass beads of rhombus shape (92LSIIM6:70).

Photo 15.13. Necklace consisting of white glass beads (84LSIM42:4).

beads and glass beads), 84LSIM45:7B (a necklace consisting of glass beads with a banded mosaic pattern), 84LSIM34:2 (a necklace consisting of glass beads in black, olive and white) and 84LSIM37:1 (a necklace consisting of glass beads). Their sizes are rather different, 0.1–1.4 cm in length and 0.2–1.2 cm in diameter. There is a glass bead in the shape of a square cylinder (84LSIM45:7B). It is 1 cm in length and 0.4 cm in width.

The other 42 glass beads are rhombus-shaped. They were discovered in tombs 92LSIIM6:70 (a necklace consisting of rhombus-shaped glass beads) (Photo 15.12), 84LSIM37:1 (a necklace consisting of glass beads) and 84LSIM42:3 (eardrops consisting of glass beads). Another glass bead is bigger and flattened rhombus-shaped, and is on a necklace consisting of glass beads (84LSIM37:1). The others are all square rhombus-shaped, and of similar sizes, 0.6–1 cm in length and 0.3–0.4 cm in width.

(f) There are 108 white glass beads. They were discovered in tombs 84LSIM37:1 (a necklace consisting of glass beads), 84LSIM34:2 (a necklace consisting of glass beads in olive, yellow and white), 84LSIM42:4 (a necklace consisting of white glass beads) (Photo 15.13) and 84LSIM30:7 (a necklace). These beads are quite small and their colors are silvery white or milky white and transparent; some are pale black. They are abacus-bead-shaped glass beads, 0.2–0.4 cm in length and 0.2–0.3 cm in diameter.

(g) There are 113 green glass beads. They were discovered in tombs 84LSIM01:c94 (a string of glass beads), 84LSIM45:7B (a necklace consisting of glass beads with a banded mosaic pattern), 92LSIIM6:164 (a string of glass beads), 84LSIM42:3 (eardrops consisting of glass beads) and 84LSIM45:6-2B (a necklace consisting of black glass beads). Their colors and shapes have some differences. They can be divided into five types: ball-shaped, abacus-bead-shaped, gourd-shaped, barrel-shaped and grape-shaped.

A ball-shaped glass bead was discovered in tomb 84LSIM01:c94. It is slightly weathered and is translucent. There are some air bubbles inside. The bead is 1.2 cm in diameter and its aperture is 0.25 cm.

A total of 104 glass beads are abacus-bead-shaped. Their sizes are nearly the same, 0.3–0.4 cm in length and 0.2–0.4 cm in diameter.

A gourd-shaped glass bead was found in tomb 84LSIM45:7B. It is 0.4 cm in length and 0.3–0.2 cm in diameter.

An eardrop ornament of grape-shaped glass bead is strung in the middle of a necklace (84LSIM42:3). One end is bigger than the other, like a grape. It is 1.2 cm in length and 0.5 cm in diameter.

The other six beads, barrel-shaped, were discovered in tomb 92LSIIM6:376. Their sizes are identical, 0.3–0.4 cm in length and 0.3 cm in diameter.

(h) There are 32 black glass beads of abacus shape. Their sizes are identical. Among them, 24 were discovered on a necklace consisting of glass beads in black, olive and white (84LSIM34:2). They are 0.3–0.5 cm in length and 0.4–0.9 cm in diameter (Photo 15.14).

(i) There are 7 glass beads of banded mosaic design. They were discovered in tombs 84LSIM45:6-2B (a necklace consisting of black glass beads), 84LSIM45:7B (a necklace consisting of glass beads with banded mosaic), 84LSIM25:8 (a necklace consisting of glass beads) and 92LSIIM6:365 (a necklace

Photo 15.14. Necklace consisting of black–olive–white glass beads (84LSIM34:2).

consisting of carbon beads and glass beads). The pattern is a banded mosaic or a combed chevron, like a cloud. Their shapes have some differences, and they can be divided into two types: abacus-bead-shaped and barrel-shaped.

The abacus-bead-shaped glass beads are four in quantity. They were discovered in tombs 84LSIM45:6-2B (a necklace consisting of black glass beads), 84LSIM45:7B (a necklace consisting of banded mosaic glass beads) and 84LSIM25:8 (a necklace consisting of glass beads). One glass bead with a white banded mosaic pattern against a blue background is on a necklace consisting of black glass beads (84LSIM45:6-2B), another glass bead with a banded mosaic pattern against a deep blue background is on a necklace consisting of glass beads (84LSIM25:8), and the other two glass beads with a banded mosaic pattern against a deep blue background are on a necklace consisting of glass beads (84LSIM45:7B). Their sizes are identical, 0.2–0.5 cm in length and 0.2–0.6 cm in diameter.

The barrel-shaped glass beads are three in quantity. They were discovered in tombs 84LSIM45:7B (a necklace consisting of glass beads with a banded mosaic design) (Photo 15.15) and 92LSIIM6:365 (a necklace consisting of carbon beads and glass beads). Two of them are with white banded mosaic decoration against a green background and are strung on a necklace (84LSIM45:7B); one is broken and the other is intact. Another glass bead with white banded mosaic decoration against a green background is strung on a necklace consisting of carbon beads and glass beads (92LSIIM6:365) (Photo 15.16). Their sizes have some differences, and are 2.4–2.9 cm in length and 0.8–1 cm in diameter.

(j) There are two swirled two-color glass beads: one is on a necklace consisting of carbon beads and glass beads (92LSIIM6:365), and the other is on a string of agate stone and glass-eye beads (92LSIIM6:368). The colors of the swirled glass beads are blue and pale yellow, but not very pure, and the interface is not clearly distinguished. Glass of two different colors was twisted and then shaped into the beads, giving the

Photo 15.15. Necklace consisting of glass beads with banded mosaic design (84LSIM45:7B).

effect of lines of different colors bending in and out in quite an irregular manner. The beads are elliptical, and 0.7–1 cm in length and 0.5–0.7 cm in diameter.

(k) There are 28 glass eye beads. Their eyes set respectively against a white background (84LSIM01:c93), a white background (84LSIM01:c111), a blue background (84LSIM02:c47) and a white background (84LSIM35:1), on a necklace consisting of white glass beads (84LSIM42:4), on a necklace consisting of the white glass beads (84LSIM44:4), against a red background (84LSIM45:2A), on a necklace consisting of glass beads with banded mosaic decoration (84LSIM45:7B), on a necklace consisting of opaque orange glass beads (84LSIM49:155L), on a string of

Photo 15.16. Necklace consisting of carbon beads and glass beads (92LSIIM6:365).

Photo 15.17. String of glass beads with eyes patterns and agate beads against a blue background (92LSIIM6:368).

glass eye beads and agate beads against a blue background (92LSIIM6:110) (Photo 2.10), on a string of green glass beads (92LSIIM6:164), on a necklace consisting of opaque orange glass beads (92LSIIM6:366) and on a string of blue glass eye beads and agate beads against a blue background (92LSIIM6:368) (Photo 15.17).

It looks like different technology was used to make the glass eye beads: one is the mosaic type, the surface of the glass beads is smoother, and the eyes have concentric rings; the other is the piled type, the surface of glass beads is uneven, and the eyes are protuberant.

Among them, seven mosaic-type glass eye beads were discovered in tombs 84LSIM01:c93 (glass eye beads against a white background), 84LSIM01:c111 (glass eye beads against a white background), 84LSIM49:155L (a necklace consisting of opaque orange glass beads) and 84LSIM45:2A (glass eye beads against a white background). Their shapes are different — most are abacus-bead-shaped and one is triangular-cylinder-shaped. The abacus-bead-shaped glass eye beads are six in quantity. Their two ends are slightly even and the walls are in the shape of an arc or a drum. Some beads are flat. Three of them were discovered in tomb 84LSIM01:c111 (on a string of eye beads): two of them have six eyes against a white background, which is in staggered distribution, and the eyes are elliptical, with four or five concentric rings. One of them is a light green background glass eye bead, and the eyes are irregular, with six concentric rings. The central eyes are blue, but the concentric rings are light green. The sizes of the beads are identical, 0.7–1.3 cm in length and 1.1–1.6 cm in diameter. Besides, in the beads with four-layered eyes against a white background, the eyes and the concentric rings are all dark blue; while in the beads with five-layered eyes against a white background, the eyes are blue and the concentric rings are light green (Photo 15.18). The glass eye bead set against a blue background on a necklace consisting of

Photo 15.18. String of eye beads (84LSIM01:c111).

opaque orange glass beads (84LSIM49:155L) is two-layered and abacus-bead-shaped. It is decorated with white concentric rings against a dark blue background, and the center is dark blue too. It is 0.6 cm in length and 0.6 cm in diameter (Photo 15.19).

The glass eye bead of cylinder shape was discovered in tomb 84LSIM45:2A. The transverse section is triangular; both sides of the wall are straight. There are three eyes with four concentric rings each, and they are decorated with white and dark reddish brown against a red background. The bead is 0.4 cm in length and 0.6 cm in diameter (Photo 15.20).

There are 21 piled glass eye beads unearthed from tombs 84LSIM45:7B (a necklace consisting of glass beads with banded mosaic patterns), 84LSIM42:4 (a necklace consisting of white glass beads), 84LSIM02:c47 (glass eye beads against a blue background) (Photo 15.21), 92LSIIM6:368 (a string of

Photo 15.19. Eye bead against a blue background (84LSIM49:155L).

Photo 15.20. Eye bead against a red background (84LSIM45:2A).

Photo 15.21. Eye bead against a blue background (84LSIM02:c47).

glass eye beads and agate beads) and 92LSIIM6:110 (a string of glass eye beads and agate beads). All of them are of abacus-bead-shaped, with light yellow protuberant dots on the blue background, and a black line is on the dots, looking like an eyeball. Their sizes are identical, 0.6 cm in length and 0.6 cm in diameter.

(l) There are four gilded glass beads. Their shapes exhibit two different types: one is gear-shaped and the other is oblate.

A gear-shaped gilded glass bead was discovered in tomb 84LSIM01:92. It has a white glass body, and its gold foil has been partially peeled off. The bead is 0.9 cm in length and 0.7 cm in diameter (Photo 15.22).

The other three oblate gilded glass beads were discovered in tomb 92LSIIM6:365. Their sizes are identical, 0.2 cm in length and 0.6 cm in diameter.

(m) There are two silver-foiled glass beads of abacus bead shape. The silver foil on one of them has been peeled off. The sizes of these two beads are identical, 0.5–0.6 cm in length and 0.6–0.7 cm in diameter.

Photo 15.22. Gear-shaped gilded glass bead (84LSIM01:92).

Photo 15.23. Glass ear pendant (84LSIM49:79).

2.2. *Glass ear pendant*

There is a glass ear pendant unearthed from tomb 84LSIM49:79. It is translucent or opaque green. It has an unsymmetrical H shape, and one side is bigger than the other. It is 1 cm in length and 0.9 cm in maximum diameter (Photo 15.23).

3. Discussion

Based on the above introduction and typological observation of ancient glassware unearthed from the Zhagunluke and Sampula cemeteries, as well as on the historical documents and chemical composition analysis of some ancient glass samples, we would like to present a brief discussion, as follows:

3.1. *Date, type and related issues*

The glassware from the Zhagunluke graveyard belongs to the second and third cultural phases. The second cultural phase dates from the Spring and Autumn Period to the Western Han Dynasty, and the third cultural phase dates from the late Eastern Han Dynasty to the Northern and Southern Dynasties. The artifacts excavated from the tombs of the second cultural phase are mainly glass beads. They have such varied colors as blue, green and white, and eye patterns against a blue background. They are abacus-bead-shaped, cube-cylinder-shaped, olive-shaped, etc. In the tombs of the third cultural phase, a glass cup and glass eye beads of rhombus or round shape set against a blue background were excavated.

The tombs of the Sampula cemetery can be divided into two cultural phases: early and late. The early phase corresponds to the period from the late Western Han Dynasty to the end of the Eastern Han Dynasty in the Central Plains of China, and the late phase to the Wei–Jin period in the Central Plains. The excavated glass artifacts are mainly beads and ornaments, and most of them were unearthed from tombs of the early period, while the glasses unearthed from tombs of the late period are few in quantity. Such a difference may be correlated with the quantity of tombs that have been excavated: more early tombs have been excavated than the late ones.

The colors of the glass beads unearthed from the early period tombs are varied, including opaque orange, brown–purple, olive–amber, blue, amber, white, green, black; and eye beads, banded mosaic beads, swirled two-color beads, silver-foiled beads, gilded beads, etc. There are many different shapes of glass beads too — abacus-bead-shaped, barrel-shaped, gourd-shaped, triangular-cylinder-shaped, oblate, rhombusic, long-grape-shaped, gear-shaped, square-tube-shaped etc. Glass eye beads were also discovered. They were made by two kinds of technology: the mosaic method and the piled method. The concentric rings of glass eye beads number up to six.

The glass artifacts unearthed from the late period tombs are of two kinds: the ear pendant and the bead ornament. Bead ornaments are dominant, including piled pattern glass eye beads with a blue background, mosaic glass eye beads with a blue background, brown–purple square-tube-shaped glass beads, opaque orange cylindrical glass beads, banded mosaic glass beads, etc.

As we can see, the Zhagunluke tombs of the second cultural phase are the earliest. It seems that the lower limit for the dates of Zhagunluke tombs is comparable to the dates of early tombs of the Sampula graveyard. However, the Zhagunluke tombs of the third cultural phase are later than the Sampula tombs. This means that the servicing time of the Zhagunluke tombs is the longest. This can also be seen from the changes of shapes and colors of the unearthed glass artifacts. For example, the glass cup was unearthed from a tomb of the third cultural phase at Zhagunluke and the glass ear pendants were unearthed from the late tombs of the Sampula graveyard, while all unearthed glass artifacts from the tombs of the second phase at Zhagunluke and from the early Sampula tombs are bead ornaments. Besides, some changes can be seen in the craft of the glass eye beads unearthed from the second phase tombs of the Zhagunluke graveyard. They were mainly made using mosaic technology, while the piled method was used to make the eye beads unearthed from the early tombs of the Sampula graveyard. This shows that mosaic glass eye beads appeared earlier.

In addition, the coloring techniques used in making glass eye beads were changing with the time. During the second phase of the Zhagunluke graveyard and the early phase of the Sampula graveyard, the coloring technique was quite simple, such as for those beads with a blue, red, white or green background. For the third phase of the Zhagunluke graveyard, we see eye beads having five-colored eyes inlaid on a blue background. The colors include yellow, deep yellow, red, white, black, etc. In general, the use of colors in the early period was relatively simple and fairly rough, but the techniques were much improved in the late period.

3.2. *Chemical composition analysis*

Five samples selected from the glass artifacts excavated from the ancient tombs of Zhagunluke and Sampula were examined by scientists of the Shanghai Institute of Optics and Fine Mechanics, CAS, by using various analytical techniques, such as the proton-induced X-ray emission technique, the energy-dispersive X-ray fluorescence method, and inductively coupled plasma atomic emissions; the analysis results are given in Table 15.1. The analysis report shows that the two glass beads in blue and light blue from tomb 98QZIM124:9 of the second phase at Zhagunluke are soda-lime-silica glass with very high K_2O and MgO. Three glass eye beads from the Sampula tombs (84LSIM49:155L, 84LSIM35:1 and an unmarked one) are of the potash glass system, with very high Al_2O_3. The scientists suggest that these beads could belong to the Roman–Sasanian glass system.

3.3. *Historical background probing*

It is generally believed that the goods called *liuli* (colored glaze) in the ancient Chinese literature before the Tang Dynasty could be

Table 15.1. Chemical composition of ancient glass samples from the Zhagunluke and Sampula cemeteries (wt%).

No.	SiO_2	Na_2O	CaO	MgO	K_2O	Al_2O_3	PbO	BaO
1	63.09	18.71	4.66	2.74	2.27	4.42	0.02	0.11
2	62.66	17.36	6.38	3.76	3.86	4.23		0.10
3	81.32		7.99		1.05	3.11		
4	78.75		9.41		0.19	2.29	0.89	
5	87.91		2.28		1.97	5.83		

No.	CuO	Fe_2O_3	TiO_2	ZnO	MnO	B_2O_3	As_2O_3	CI	P_2O_5
1	1.33	0.67	0.17	0.05	0.05	0.13	0.01	1.11	0.41
2	0.56	0.80	0.16	0.01	0.06	0.06			
3	0.48	1.26	0.09		1.88		0.25	1.03	0.21
4	0.29	1.84	0.13		0.23		0.10	2.04	0.71
5	1.53	0.5							

1(XJ6A), 2(XJ6B): Blue and light blue glass beads unearthed at Zhagunluke (98QZIM124:9). 3,4(XJ7A,XJ7B): Eye beads of the inlaid type unearthed at the Sampula cemetery (84LSIM49:155L, 84LSIM35:1). 5: Glass eye beads of the piled type unearthed at the Sampula cemetery.

glass. Searching through the historical documents related to the Western Regions (now the Xinjiang area) indicates that glass production did not exist in the ancient kingdoms of Xinjiang before the Tang Dynasty. All glass artifacts were produced and imported from Western countries. According to the record in *Houhanshu — Xiyuzhuan* (*History of the Eastern Han Dynasty — Memoir of the Western Regions*), there were a great deal of gold, silver and other wonderful treasures in the Grand Qin Empire, including phosphorescent jade disks, bright pearls, *haijixi*, coral, amber, *liuli*, *langgan* (pearl-like stone), bright red and blue–green jade. Here the Grand Qin Empire refers to the land of the Roman Empire, while *liuli* and *langgan* may refer to the ancient glass. In the book *Wei Lue* (*History of the Wei Dynasty*), it is recorded that the Grand Qin Empire still produced ten colors of *liuli*, including red, white, black, yellow, blue, green, white–blue, crimson, red–black, purple, etc. This suggests that glass manufacturing was extremely skilled at that time. Then the commerce of the Grand Qin Empire became more developed and its products were exported to all parts of the world through the Mediterranean Sea, the Baltic Sea, the Black Sea, the Red Sea, the Indian Ocean, etc. It cannot be ruled out that glassware was included in the traded goods. It is recorded in *Weishu — Xiyuzhuan* (*History of the Wei Dynasty — Memoir of the Western Regions*) that the Persian Empire produced *liuli*, and the people of the Grand Yen Chin went to the capital of the Central Plains of China and engaged in trade. Some of these people claimed that they could make five-colored *liuli* using stone. They thus obtained materials from the mountains and transported them to the capital (now Luoyang, Henan province) to produce the five-colored *liuli*. They were successful and their products were even more beautiful than those imported from the West. Here five-colored *liuli* should refer to the ancient glass; the Persian Empire is today's Iran, and the Grand Yen Chin lies in today's Amu River area. From these records we can have some idea of how the glass technology was disseminated from the West to the East.

The Zhagunluke cemetery is a relatively large public cemetery of the ancient Charchan Kingdom, while the Sampula cemetery is a relatively large public cemetery of the ancient Khotan Kingdom. Both Charchan and Khotan were ancient city kingdoms along the route to the south of the Silk Road. The cultural relics unearthed from the tombs provided some evidence reflecting the development of the civilizations in this region. The chemical compositions of the glassware unearthed from Zhagunluke and Sampula tombs show some differences between these two places. The Zhagunluke glassware belongs to the Na_2O–CaO–SiO_2 glass system, while the Sampula glassware can be included to the K_2O–CaO–SiO_2 glass system. These compositional characteristics indicate that they all belong to the Roman–Sasanian glass system. It seems that the glass artifacts of the Roman–Sasanian glass system were transported to today's Hetian and Qiemo areas of Xinjiang along the Silk Road during the Spring and Autumn period and the Warring States period. In the early period, the imported glass wares were mainly bead ornaments. Later other, bigger glass artifacts were introduced. This is evidenced by the pale blue glass cup with elliptical facet decoration (96QZIM49:4), which is extremely similar to the glass cups of typical Roman–Sasanian glass system.

In addition, the glass ear pendant (84LSIM49:79) from the Sampula cemetery is an extremely significant find. The ear pendant is a kind of ear ornament and is also called *erci*. According to the description in the book *History of the Eastern Han Dynasty*, the hairpin, the jade or pearl earring and the ear pendant were hanging ornaments. It is generally thought that the ear pendant of the Han Dynasty was in the shape of H, with one side bigger than the other. Such a shape is extremely similar to that of the ear pendant unearthed at the Sampula cemetery. As we can see from other archeological evidence available so far, the earliest ear pendants were found in the Central Plains of China. The pottery ear pendants were discovered at the Hemudu site of Zhejiang and the Baojing site of Guangdong. Besides, ear pendants made of bone, amber beads, agate, crystal marble, gold etc. were discovered in some other places of China. Glass ear pendants had appeared since the Warring States period, and during the Han Dynasty they were distributed widely in many provinces, such as Gansu, Ningxia, Henan, Yunnan, Guizhou, Guangxi and Guangdong. So, in terms of typology, the glass ear pendant should belong to the cultural tradition of the Central Plains of China. It could be transported westward to today's Hetian area of Xinjiang from the East along the Silk Road.

References

1. F. X. Gan, Research on the early glass beads unearthed in the Baicheng and Tacheng regions, Xinjiang, *J. Chin. Ceram. Soc.* (in Chinese) **31**, 663–668 (2003). F.X. Gan (ed.), *Research on the Ancient Glass Unearthed from the Southern Part of China: Several Views about Research on the Ancient Glass of China* (Shanghai Science and Technology Publishers), in Chinese, pp. 1–9.

2. Museum of Xinjiang Uygur Autonomous Region, CPAM of Bazhou and CPAM of Charchan County, Excavation of Graveyard No. 1 at Zagunluk in Charchan, Xinjiang, *J. Archaeol.* **1**, 39–136 (2003). Xinjiang Uygur Autonomous Region Museum and Xinjiang Institute of Archeology (ed.), *Sampula in Xingjiang of China: Revelation and Study of Ancient Khotan Civilization* (in Chinese) (Xinjiang People's Publishing House, 2001).
3. Xinjiang Uygur Autonomous Region Museum and Xinjiang Institute of Archeology (ed.), *Sampula in Xingjiang of China: Revelation and Study of Ancient Khotan Civilization* (in Chinese) (Xinjiang People's Publishing House, 2001).

Chapter 16 第十六章

Chemical Composition Analyses of Early Glasses of Different Historical Periods Found in Xinjiang, China

Li Qinghui and Xu Yongchun

Shanghai Institute of Optics and Fine Mechanics,
Chinese Academy of Sciences, Shanghai 201800, China

Zhang Ping

Xinjiang Institute of Archeology, Urumchi 830011, China

Gan Fuxi

Shanghai Institute of Optics and Fine Mechanics,
Chinese Academy of Sciences, Shanghai 201800, China

Fudan University, Shanghai 200433, China

Cheng Huansheng

Institute of Modern Physics, Fudan University,
Shanghai 200433, China

1. Introduction

Spurred on by renewed archeological excavations, many ancient glasses have been found in different areas of China.[1–4] This has built a physical foundation for further study. Scholars both in China and abroad have paid great attention to the research on ancient Chinese glasses.[5–11]

To study the chemical and physical properties by the modern technical means is one of the important aspects of research on ancient glass, which could provide scientific evidence for the archeologist. This is useful for confirmation of the date, making technology and technical origin of the ancient glass. Since the 1920s, scholars both in China and abroad have carried out some work in succession.[12–18] The authors have also obtained new results in recent years.[19–22]

Xinjiang Uygur Autonomous Region, located in the northwest of China, is one of the most significant regions along the desert route of the ancient Silk Roads. In history, Xinjiang has played a very important role in the cultural and technical exchanges between East and West. In the 1980s, some work on the ancient glasses found in Xinjiang was done.[12, 13, 23] There are many questions about the history of glass and glassmaking in Xinjiang, and they remain either unresolved or partially resolved. In this article, we will report the chemical composition of the 65 early glasses found in Xinjiang.

2. Samples and Experiment

All the ancient glass samples were kindly provided by the Xinjiang Institute of Archeology. They include colored beads, eye beads, tubes, glass rings and different fragments of glass vessels. They were either unearthed from cemeteries or collected from historical sites, mostly around the Takelamagan (Taklamakan) desert. The dates of these glass samples are from the Western Zhou to the Song Dynasty (about 1100 BC–1279 AD). The total number of samples is 65. More details of the samples are shown in the Appendix, according to the time sequence.

Before determination of the chemical composition, the structure state of the samples, which were hard to discriminate with the naked eye, was first analyzed with a D/Max 2550V type X-ray diffractometer. Figure 16.1 shows the X-ray diffraction spectra for some samples. It could be found that there were only a few diffraction peaks in samples XJ-1B and XJ-3B, which indicated that

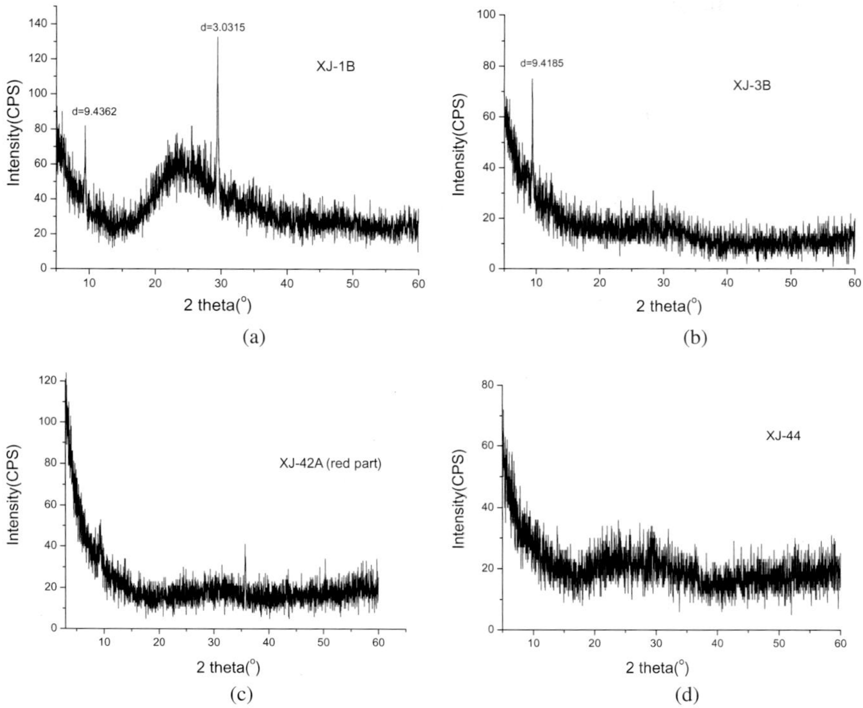

Fig. 16.1. X-ray diffraction spectra of some samples: (a) spectra of a blue fragment unearthed from the tomb at Emin Tiechanggou-Wanquan; (b) spectra of a blue eye bead glass unearthed from the tomb at Kiziltur; (c) spectra of a blue glass fragment unearthed from the tomb at Kiziltur; (d) spectra of the red part of an eye bead unearthed from ancient town Akespili.

these samples contained some crystalline grains. But the main body of these samples was in the noncrystalline state.

Quantitative chemical analyses have been carried out on 48 specimens of early glasses of Xinjiang using a combination of the external-beam PIXE technique, ICP-AES (type of setup: IRIS INTREPID), energy-dispersive X-ray fluorescence spectrometry (EDXRF; type of setup — EDAX Eagle III) and Rutherford backscattering spectroscopy (RBS). We have used the external-beam PIXE technique to analyze the chemical composition of the early glasses in southern and southwestern China. To deduce the influence of the air absorption on the detection of the typical X-ray fluorescence of Na and Mg, a modified external-beam PIXE technique was used in this work. During the experiment, the flowing He gas was infused between the sample and the detector. So the lower mass elements such as Na and Mg could be successfully detected. The details of the method can be found in the article by Cheng Huansheng *et al.* in Ref. 24.

Because of the limited availability of the samples suitable for analysis, the used method was determined by the condition of the samples. The morphology of the specimens was determined with an EPMA 8705 QH2 type scanning electron microscope (SEM). More information about these methods can be found in the literature.[20, 24, 25]

3. Results

The analytic results on the chemical composition of the studied glass samples are shown in Table 16.1. At present, we just classify these samples according to compositional similarities. This is the first step toward further study.

3.1. *From the Western Zhou to the Spring and Autumn Period (about 1100–500 BC)*

For this period, glass samples were unearthed from reservoir tombs at Kiziltur, Baicheng,[26] and the ironworks-chimb to the Wanquan tomb.[27] Kiziltur was one of the important prehistoric sites of the Bronze Age to the early Iron Age in the ancient Quzi Kingdom. It has large-scale residential ruins, bronze metallurgy ruins and public cemeteries.[28] The cemeteries lie on the eastern and western mesas of the Kizil River. According to ^{14}C isotope analysis and emendation of the wood's annual rings (tree-ring calibration), these cemeteries date back to 1100–700 BC, corresponding to the period from the Western Zhou to the Spring and Autumn Period in central China.[29] The ironworks-chimb to the Wanquan tomb was the graveyard of the early nomadic people, dating back to 700–500 BC.

Previously, some work had been done on a few samples from tombs M21, M26 and M37.[30] For this article, the samples chosen were

Table 16.1. Chemical compositions of the early glasses unearthed and collected in Xinjiang (wt%).

Sample	SiO_2	Na_2O	CaO	MgO	K_2O	Al_2O_3	PbO	BaO	CuO	Fe_2O_3	TiO_2	ZnO	MnO	Cl	P_2O_5	Others	Means
XJ-1A	62.50	18.27	5.88	5.20	2.57	1.12	0.09	0.02	0.79	0.57	0.07	0.05	0.04	0.68	1.01		ICP, PIXE
XJ-1B (green)	73.99		4.68	5.6		1.6	0.09		1.2	0.71	0.09				10.79	SrO 0.04	EDXRF
XJ-1B (green)	83.74		5.92		0.45	0.33	0.31		0.94	1.23	0.18		0.06	0.82	4.84	SO_3 1.15	PIXE
XJ-1B (white)	22.30		60.36		0.51	3.32	3.33		0.19	2.29	0.06		0.10	0.95	3.06	SO_3 3.53	PIXE
XJ-1C (blue)	72.81	2.10	5.18	7.01	1.25	2.32					0.13				0.92		EDXRF
XJ-1C (blue)	77.16	2.49	4.12	6.06	1.04	2.85			1.46	0.92	0.17		0.74		1.74		M-PIXE
XJ-1C (white)	69.14	1.10	15.79	4.59	0.43	3.27	4.40		0.21	0.68	0.13		0.10	0.12			M-PIXE
XJ-1D	81.07	1.96	3.28	4.77	1.37	2.70			0.61	1.04	0.20		0.03	0.73	1.88		M-PIXE
XJ-1M	50.24	2.89	38.08	2.37	0.24	1.27			1.47	0.54	0.11		0.33	0.32	1.33		M-PIXE
XJ-2A	64.54	11.54	8.88	5.02	1.59	1.99	1.93	0.01	0.01	1.03	0.02	0.02	0.04	0.68	0.13	Sb_2O_5 0.72	ICP, PIXE
XJ-2B	64.31	12.05	4.80	2.67	2.42	1.36	9.01	0.008		1.10	0.07	0.02	0.02		0.34	Sb_2O_5 1.60	ICP
XJ-2C	66.42	13.06	7.55	5.03	1.11	1.76	3.55		0.02	0.96	0.04	0.01	0.02				M-PIXE
XJ-3A	64.14	15.27	6.65	3.66	2.93	1.44	0.02	0.02	0.76	0.86	0.13	0.06	0.03	1.33	2.53		ICP, PIXE
XJ-3B	71.37	14.50	6.55	3.10	1.65	0.53	0.30	0.01	1.16	0.66	0.03	0.05	0.03				ICP
XJ-4A	66.11	14.29	6.61	4.58	2.19	1.89	0.62	0.01	0.90	1.07	0.17	0.05	0.03			Sb_2O_5 1.44	ICP

(*Continued*)

Table 16.1. *(Continued)*

Sample	SiO_2	Na_2O	CaO	MgO	K_2O	Al_2O_3	PbO	BaO	CuO	Fe_2O_3	TiO_2	ZnO	MnO	Cl	P_2O_5	Others	Means
XJ-4B	78.44		10.06		4.55	0.57			1.97	0.95	0.06		0.05	1.23	0.91	SO_3 1.14	PIXE
XJ-4M	69.98	3.16	4.40	2.61	1.49	7.72	2.34		0.46	1.92	0.34	0.04	0.17	0.60	2.76	SO_3 1.92	M-PIXE
XJ-30	75.44	9.08	7.74	3.35	1.51	1.43	0.02	0.01	0.56	0.34	0.11	0.11	0.02			Sb_2O_5 0.03	ICP
XJ-30M	71.45	9.00	5.71	5.14	1.58	2.95			0.95	0.56	0.15	0.06	0.03	0.41	0.47	SO_3 1.55	M-PIXE
XJ-32	87.52	0.99	5.35	2.52	0.22	1.23	0.06	0.01	1.35	0.48	0.14	0.06	0.03			Sb_2O_5 0.01	ICP
XJ-33	75.03	10.22	4.17	3.17	1.91	1.49	0.03	0.03	1.68	0.74	0.13	0.05	0.72		0.32	Sb_2O_5 0.28	ICP
XJ-33M	71.45	9.00	5.71	5.14	1.58	2.95			0.95	0.56	0.15	0.06	0.03	0.41	0.47		M-PIXE
XJ05-10	68.58	1.96	5.95	19.16	0.54	0.93			0.72	0.52	0.21		0.03	0.43	0.49		M-PIXE
XJ05-11	69.34	5.92	11.44	4.57	1.73	2.75			0.95	0.89	0.49		0.03	0.39	0.53		M-PIXE
XJ05-12	80.49	1.07	6.27	3.50	0.46	2.76			2.40	1.13	0.20		0.16	0.31	0.77		M-PIXE
XJ-44	73.83	9.17	5.19	3.87	1.96	3.00	0.09	0.01	1.04	0.59	0.05	0.03	0.07		0.37	0.05	EDXRF
XJ-44	68.88	15.93	6.11	4.03	2.20	0.87	0.02	0.01	1.10	0.56	0.04	0.05	0.08			Sb_2O_5 0.01	ICP
XJ-5A	77.92	0.82	1.97	0.36	15.60	1.63	0.005	0.03	0.91	0.57	0.12	0.02	0.01				ICP
XJ-5B	78.71	0.81	2.36	0.47	14.18	1.63	0.005	0.02	1.19	0.43	0.06	0.06	0.02			Sb_2O_5 0.02	ICP
XJ-6A	63.09	18.71	4.66	2.74	2.27	4.42	0.02	0.11	1.33	0.67	0.17	0.05	0.05	1.11	0.41	Sb_2O_5 0.05	ICP, PIXE

(Continued)

Table 16.1. *(Continued)*

Sample	SiO_2	Na_2O	CaO	MgO	K_2O	Al_2O_3	PbO	BaO	CuO	Fe_2O_3	TiO_2	ZnO	MnO	Cl	P_2O_5	Others	Means
XJ-6B	62.66	17.36	6.38	3.76	3.86	4.23		0.10	0.56	0.80	0.16	0.01	0.06				ICP
XJ-6C	72.41	3.53	4.81	3.57	4.62	6.62			0.37	0.96	0.25		0.06	1.51	0.75		M-PIXE
XJ-46	20.18	2.20	1.90	0.28	0.36	1.00	47.14	14.62	0.06	12.03	0.05		0.01		0.11	Sb_2O_5 0.04	ICP
XJ05-2	70.03	3.78	5.66	3.16	4.41	5.57			2.92	1.97	0.40		0.31	1.00	0.44		M-PIXE
XJ05-5A	74.69	3.03	2.69	1.00	7.68	8.04			0.02	0.87	0.12		0.00	0.57	0.91		M-PIXE
XJ05-8	77.71	0.12	13.75	0.84	1.01	2.83			1.31	0.86	0.15		0.00	0.21	0.69		M-PIXE
XJ-7A (blue)	81.32		7.99		1.05	3.11			0.48	1.26	0.09		1.88	1.03	0.21	CoO 0.25	PIXE
XJ-7A (white)	78.75		9.41		0.19	2.29	0.89		0.29	1.84	0.13		0.23	2.04	0.71	CoO 0.10	PIXE
XJ-40 (blue)	79.19	1.80	2.98	2.02	2.75	5.97			2.22	0.68	0.16		0.03	0.70	0.83		M-PIXE
XJ-40 (red)	65.15	4.41	6.45	4.19	5.15	7.64	3.73		0.91	1.17	0.21		0.09	0.75			M-PIXE
XJ-40 (white)	56.69	4.63	6.25	4.79	5.53	8.72	10.49		0.19	1.36	0.13		0.06	0.88			M-PIXE
XJ-8	77.28	2.44	2.69	2.22	2.20	7.86			1.39	1.29	0.26		0.09	0.35	1.28		M-PIXE
XJ05-1A	57.97	12.00	6.98	3.86	6.60	7.92			0.03	1.70	0.16		1.38	0.30	0.64		M-PIXE
XJ05-1B	52.20	8.62	6.42	4.51	5.62	8.97	8.28		0.15	1.73	0.10		0.46	0.89	2.05		M-PIXE
XJ05-3A	91.24	0.99	0.65	0.26	1.05	4.59			0.19	0.39	0.16		0.00	0.19	0.31		M-PIXE
XJ05-3A (side)	88.06	0.66	2.84	1.33	0.63	4.59			0.54	0.75	0.10		0.00	0.28	0.14		M-PIXE
XJ05-3B	82.80	1.91	2.69	1.59	1.10	5.49			1.80	1.31	0.16		0.03	0.48	0.22		M-PIXE

(Continued)

Table 16.1. *(Continued)*

Sample	SiO_2	Na_2O	CaO	MgO	K_2O	Al_2O_3	PbO	BaO	CuO	Fe_2O_3	TiO_2	ZnO	MnO	Cl	P_2O_5	Others	Means
XJ05-3C	72.84	3.53	9.30	2.87	1.30	6.98			0.00	1.44	0.20		0.03	0.82	0.22		M-PIXE
XJS05-3d	61.61	6.70	6.36	4.58	3.97	6.78	5.57		0.06	1.23	0.11		0.47	0.91	1.65		M-PIXE
XJ05-6A	63.41	11.58	6.65	6.10	4.01	5.22			0.11	1.16	0.11		0.06	0.61	0.60		M-PIXE
XJ05-6B	72.88	1.63	3.33	1.22	12.16	2.79			3.36	1.20	0.11		0.09	0.37	0.62		M-PIXE
XJ05-6C	64.59	8.72	8.02	5.65	3.83	5.44			0.23	1.38	0.19		0.06	0.92	0.53		M-PIXE
XJ05-6D (black)	60.68	7.34	7.24	3.38	5.77	7.10			0.07	6.13	0.20		0.61	0.41	0.47		M-PIXE
XJ05-6D (eye part)	62.18	7.18	6.34	4.15	5.32	7.01	0.60		0.07	5.55	0.10		0.42	0.41	0.67		M-PIXE
XJ-45	72.39	18.91	5.35	0.41	0.40	1.57		0.02		0.28	0.05	0.01	0.01		0.05	Sb_2O_5 0.49	ICP
XJ-9A	60.36	17.41	7.76	4.48	2.75	3.23	0.69	0.07	0.21	1.40	0.17	0.05	0.12	1.00	0.05		ICP, PIXE
XJ-10M	59.34	22.98	4.96	3.27	3.05	2.78			0.15	1.19	0.11	0.09		1.20	0.67		M-PIXE
XJ-10A	45.05	30.77	6.73	3.86	7.11	3.29		0.07	0.28	1.40	0.28	0.01	0.07	0.61			ICP, PIXE
XJ-12A	67.58	15.18	8.12	3.55	2.23	1.95	0.02	0.03		0.72	0.42	0.04	0.03				ICP
XJ-13A	60.73	9.94	9.06	2.82	8.76	4.25	0.07	0.52	1.78	1.29	0.65	0.05	0.03			Sb_2O_5 0.02	ICP
XJ-13B	66.33	12.80	7.22	4.55	4.69	1.98		0.06		0.64	0.06	0.01	1.35		0.23	Sb_2O_5 0.02	ICP
XJ-14A	70.15	17.98	7.70	0.46	0.53	1.66	0.01	0.04		0.59	0.12	0.01	0.63		0.09		ICP
XJ-15A	58.32	15.62	7.45	4.36	5.82	5.50		0.04	0.01	1.26	0.03	0.02	1.50				ICP
XJ-16A	72.47		7.74		4.20	4.67	0.21	0.03	2.44	2.37	0.25		0.13	2.22	0.74	SO_3 2.34	PIXE

(Continued)

Table 16.1. *(Continued)*

Sample	SiO_2	Na_2O	CaO	MgO	K_2O	Al_2O_3	PbO	BaO	CuO	Fe_2O_3	TiO_2	ZnO	MnO	Cl	P_2O_5	Others	Means
XJ-16B	65.61	18.66	4.48	2.52	3.34	2.17	0.90	0.18	0.48	0.45	0.11	0.02	0.06				ICP
XJ-17A	63.25	13.92	3.33	1.13	7.50	8.56		0.11	0.01	1.46	0.42	0.01	0.03		0.23		ICP
XJ-18A	54.69		17.19		4.95	14.48	0.90	0.34	0.01	0.60	1.39		0.03	0.30	0.03	SO_3 5.08	PIXE
XJ-18A	57.25		21.94		4.01	10.75	0.45	0.18		0.45	1.24		0.02	0.20		SO_3 3.45	PIXE
XJ-18B	69.28	6.89	6.21	3.96	5.15	5.71				0.81	0.13		0.47	0.61	0.44		
XJ-43A	68.00	15.55	8.43	3.21	3.3	1.02				0.48							RBS
XJ-43A	67.77	15.57	8.12	3.30	3.41	0.96		0.02		0.46	0.03	0.03	0.03		0.14		ICP
XJ-43B	64.69	9.1	7.14	3.71	0.69	2.10				12.55							RBS
XJ-19A	54.35	15.81	7.50	3.37	8.23	8.66	0.02	0.08		1.00	0.10	0.01	0.04		0.80		ICP
XJ-20A	59.65	15.36	11.93	4.03	3.28	3.31	0.09	0.06		1.32	0.15	0.01	0.05		0.73		ICP
XJ-21A	71.33	7.01	5.22	5.55	2.93	4.07	1.07		1.24	0.80	0.13			0.63			M-PIXE
XJ-22	66.90	15.77	8.37	4.81	1.92	1.48		0.05		0.43	0.12	0.02	0.025				ICP
XJ-35	66.67	16.19	5.94	3.56	3.79	2.21	0.04	0.01	0.28	0.97	0.08	0.04	0.08				ICP
XJ-35	67.18	16.27	6.21	3.63	3.79	2.06				0.84							RBS
XJ-24	44.18	30.11	7.64	4.31	10.21	1.36	0.11	0.54	0.06	0.55	0.10	0.06	0.72				ICP
XJ-25	69.75	19.35	6.15	0.70	0.50	2.15		0.02		0.63	0.10	0.01	0.22		0.07	Sb_2O_5 0.32	ICP
XJ-27	78.49		8.00		4.64	3.09	0.21			1.22	0.29		0.32	1.77			PIXE
XJ-34A	64.00	16.19	8.02	3.62	4.94	2.05				1.18							RBS
XJ-41A	71.79		10.76		5.22	8.35		0.12	0.01	2.91	0.29		0.09				PIXE
XJ-42A (red)	71.63		9.25		4.04	6.50	0.66	0.16	3.79	2.48	0.33		0.08		0.45		PIXE

(Continued)

Table 16.1. *(Continued)*

Sample	SiO_2	Na_2O	CaO	MgO	K_2O	Al_2O_3	PbO	BaO	CuO	Fe_2O_3	TiO_2	ZnO	MnO	Cl	P_2O_5	Others	Means
XJ-42A (black)	73.12		12.52		3.91	5.89	0.02	0.06	0.37	2.51	0.48		0.10		0.12		PIXE
XJ-42A (white)	73.22		11.42		3.87	6.05	0.65	0.08	0.28	2.29	0.30		0.07		0.19		PIXE
XJ-42A (yellow)	66.17		9.35		3.81	4.86	13.07	0.09	0.53	1.57	0.16		0.06		0.32		PIXE
XJ-42A (green)	73.51		10.15		4.40	5.39	1.32	0.01	1.99	2.12	0.27		0.07		0.23		PIXE
XJ-42A (green)	70.66		10.44		3.85	6.06	2.51		2.54	2.89	0.29		0.08			SrO 0.07	EDXRF
XJ-42B (black)	56.98	3.36	1.13	0.16	0.44	2.39	23.7	1.34	2.86	4.44					0.54	Ce_2O_3 0.13	EDXRF
XJ-42B (white)	67.64		0.72		0.69	1.94	24.79	1.27	0.47	1.48					0.99		PIXE
XJ-42B (green)	57.04		3.41		0.51	6.94	23.77	5.42	1.08	0.84					0.98		PIXE
XJ-42B (green)	49.35		2.52		0.37	3.19	34.24	7.87	0.64	0.82			0.01		0.97		PIXE
XJ-42B (black)	43.52		1.49		0.76	3.65	31.28	1.43	5.28	9.96	0.06		0.03		2.51		PIXE
XJ-28	52.72	17.81	8.68	4.49	5.88	9.12	0.01	0.05		1.01	0.02	0.04	0.04				ICP

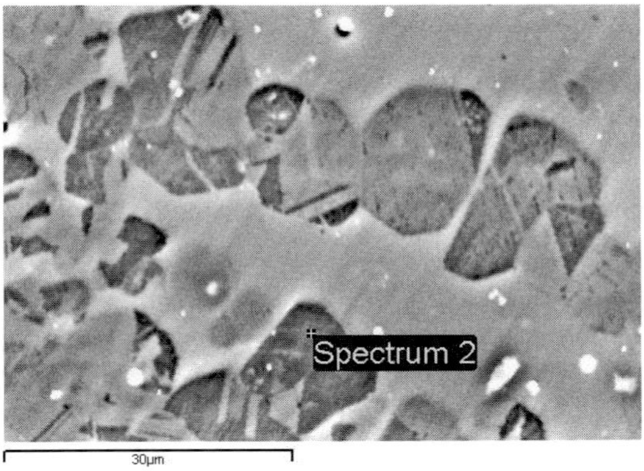

Fig. 16.2. SEM image of sample XJ-2A.

from tombs M11, M26, M27, M60, M61 and M75. The total number was 21. Most of the samples were single-colored, such as blue, light yellow and green. The glass body was weathered at different levels and contained air bubbles and unmelted grains. The body surface was covered with some white or brown material. Most of the samples were opaque. According to X-ray diffraction analysis, part of the material was crystalline $CaCO_3$. Figure 16.2 is the SEM image of sample XJ-2A. In the figure, the unmelted grains can be seen clearly. According to EDX spectroscopy, the chemical composition of the indicated grains was Na 24.7%, Si 43.4%, Ca 31.2%, Mg 0.7%.

Three eye bead samples, XJ-1B, XJ-1C and XJ-1M, were found. Each had two eyes with white rings inlaid in the glass body. Sample XJ-1 and one of the eye beads, XJ-1C, are shown in Fig. 16.3. It was found that there was some difference between the analytical results determined by different means, which was correlated with the unevenness of the samples, system errors and different characteristics of the means.

Based on the analytical results, samples of this period are mainly of four types.

(1) Na_2O–CaO–SiO_2 glass

About ten samples belong to this kind of glass, which occupied about 48% of the analyzed samples of this period. These samples

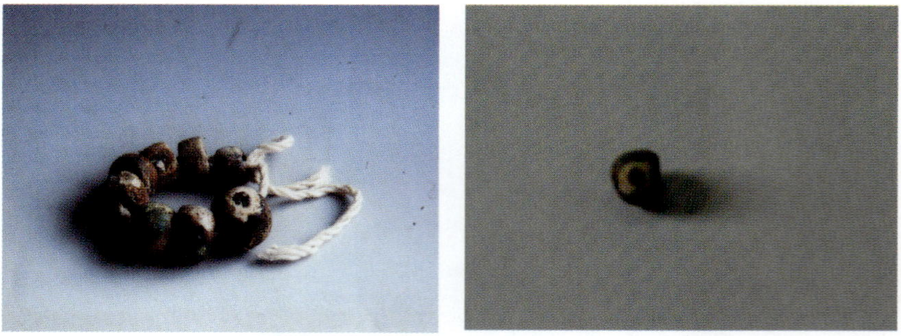

(a) (b)

Fig. 16.3. (a) Sample XJ-1 and (b) one of the eye beads, XJ-1C.

included XJ-1A, XJ-3A, XJ-3B, XJ-4A, XJ-30, XJ-33, XJ-33M and XJ-44. They covered nearly all the sampled tombs. All of them are similar in chemical composition, the Na_2O content being 10%–14%, the CaO content 5%–8%, and the K_2O and Al_2O_3 contents less than 3%. The alkali ratio (Na_2O/K_2O) is between 3 and 5.

(2) Na_2O–CaO–PbO–SiO_2 glass

Samples XJ-2A, XJ-2B, XJ-2C and XJ-4M belong to this kind of glass, which occupied about 19% of the analyzed samples. One of the characteristics of the samples is that they have a high PbO content. The PbO content of the samples is 1.93%, 9.01%, 3.55% and 2.34% respectively. These samples are mainly from tombs M26 and M11, and all are yellow or yellow–green in color. The Na_2O content of sample XJ-4M is only 3.16%, but that of Al_2O_3 is up to 7.72% and so it is unique.

(3) CaO–MgO(PbO)–SiO_2 glass

This kind of glass used CaO and MgO as the main flux, with a low Na_2O content (less than 2.5%) or undetected. Five samples, XJ-1B, XJ-1D, XJ-32, XJ05-10 and XJ05-12, belong to this kind of glass, which occupied about 29% of the analyzed samples. They are mainly from tombs M26, M9 and M61. For the glass eye beads, there is an obvious difference between the chemical composition of the white eye parts and of the glass bodies. The white rings of the eyes have higher CaO and PbO contents than the body.

(4) Others

The contents of Na_2O and MgO in XJ-4B were not determined. Maybe they belong to Na_2O–CaO–SiO_2 or CaO–MgO–SiO_2 glass.

Generally, the samples unearthed from Baicheng and Tacheng are characterized by a high content of MgO and a low content of Al_2O_3, and some samples have a higher PbO content. For XJ-2A, XJ-2B and XJ-4A, there is a high Sb_2O_3 content. It should be pointed out that similar Na_2O–CaO–PbO–SiO_2 glass beads were found in Marquis Zeng mausoleum in Sui county of Hubei province.[31]

The glass beads unearthed from the Kiziltur cemetery are the earliest glasses found in China and were analyzed through technical methods. They are mostly single-colored and are weathered at different levels. Only three eye beads are found now, and they have only two eyes. The shape of the beads is irregular. The making technology for these eye beads was obviously backward, compared to that of the Middle and East Asian areas.[32] The results of X-ray diffraction and SEM analysis show that these beads contain some little crystalline grains and air bubbles. All these indicate that the glassmaking was immature. The obvious difference between the composition of the white part and that of the green body (XJ-1B) shows that the craftsman was familiar with some properties of the raw materials, such as lead ores.

The compositions of the Na_2O–CaO–SiO_2 glass samples from Kiziltur and Baicheng are very similar to those of glasses from Mesopotamia (approximately 800 BC) and the Persian Kingdom.[21] But the other kinds of glass are rarely found elsewhere. From the Western Zhou to the Spring and Autumn Period, the metallurgy was fully developed in the prehistoric

Qiuzi Kingdom. Once, a bronze scoop with a lead (Pb) content higher than 10% was excavated from the reservoir tombs at Kiziltur.[33] Possibly, the Na_2O–CaO–SiO_2 glass beads from Kiziltur were either imported from Middle and West Asia, or made locally after absorbing the imported glassmaking technology. The other glass beads should have been made locally using the native raw materials. The glassmaking had a close relationship with the local metallurgy of the bronze and lead ores. The nomads played a very important role in the glass trade and the spread of glassmaking.

3.2. *The Warring States Period (475–221 BC)*

The six samples of this period were unearthed at Wensu, Qiemo and Hami. All of them have accurate provenances and dates. They can be classified into three groups: (1) samples XJ-5A and XJ-5B belong to the K_2O–SiO_2 system glass; their K_2O content is 15.60% and 14.18% respectively, and their Na_2O content is less than 1%; (2) samples XJ-6A and XJ-6B belong to the Na_2O–CaO–SiO_2 system glass and have a high MgO content (about 3%); (3) sample XJ-46 (Fig. 16.4) belongs to the PbO–BaO–SiO_2 system glass and contains 47.14% PbO, 14.26% BaO and 12.03% Fe_2O_3.

Fig. 16.4. Glass beads XJ-46.

PbO–BaO–SiO_2 glasses of the Warring States period were widely excavated in the valleys of the Yellow River and the Yangtze River. In Hunan province, K_2O–SiO_2 glass of the Warring States period was excavated, which coexisted with PbO–BaO–SiO_2 glass.[13] Besides Wensu county, K_2O–SiO_2 glass of the Warring States period was found in Jiangchuan county, Yunnan province.[34] Later, large numbers of K_2O–SiO_2 glass relics were excavated from the tombs of the Han Dynasty (206 BC – 220 AD) located in Guangxi, Guangdong, Yunnan, Guizhou and other provinces.[8,14] The archeological research results about the corals and seashells unearthed in Xinjiang, and about the Hetian nephrite unearthed in central and southwestern China, revealed that these areas had direct or indirect cultural and technical exchange as early as the Shang Dynasty and the Western Zhou.[35] The results of this article show that the exchange included the glassmaking and artifacts between Xinjiang and the other areas of China.

3.3. *From the Western Han Dynasty to the Eastern Han Dynasty (206 BC–220 AD)*

Samples dated after the Warring States period were mainly gathered from the historical sites. Some sites lasted for several hundred years. Only three samples were confirmed to be of this period. They are XJ05-2, XJ05-5A and XJ05-8, from Luntai county, Yili city and Kuche county respectively, in the north of the Tekelmakan Desert. The chemical compositions of XJ05-2 and XJ05-5A are similar to that of XJ-6C from Qiemo county, which belongs to mixed alkali silicate glass. XJ05-2 and XJ05-5A have a high Al_2O_3 content, 5.57% and 8.04% respectively; while the chemical composition of XJ05-8 is different, with CaO (13.75%) as the main flux and lower K_2O (1.01%) and Na_2O (0.12%).

3.4. *From the Han Dynasty to the Song Dynasty (206 BC–1279 AD)*

Most of the samples of this period were collected from different historical sites at Ruoqiang, Minfeng, Cele, Hetian, Pishan, Shufu,

Fig. 16.5. Glass bead XJ-7A.

Fig. 16.6. Glass beads XJ-42A (*left*) and XJ-42B (*right*).

Keping, Kuche and other places, along the southern route of the desert Silk Road. During this period, the types of glass were obviously increased, including glass eye beads, glass rings, and fragments of glass vessels like bottles and cups. The glass quality also improved. The making techniques included molding, pressing, blowing, twisting, cold processing, etc. The total number of samples is 43. Figures 16.5 and 16.6 are photographs of some glass eye beads, such as XJ-7A, XJ-42A and XJ-42B. It can be seen that these eye beads are very fine and that the making technology had improved greatly from those at Kizil, Baicheng. The samples can be divided into five groups:

(1) Na_2O–CaO–SiO_2 glass

About 15 samples (XJ-45, XJ-9A, XJ-10M, XJ-12A, XJ-13B, XJ-14A, XJ-16B, XJ-43A, XJ-43B, XJ-20A, XJ-22, XJ-35, XJ-24, XJ-25, XJ-34A) belong to this kind of glass, which occupied about 35% of the analyzed samples. Besides XJ-43B, the Na_2O content of these samples is between 15% and 20%. The Na_2O–CaO–SiO_2 glass is similar to the following Na_2O–K_2O–CaO–SiO_2 glass, but has a lower content of K_2O and Al_2O_3 (both less than 4%) except for XJ-34A (4.94% K_2O) and XJ-24 (10.21% K_2O).

This kind of glass can be divided into two types, depending on the contents of K_2O and MgO. One type is the Roman glass characterized by low contents of K_2O and MgO (<1%), such as samples XJ-14A, XJ-7A, XJ-25 and XJ-45. The other type is the Sasanian glass having higher contents of K_2O and MgO (>3%) than the Roman glass, such as samples XJ-35 and XJ-43A. The two types of glasses were imported into China through the Desert Silk Road.

(2) Na_2O–K_2O–CaO–SiO_2 glass

This kind of glass is mixed alkali silicate glass characterized by higher contents of K_2O and Al_2O_3. The K_2O content of these samples is mostly higher than 7%. The Al_2O_3 content is also very high, and the value is about 4%–9%. According to the Na_2O content, this kind of glass can be divided into two types. One type has a Na_2O content that is less than 10%, which occupied 23% of the samples for this period, such as samples XJ-40, XJ-8, XJ05-1B, XJ05-3C, XJ05-3D, XJ05-6C, XJ05-6D, XJ-13A, XJ-18B and XJ-21A. Among them, sample XJ-21A contains some PbO (1.07%), and the content of K_2O is also lower than for the others. The other type has a Na_2O content that is higher than 10%, such as samples XJ05-1A, XJ05-6A, XJ-10A, XJ-15A, XJ-17A, XJ-19A and XJ-28.

For the period from the Jin Dynasty through the Song Dynasty (about the 3rd–13th centuries), the Na_2O–K_2O–CaO–SiO_2 glass was widely found and lasted for a long time. We think that this kind of glass was a native product and saltpeter was possibly the melting flux.

(3) PbO–BaO–SiO_2 and $Na_2O(K_2O)$–CaO–PbO–SiO_2 glass

Three glass eye beads belong to this kind of glass. Sample XJ-42B is PbO–BaO–SiO_2 glass, with PbO 24%–34% and BaO 1%–8%. The yellow part of sample XJ-42A and the red and white parts of sample XJ-40 are $Na_2O(K_2O)$–CaO–PbO–SiO_2 glass, with the content of PbO being 13.07% and 10.49% respectively, and the content of Al_2O_3 is higher than 5%.

(4) K_2O–SiO_2 glass

Only one sample, XJ05-6B, belongs to this kind of glass, with K_2O (12.16%) as the main flux and 3.33% CaO and 2.79% Al_2O_3.

(5) Faience and others

According to the SEM and XRD analytical results (not provided here), samples XJ05-3A and XJ05-3B from the Niya sites belong to faience beads. These two samples were colorized by Fe and Cu elements, and contain many crystalline α-quartz grains and little K_2O, Na_2O, MgO and CaO.

There are some samples whose system cannot be clearly determined now. According to the present results, samples XJ-7A, XJ-16A and XJ-27 are probably Na_2O–CaO–SiO_2 glass, and samples XJ-18A and XJ-41A are probably Na_2O–K_2O–CaO–SiO_2 glass.

4. Discussion

Based on the experimental results, we can see that the chemical composition of the early glasses in Xinjiang has special characteristics.

Although other kinds of glass also exist, they are mainly Na_2O–CaO–SiO_2 and Na_2O–K_2O–CaO–SiO_2 glasses. The ratio of the PbO–BaO–SiO_2 glasses is low; perhaps they were introduced from the central areas of China. The glassmaking in Xinjiang was influenced by both the West and the central China areas and had a characteristic origin. Its technical development was also different from that in central China. Combining our analytic results for the early glasses of the Han Dynasty and the Warring States Period which were unearthed in Gansu province and the literature, we can confirm that the Northwest Silk Road was one of the routes along which the glass trade and glassmaking exchange took place. As early as the Spring and Autumn and Warring States periods (770–221 BC), there was already intercourse, either direct or indirect, between Xinjiang and the West through this route. Further work is ongoing.

Appendix: Sample Descriptions

1. Samples from the Western Zhou to the Spring and Autumn Period (about 1100–500 BC).

Samples	Site and others
XJ-1A	Reservoir tomb 90BKKM26:6 at Kiziltur, Baicheng. Glaucous bead, outer diameter 1.5 cm.
XJ-1B	Reservoir tomb 90BKKM26:6 at Kiziltur, Baicheng. Eye bead, green glass with two green-and-blue eyes, outer diameter 1.1 cm, internal diameter 0.5 cm, height 0.5–1.0 cm, opaque.
XJ-1C	Reservoir tomb 90BKKM26:6 at Kiziltur, Baicheng. Eye bead, turbid blue glass with two blue-and-white eyes, outer diameter 1.5 cm, internal diameter 0.4 cm, height 0.8–1.4 cm, opaque.
XJ-1D	Reservoir 90BKKM26:6 at Kiziltur, Baicheng. Green glass bead, outer diameter 1.3 cm, height 0.9 cm, opaque.
XJ-1M	Reservoir tomb 90BKKM26:6 at Kiziltur, Baicheng. Uneven green glass bead with air bubble, opaque, remaining white ring inlay in green body.
XJ-2A	Reservoir tomb 90BKKM26:6 at Kiziltur, Baicheng. Light yellow bead, outer diameter 1.2 cm, opaque.

(*Continued*)

(*Continued*)

Samples	Site and others
XJ-2B	Reservoir tomb 90BKKM26:6 at Kiziltur, Baicheng. Light yellow bead with break, outer diameter 0.8 cm, internal diameter 0.4 cm, height 0.6 cm, opaque.
XJ-2C	Reservoir tomb 90BKKM26:6 at Kiziltur, Baicheng. Yellow bead fragment.
XJ-3A	Reservoir tomb 90BKKM4:7 at Kiziltur, Baicheng. Green bead fragment, opaque.
XJ-3B	Reservoir tomb 90BKKM4:7 at Kiziltur, Baicheng. Blue bead, semitransparent, outer diameter 1.0 cm.
XJ-4A	Reservoir tomb 90BKKM11 at Kiziltur, Baicheng. Light green bead fragment, opaque.
XJ-4B	Reservoir tomb 90BKKM11 at Kiziltur, Baicheng. Blue bead fragment, opaque.
XJ-4M	Reservoir tomb 90BKKM11 at Kiziltur, Baicheng. Yellow–greenish bead fragment, opaque.
XJ-30	Reservoir tomb 91BKKM3:9 at Kiziltur, Baicheng. Green bead fragment, opaque.
XJ-30M	Reservoir tomb 91BKKM3:9 at Kiziltur, Baicheng. Green bead fragment, opaque.
XJ-32	Reservoir tomb 91BKKM9:4 at Kiziltur, Baicheng. Green bead fragment, opaque.
XJ-33	Reservoir tomb 91BKKM27 at Kiziltur, Baicheng. Green bead fragment, opaque.
XJ-33M	Reservoir tomb 91BKKM27 at Kiziltur, Baicheng. Green bead fragment, opaque.
XJ05-10	Reservoir tomb M60:1 at Kiziltur, Baicheng. Light blue bead fragment.
XJ05-11	Reservoir tomb M75 at Kiziltur, Baicheng. Blue bead fragment, opaque, with air bubble and unmelted part.
XJ05-12	Reservoir tomb M61 at Kiziltur, Baicheng. Light yellow bead fragment, with air bubble.
XJ-44	From ironworks-chimb to Wanquan tomb No. 1 at E'min, Tacheng. Blue bead fragment, opaque.

2. Samples after the Warring States (475–221 BC).

Samples	Site and others
XJ-5A	Tomb 83WBM41 at Bao-Zi-Dong, Wensu. Blue-and-green bead fragment, opaque.
XJ-5B	Tomb 83WBM41 at Bao-Zi-Dong, Wensu. Blue bead fragment, opaque.
XJ-6A	Tomb 98QZIM1249:9 at Zha-Gun-Lu-Ke, Qiemo. Light blue bead, outer diameter 0.4 cm, semitransparent.
XJ-6B	Tomb 98QZIM1249:9 at Zha-Gun-Lu-Ke, Qiemo. Blue bead, height 0.3 cm, opaque.
XJ-6C	Tomb 98QZIM1249:9 at Zha-Gun-Lu-Ke, Qiemo. Light green bead fragment.
XJ-46	Tomb 96HHSHM14 at Hami. Blue pumpkin-shaped bead fragment.

3. Samples after the Han Dynasty (206 BC–220 AD).

Samples	Site and others
XJ05-2	Historical site of Ke-You-Ke-Qin castle in Luntai county. Blue bead, outer diameter about 0.6 cm, opaque.
XJ05-5A	Cemetery of Red Flag factory in Gongliu county, Yili city. Light green semitransparent bead with hexagonal prism figure. Body length about 0.9 cm, arris length of the bottom surface about 0.5 cm.
XJ05-8	Ma-Zha-Pu-Tang cemetery in Kuche county. Light blue bead fragment.

4. Samples after the Han Dynasty (220–1271 AD).

Sample	Time	Site and others
XJ-7A	3rd–4th cent. AD	Shan-Pu-La cemetery 84HLSM49:155 or 84HLSM35:1, Luopu county. Glass eye bead with dark blue body inlaid with several white-and-blue rings. Inner diameter about 0.6 cm outer diameter about 1 cm.

(*Continued*)

(*Continued*)

Sample	Time	Site and others
XJ-40	3rd–4th cent. AD	Shan-Pu-La cemetery, Luopu county. Eye bead fragment with blue body inlaid with black–yellow–brown parts.
XJ-8	2nd cent. BC to 4th cent. AD	Niya historical site, Minfeng county. Blue bead with many faces.
XJ05-3	2nd cent. BC to 4th cent. AD	Four samples from Niya historical site, Minfeng county. XJ05-3A: milky–light green faience bead with nipple-shaped decoration, outer diameter about 1 cm. XJ05-3B: green faience bead with irregular shape, length about 0.8 cm. XJ05-3C: three-linked yellow glass beads, length about 1.1 cm. XJ05-3D: glass eye bead fragment with black body inlaid with white-and-black eye parts.
XJ05-1	2nd cent. BC to Tang Dynasty	Two samples from Tu-Mu-Xiu-Ke historical site, Bachu county. XJ05-1A: light yellow glass fragment with two yellow-and-green striae on the surface. XJ05-1B: funnel-shaped bead with black-and-white alternate-stria body, one white ribbon around the waist, two end surfaces (diameter 6 mm and 3 mm).
XJ05-6	2nd cent. BC to Tang Dynasty	Four samples from Da-Wang-Ku-Mu historical site, Xinhe county. XJ05-6A: yellow glass fragment containing air bubbles. XJ05-6B: celeste semitransparent glass bead, length about 1 cm, diameter about 0.6 cm. XJ05-6C: black bead fragment, diameter about 1.2 cm. XJ05-6D: glass eye bead with black body inlaid with five white-and-blue double-layered eye parts.

(*Continued*)

(Continued)

Sample	Time	Site and others
XJ-18	2nd cent. BC to Tang Dynasty	Two samples from Pishan county, Hetian city. XJ-18A: celeste quadrate glass tube, opaque, length about 0.4 cm, width about 0.3 cm. XJ-18B: light yellow glass fragment.
XJ-43	Han Dynasty to Tang Dynasty, 206 BC–907 AD	Two samples from ancient An-Di'er castle site, Minfeng county. XJ-43A: light green–yellow glass vessel fragment with elliptic pattern, transparent. XJ-43B: yellow–greenish glass vessel fragment with raised stria.
XJ-45	2nd cent. BC to 4th cent. AD	Tomb 80LBC:180A, Loulan historical site, Ruoqiang county. White fragment of glass cup, semitransparent, molding.
XJ-9A	4th–8th cent. AD	Da-Ma Gorge site, Ce-Le county. Navy-blue glass fragment with mouth-shaped part, semitransparent to transparent, containing air bubbles.
XJ-10A	4th–8th cent. AD	Above sited navy-blue opaque glass fragment, containing air bubbles.
XJ-10M	4th–8th cent. AD	Above sited green hexagonal prism glass bead fragment.
XJ-13	Southern and Northern Dynasties to Tang Dynasty	Two samples from Ka-La-Ke'er historical site, Luopu county. XJ-13A: bottle-green opaque glass fragment. XJ-13B: light purple semitransparent glass fragment.
XJ-14A	Tang Dynasty to Northern Song Dynasty	Mai'e-Pur mausoleum site. White transparent glass fragment containing air bubbles.
XJ-15A	6th–8th cent. AD	Tuo-Mu-Li-Ke castle site, Keping county. Yellow fastener-shaped glass fragment, semitransparent.

(Continued)

(*Continued*)

Sample	Time	Site and others
XJ-16	6th–8th cent. AD	Two samples from Qiong-Ti-Mu historical site, Keping county. XJ-16A: blue–greenish glass tube, length about 1.2 cm, outer diameter about 0.3 cm. XJ-16B: green glass fragment.
XJ-17A	Tang Dynasty	Bo-Xi-Ke-Re historical site, Shufu county. Green transparent glass vessel fragment.
XJ-19A	Tang Dynasty to Song Dynasty	Mo'er stupa site, Shufu county. Light green transparent glass fragment, containing many little air bubbles.
XJ-20A	Tang Dynasty	La-Yi-Su watch tower site, Luntai county. Blue glass fragment.
XJ-21A	Tang Dynasty	Tuo-Gai-Ta-Mu castle site, Kuche county. Blue semitransparent glass bead, outer diameter 0.6 cm, height about 0.3 cm.
XJ-22	Tang Dynasty	Ku-Mu-Tu-La historical site, Kuche county. Light green semitransparent glass fragment.
XJ-35	Tang Dynasty	Ku-Mu-Tu-La historical site, Kuche county. Light blue glass vessel fragment, surface covered with polychrome layer, containing holes and air bubbles.
XJ-24A	Tang Dynasty	Wu-Jia-Bi'e-Mu mausoleum site, Hetian city. Light purple glass fragment, weathered surface, coexisting with copper coin of the Tang Dynasty (Kai-Yuan period).
XJ-25	Southern and Northern Dynasties to Tang Dynasty	Stupa site at Bu-Gai-Wu-Yi-Li-Ke, Moyu county. Colorless transparent glass fragment.
XJ-27	Tang Dynasty	Da-Ma Gorge site, Ce-Le county. White transparent glass ring, inner diameter 1.8 cm, outer diameter 2.0 cm, with octagonal shape on the outside.

(*Continued*)

(*Continued*)

Sample	Time	Site and others
XJ-34A	Southern and Northern Dynasties to Tang Dynasty	Sen-Mu-Sai-Mu grotto site, Kuche county. Green glass fragment with raised arris.
XJ-41A	4th–8th cent. AD	Mai-Li-Ke-Ga-Wa-Ti site, Hetian city. Green glass tube, length about 1.6 cm, aperture about 0.2 cm.
XJ-42	Tang Dynasty to Song Dynasty, maybe Han Dynasty	Two samples from ancient A'Ke-Si-Pi-Li castle ruins. XJ-42A: multi-colored (red, yellow, green, white and black) glass eye bead fragment. XJ-42B: Big eye bead with green-and-white eyes, black glass, opaque, outer diameter 1.9 cm, height 1.6 cm.
XJ-28	Song Dynasty to early Yuan Dynasty	Wa-Shi Gorge site, Ruoqiang county. Light green oral fragment of glass bottle, semitransparent, containing air bubbles, irregular shape.

Acknowledgments

This research is supported by the National Natural Science Foundation of China with Grant No. 50672106 and the Intellectual Innovation Project of the Chinese Academy of Sciences ("Technical Research of the Ancient Chinese Glass and Jade"). The authors are grateful to Profs. Jiazhi Li and Juan Wu of the Shanghai Institute of Ceramics, Dr Bin Zhang of Fudan University and Mr Yongchun Xu of the Shanghai Institute of Optics and Fine Mechanics for their help in the related experiments.

References

1. B. D. Yang, *Palace Museum Journal* (in Chinese) **2**, 14–24 (1980).
2. Z. X. Gao, *Cultural Relics* (in Chinese) **12**, 54–65 (1984).
3. J. Y. An, *Acta Archaeologica Sinica* (in Chinese) **4**, 413–448 (1984).

4. F. X. Gan (ed.), *Scientific Research on Early Chinese Glass — Proceedings of the Archeometry of Glass Session of the International Symposium on Glass*. Beijing, 1984; (in Chinese) (China Architecture & Building Press, Beijing, 1986).
5. F. X. Gan, Z. F. Huang and B. R. Xiao, *J. Chin. Ceram. Soc.* (in Chinese) **6**, 99–104 (1978).
6. B. D. Yang, *Palace Museum Journal* (in Chinese) **5**, 76–78 (1979).
7. D. J. Hou, *The Mining, Metallurgy, Lacquer and Glass Making in the Chu Area* (Hubei Education Press, Hankou, 1995), in Chinese.
8. Q. S. Huang, *Archeology* (in Chinese) **3**, 264–276 (1988).
9. H. C. Beck and C. G. Seligman, Barium in ancient glass, *Nature* **133**(6), 982 (1934).
10. C. G. Seligman, P. D. Ritchie and H. C. Beck, Early Chinese glass from pre-Han to Tang times, *Nature* **138**, 721 (1936).
11. K. H. Zhao, *Stud. Hist. Natural Sci.* (in Chinese) **10**(2), 145–156 (1991).
12. Research Institute of Building Materials, Qinghua University, Institute of Archeology, CASS, *Acta Archaeologica Sinica* (in Chinese) **4**, 449–457 (1984).
13. F. K. Zhang, Z. H. Cheng and Z. G. Zhang, *J. Chin. Ceram. Soc.* (in Chinese) **11**(1), 67–75 (1983).
14. M. G. Shi, O. L. He and F. Z. Zhou, *J. Chin. Ceram. Soc.* (in Chinese) **14**(3), 307–313 (1986).
15. S. Kwan, *Early Chinese Glass* [M] (Art Museum, Chinese University of Hong Kong, 2001), in Chinese.
16. M. G. Shi and F. Z. Zhou, Some Chinese glasses of the Qing Dynasty, *J. Glass Studies* **3**, 102–105 (1993).
17. J. Z. Li and X. Q. Chen, *J. Chin. Ceram. Soc.* (in Chinese) **14**(3), 293–296 (1986).
18. R. H. Brill and S. Hiroshi, Lead-isotope analysis of some Asian glasses, in *Proceedings of the XVII International Congress on Glass* (Beijing, 1995) (Chinese Ceramic Society), Vol. 6, pp. 491–496.
19. F. X. Gan (chief ed.), *Study on Ancient Glasses in Southern China — Proceedings of the 2002 Nanning Symposium on Ancient Glasses in Southern China* (Shanghai Science and Technology Publishers, 2003), in Chinese.
20. Q. H. Li, B. Zhang, H. S. Cheng *et al. J. Chin. Ceram. Soc.* (in Chinese) **31**(10), 950–954 (2003).

21. Q. H. Li, B. Zhang, H. S. Cheng *et al.*, *J. Chin. Ceram. Soc.* (in Chinese) **31**(7), 663–668 (2003).
22. Q. H. Li, D. H. Gu, F. X. Gan *et al.*, *Nuclear Techniques* (in Chinese) **26**(12), 922–925 (2003).
23. J. Y. An, Ancient glasses found in Xinjiang, in *Proceedings of the International Congress on Glass* (Edinburgh, 2001), Vol. 2. Extended Abstract, 13.
24. H. X. Zhu, H. S. Cheng, F. J. Yang *et al.*, *Nuclear Techniques* (in Chinese) **24**(2), 149–153 (2001).
25. Q. G. Li, Archeometry and analyzing techniques of the ancient glasses. In: Gan Fuxi (chief ed.), *Study on Ancient Glasses in Southern China — Proceedings of the 2002 Nanning Symposium on Ancient Glasses in Southern China* (Shanghai Science and Technology Publishers, 2003), in Chinese pp. 65–75.
26. Xinjiang Institute of Archeology, *Archeology* (in Chinese) **6**, 14–28 (2002).
27. X. T. Liu and T. H. Ti, *China Cultural Relic News* (in Chinese) **19**(1), 2002–07.
28. P. Zhang, *Cultural Relics of Xinjiang* (in Chinese) **2**, 59–65 (1999).
29. C. W. Ruan and J. Y. Liu, *Cultural Relics of Xinjiang* (in Chinese) **3–4**, 114–115 (1999).
30. W. Qian, P. Zhang and Q. M. Li, In: F. B. Wan and E. Bamo (eds.), *Proceedings of the 5th International Conference of Chinese Minorities Science and Technology History* (Xichang, 2000) (Guangxi Minzu Press, 2001), in Chinese, pp. 138–145.
31. Hubei Provincial Museum, *Marquis Zeng Mausoleum of Sui Country* (Wenwu Press, Beijing, 1989), in Chinese, pp. 423–425.
32. J. Y. An, Glass eye beads in China. In: UNESCO, Institute of Archeology, Chinese Academy of Social Sciences (eds.), *Land Routes of the Silk Roads and the Cultural Exchanges Between the East and West Before the 10th Century* (New World Press, Beijing, 1996), in Chinese, pp. 351–367.
33. I. A. Rasul, P. Zhang and W. Qian, *Cultural Relics of Xinjiang* (in Chinese) **1**, 53–59 (2002).
34. Yunnan Provincial Museum, *Acta Archaeologica Sinica* (in Chinese) **2**, 97–156 (1975).
35. Y. Z. Zhang, *Cultural Relics of Xinjiang* (in Chinese) **3**, 27–31 (1996).
36. P. Zhang, *Xinjiang Soc. Sci.* (in Chinese) **3**, 87–92 (1986).

Chapter 17 第十七章

Glass Materials Excavated from the Kiln Site of Tricolor Glazed Pottery at Liquanfang in Chang An City of the Tang Dynasty

Jiang Jie

Famen Temple Museum, Shaanxi, 722201, China

1. Introduction

The Shaanxi Provincial Institute of Archeology (SPIA) found a tricolor glazed ceramics kiln site of the Tang Dynasty and did rescue work in summer 1999. Four kiln sites, ten ash pits and one modern tomb were excavated, covering 140 square meters. More than 10,000 different kinds of tricolor glazed pottery fragments and porcelain fragments were unearthed, including some residual glass fragments and pieces of bone tools, etc.

The center of the kiln site is located on Xiguanzheng Street, outside of Ximen ("West Gate"), south of Fenggao Road, west of Caoyang village and South Laodong Road, and northeast of the former Xi'an Airport runway. Now it is a dormitory building of the Northwest Aviation Administration Bureau. This site is 3.5 km from the Bell Tower of the city center, with plane topography and 450 m above sea level, at 108°53′30″ east longitude and 30°15′10″ north latitude.

According to the research done by Mr Shi Nianhai, this site was Liquanfang ("*fang*" means "lane" or "workshop" in Chinese) of the ancient Chang An city during the Sui and Tang Dynasties. Liquanfang was one of the largest communities in Chang An city during that period. It was not only near the West Market, but also near the central government and the imperial palace. Shi pointed out that "Liquanfang was still the top prosperous place even after the political center moved to Daming Palace in the east of the city." The Chunmingmen–Jinguangmen streets passed through Liquanfang and the West Market, which used to be an important labor and commodity trade center and economic artery of Chang An city.

From investigation it has been inferred that Liquanfang had a four-side wall built with rammed earth. There was a gate on each side of the wall in the area of Liquanfang, 838 m from south to north and 1032 m from east to west, respectively. There was a cross street, 15 m wide, through the gate. According to the books *New Notes of Two Capitals* (by Wei Shu) and *Chang An's History* (by Song Minqiu), there were four cross streets in the directions of east, west, south and north, set within three *fang* (lanes) which had communities at the palace and imperial city of Chang An. The cross streets divided the area of *fang* into four sections — northeast, southeast, southwest and northwest. The layout of Liquanfang was established by this mode.

From the historical literature, the activities of ancient people at Liquangfang in the Tang Dynasty covered three places — the northwest, southwest and southeast parts; only the northeast part is not mentioned in writings. It is interesting that the tricolor glazed ceramics kiln was just located at this place. We believe that the kiln site, adjacent to Liquan Temple separated by North Street, is in the northeast of Liquanfang if the document "Liquan Temple is in the Northwest of the Cross Street" is correct. This is because the tricolor kiln site is located 300 m from the east of the Liquanfang site found in 1986. That is to say, the position of the kiln site should have been in the northeast of the Cross Street.

Based on our research, this kiln was used during the reign of Emperor Xuan Zong of the Tang Dynasty, from the year of late

Kai Yuan to Tian Bao, namely in the 20s to late 50s of the 8th century AD, spanning not more than 30 years. The additional important evidence supports this judgment. Before the kiln site was unearthed, some tricolor ceramic fragments and pottery figure fragments had been found in the bottom section of a ditch near the No. 3 and No. 4 kilns. They were heaps of residual and defective products from the kilns. One piece of a red pottery figure, an animal patron, has the inscription "... the fourth year of Tian Bao...Zuming," dating back to 745 AD. In the history of China, the term "year" was replaced by "*zai*" from the third year of Tian Bao, i.e. 744 AD. From the archeological study of the Tang Dynasty, this period was a transitional time after the peak of the flourishing Tang Dynasty. It was during this time that Tang tricolor pottery was not used as decorative sacrificial objects in fashion in the capital area. Most of the unearthed tricolor wares are utensils and religious figures from the kiln site of this period, which reveals that the function of Tang tricolor figures had changed greatly. This was an important specific characteristic of Tang culture during its post-prosperity days.

2. Unearthed Glass Materials

2.1. *Glasses*

There are 17 glass fragments unearthed from one ash pit, No. T2H1, 5 m from the east of the kiln site. The ash pit exists in a cultural stratum of the Tang Dynasty, the same stratum as the kiln site. The mouth of the ash pit is round, and its bottom looks like a cauldron, with a diameter of 3.60 m and a depth of 2.20 m. Residual pieces of monochrome plate-mouth-shaped vases, bowls and dishes have been unearthed from this ash pit, as well as pieces of green glaze tube-shaped tiles and tricolor spittoons. Also, pottery mold fragments, trifoot supporting tacks, spacers, and many pieces of white glaze bowls, small mouth bowls with black-outside and white-inside glaze, and pieces of glass materials etc. have been uncovered.

These 17 samples of glass material fragments, serial Nos. T2H1:54 — 1–17, can be divided into three categories: glass fragments, unfinished or defective products, and glass raw material fragments.

2.1.1. *Glass fragments*

There are a total of seven glass fragments, with serial numbers A1–A7 (sample T2H1:54 — 1–7), showing irregular shapes: (A1) brown–yellow, transparent, 3.5 cm × 3.7 cm; (A2) light orange, transparent, 3.2 cm × 2.6 cm; (A3) light purple, transparent, 4.1 cm × 2.6 cm; (A4) brownish yellow, transparent, 3.5 cm × 2.8 cm; (A5) purple, transparent, 3.8 cm × 2.5 cm; (A6) blue–green, transparent, 4 cm × 3.5 cm; (A7) light green, transparent, 7.2 cm × 6.8 cm.

2.1.2. *Unfinished and defective product fragments*

There are six pieces with man-made shapes, with serial numbers B1–B6 (sample T2H1:54 — 8–13): (B1) light green, transparent and unsuccessful blowing-bottle-like ware; round lips; flattened by pressing and bonding; 3.9 cm high and 3.1 cm wide; (B2) light orange–red, transparent and long strip; some little bubbles in it; the pad cohered spots on a flat bottom; the strip is 7 cm long and 2.6 cm wide; (B3) green, transparent fragment of round bottle; mouth diameter 2.4 cm and lip width 0.45 cm; (B4) round ring; white, transparent; V shape; outer diameter 2.4 cm and inner diameter 1 cm; (B5) round lid with button; light gray, transparent; diameter 1.8 cm and thickness 1.1 cm; (B6) defective long strip; same color as B5; 5 cm long, 2.5 cm wide and 0.8 cm high.

2.1.3. *Incompletely melted glass material ingots*

There are four pieces, with serial numbers C1–C4 (sample T2H1:54 — 14–17), showing characteristics of a combination of green glassy material and white crystal material: (C1) 9.5 cm × 9.2 cm, (C2) 9.4 cm × 6.5 cm, (C3) 3.8 cm × 3 cm, (C4) 7.1 cm × 5.2 cm.

2.2. *Mineral fragments*

A total of 11 mineral fragments were discovered, with 2 pieces from the fire door of the No. 2 kiln and the others from ash pit T1H2. These fragments are deep brown and deep green. Some of them display layer structure and rust. The biggest one is 16.5 cm × 10.5 cm. In general, the fragments are between 2.5 cm × 3 cm and 4.6 cm × 8.2 cm (sample Y2:36 — 1–2 and sample T1H2: 34 — 1).

3. Discussion

Altogether 17 glass material fragments have been found, which can be divided into three categories: nonproduct (ingot) fragments, defective and residual fragments or unfinished products, and incompletely melted material pieces. Most of the glasses previously excavated at domestic sites show complete or broken shapes of the finished products. However, these three kinds of glass fragments reveal important information on the glass-producing process, particularly the space, time and elements involved in the whole process at the kiln site; so this is rare and precious.

These kinds of glass fragments are transparent and greenish, blue–green, orange, brown–yellow, light purple or purple, and some of the fragments show unsuccessful blowing shapes and states, or unfinished and deformed products, or defective and residual products. Meanwhile, chemical analysis of the samples has revealed that the glass fragments are Na–Ca glass and there is no lead oxide in them.[1]

First, these characteristics exclude the possibility of glass slag being produced by the metallurgical process. It is easy to form glassy residues containing lead at a low melting point in the process of metal production. However, the unearthed glasses do not contain lead. And the excavated incompletely melted glass material is mainly the sintering consisting of crystal quartz powders and inorganic glass, and has no relationship with glassy residuals from metal melting.

Second, these Na–Ca glass fragments are very different in composition and technology from the tricolor wares, in which low temperature lead glaze was used. This shows a different usage.

Obviously, these fragments were the materials or unfinished products used for producing glassware. It is interesting that the glass materials were unearthed from the kiln which was used for making tricolor pottery. This finding shows the diversified production structure of the kiln and the related location of *fang*. But the origin of glass raw materials would constitute a new research project.

From the viewpoint of archeology, research on glass artifacts is generally focused on comparative study of the technological process, shape, decorative pattern, chemical composition, etc. However, these glass fragments show an interrupting and discarding state in the production process because the shape-forming, technology and decoration-making are not complete; namely, they do not exceed the concept of material in the broad sense. It is believable that these glass fragments have three features. First, they were specially used for glassware. Second, they belong to the Tang Dynasty, because the inscription "… the fourth year in Tian Bao" is engraved on the pieces of pottery in form of animal patron (called Zu Ming), which were unearthed from the kiln. Third, the samples of fragments all contain Na_2O, MgO, SiO_2, CaO, ZnO etc. by chemical analysis. And some samples also contain MnO, Fe_2O3, Al_2O_3, $Zr\,O_2$ etc., but without elements of Pb and Ba. Hence, these fragments belong to the Na–Ca glass system, not Pb glass.

The history of Na–Ca glass is longer than that of Pb glass. The early Na–Ca glass was produced in ancient Egypt and in the valley of the Tigris and Euphrates Rivers. There is not enough evidence to prove that Na–Ca glass could be produced until the fourth century.[2] It is generally accepted that the production of Pb glass continued to develop after the fourth century. Na–Ca glasses have been excavated from some tombs of the Wei Dynasty, the Three Kingdoms, the Jing Dynasty, and the Southern and Northern Dynasties. For example, there are three glass fragments unearthed from the tomb of the Southern and Northern Dynasties in Nanjing; a cameo glass was unearthed from the tomb of the Western Jin

Dynasty at Ercheng, Hubei province; a glass alms bowl was unearthed from the Fengsufu tomb of the Northern Yan Dynasty at Beipiao, Liaoning province; glass bowls and cups were unearthed from the tombs of the Feng clan of the Northern Wei Dynasty at Jingxian, Hebei province; etc. An Jiayao pointed out, based on her research, that "... these glass wares are not the typical Chinese wares. Similar products can be found in the foreign production center of glass, and the composition is approximately the same as that of the glass made in Rome, possibly imported from outside."[3]

Pb glass and Na–Ca glass have been unearthed together from some tombs of the Sui and Tang Dynasties, such as glass wares with different compositions from the tombs of Li Jingxun of the Sui Dynasty in Xi'an and from the Li Tai tomb of the Tang Dynasty at Yunxian, Hubei province. Regarding the Na–Ca glass in this period, Mr Su Bai verified that this kind of glass could be produced domestically in China.[4] The glass fragments unearthed from the tricolor kiln at Liquanfang provided new evidence to support his viewpoint.

The raw material resources for making Na–Ca glass were rather scarce in China in ancient times. Hence, the production of high Pb glass prevailed over that of other glass systems. It seems that, owing to a special background in history, a number of Na–Ca glasses could be produced domestically in the Sui and Tang Dynasties. Liquanfang is located at a central place of the city where foreigners gathered and lived. They might have brought Na–Ca glass materials and produced glass artifacts locally.

4. Conclusion

The discovery of the tricolor kiln site at Liquanfang of the Tang Dynasty has confirmed a long-time guess in academic circles, i.e. there was a tricolor ceramics workshop in Chang An city. This is an important discovery, after the big and small tricolor kilns of the Tang Dynasty found at Huangyi of Gongyi county, Henan province, and the Huangbu kiln of the Tang Dynasty in Shaanxi province. The rich archeological materials of the kiln provide verification for examining the features of the kiln site and changes of the product structure. Plenty of unearthed artifacts exceed the range of tricolor pottery; particularly, the finding of glass materials is rare and valuable. This gives us a new direction of research and a new insight into some functions and activities at *fang* around Chang An city. Meanwhile, it is valuable for understanding the different types of products and the scale of the workshop within Chang An city of the Tang Dynasty.

References

1. *The X Fluorescence Analysis of Energy Chromatic Dispersion on Glass Unearthed at the Tricolor Kiln Site at Liquanfang, Chang An City, Tang Dynasty; The Excavation Report on the Tricolor Kiln Site at Liquanfang in Chang An City, Tang Dynasty* (Cultural Relic Press, Beijing, 2006) in Chinese.
2. *Some Problems Regarding Research on Glass in Ancient Times; The Memorial Collections of Mr Xia Nai's Fifty-Year Work on Archeology; The Study of China Archeology* (Cultural Relic Press, Beijing, 1986) in Chinese, pp. 337–345.
3. Su Bai, *The Golden and Silver, and Glass Wares in Ancient China* (Su Bai, China Relic Newspaper), 3rd edn. (1992) in Chinese.
4. Han Xiang, *The Research on the Gathering and Culture of Middle Asia in Chang An City of the Tang Dynasty* (Nationality Research), 3rd edn. (2003) in Chinese, pp. 63–72.

Chapter 18 | 第十八章

Ancient Glass in the Grassland of Inner Mongolia

Huang Xueyin

The Capital Museum, Beijing 100045, China

1. Introduction

Glass is "a noncrystalline solid formed by cooled melt while maintaining the amorphous liquid microstructure at room temperature." There are natural glasses, like obsidian, which is produced when felsitic lava erupts from a volcano and cools rapidly and, tektite, which was formed under high temperature and pressure after stars entered the atmosphere and crashed into Earth. In this article, "ancient glass" refers to the man-made material in the form of glass microstructure.

Due to the limited quantity and quality of ancient glasses unearthed in Inner Mongolia, there have been few studies on the glass and glass-made objects found there. There are neither many excavation reports on nor any detailed descriptions of the few glass objects found in the grassland. In recent years, however, excavations and researches on ancient aircraft have developed rapidly and many glass objects have been uncovered from the excavation sites and have eventually caught the attention of academic researchers. These glass objects contain many cultural elements and provide us with rare material to learn about the grassland culture and its relationship with the outside world. I hope my article will provide useful information for the researchers who are interested in the subject.

Located in the northern part of China, Inner Mongolia is characterized by a vast grassland, where many nomadic minorities settled throughout the long history of China. As they were nomads and their lives depended on the search for grass and water, their daily wares and ornaments were mostly made of minerals (such as gold, silver, jade and stone), plants (such as maple tree bark) and animal skin, but very few glass wares were available to them then. Some early researchers referred to the ancient glasses in their reports as *liaoqi* ("glass").

Throughout the history of China, many words have been used to refer to the material we now know as glass. Ancient historical documents record it as *liuling, liuli, biliuli, boli,* etc., and it was finalized as *liuli* in ancient times. After the Song Dynasty (960–1279 AD), however, the word "*liuli*" was used for glazed pottery only. It was in the early Qing Dynasty (1644–1911 AD), during the Kangxi reign (1662–1722 AD), that the word "*bolichang*" ("glass factory") was formally introduced to differentiate *liulichang* ("glazed-tiles factory"), when the emperor ordered the building of a glass factory inside the Forbidden City under internal administration. According to the historical records China developed the technology of glassmaking as early as 2000 years ago, and the earliest kiln site found so far is the Boshang Glass Factory of the Yuan Dynasty (1279–1368 AD). In the context of the world, based on the historical relics and documents, the earliest ancient glass dates back to 2500 BC and was from the region of the Euphrates River and the Tigris River (Mesopotamia). Around the 15th century BC Egyptian glass technology was developed, and around the 10th century BC ancient Greece and Rome became the centers of glass production. The technology of glassmaking in China was about 1000 years behind that of the West, and it developed and matured due to the movements of nomadic tribes and trading along the ancient Silk Road through the grassland or over the sea. A good example is a belt ornament with glass inlays of the Hun (Xiongnu) styles excavated from the tomb of the king of Southern Yue State in Guangzhou.

2. Glassware Discovered in the Grassland

(1) Based on the current glass artifacts excavated in Inner Mongolia, the earliest glassware found at Lui Cheng of Erjinaqi dates from the Western Zhou Dynasty.

(2) Glassware became more popular from the Warring States Period to the Han Dynasty in the grassland, where the Hun, Xianbei and Wuhuan inhabitants started using small glass beads as ornaments. The major discoveries include the glass beads[1] found in 1972 at a group of tombs called Taohongbala at the village of Sawuozhi ("sand dump") at Hangjinqi, Erdos city, Inner Mongolia[2] (Photo 2.1). In 1979, a group of important and valuable cultural relics were found at Xigoupan, Zhonggeerqi, Erdos city, Inner Mongolia. The majority of them are gold and silver wares, but 155 glass beads were excavated from the M4 site, which are light yellow, blue and light blue in color and balls, cylinders and ovals in shape. All the beads were sent to the Shanghai Institute of Optics and Fine Mechanics for measurement (Photo 2.2).

(3) The glassware uncovered from the Han Dynasty tombs in the southwest of Inner Mongolia comprises:

Two pieces of earrings (without holes) found at M9 of the later period of the Western Han Dynasty at Liangcheng county, Wulanchabu city.[3]

Glass fills (called *zhen* in ancient Chinese) were found at M2 of the period from the Middle Western Han Dynasty to the early Eastern Han Dynasty, at Sanduandi.[4]

Glass earrings, nose fills and glass fills were found at a group of Han tombs at Zhaowan.[5]

Glass beads, glass dragonfly eye beads and glass fills were found at a tomb of the later Western Han Dynasty at Nalintaohai, Bayanor city.[6]

Glass fills were found at a tomb of the later Western Han to the early Eastern Han Dynasty at Shajingtaoxi, Banyanor city.[7]

Two glass ear ornaments were found at M9 of a group of tombs of the later Han Dynasty at Beiyinzhi, Liangcheng county, Wulanchabu city.[8]

One glass goat[9] was found at M8 during the excavation of a Han Dynasty tomb at Wulantaogai, Hangjiqi, Erdos city.

One glass button and one earring were found at M48 of the Han Dynasty at Zhaowan of the Baotou suburb.[10]

(4) More glass artifacts of the Wei, Jing, Southern and Northern Dynasties were found. During this period, the Xianbei and Wuhuan were nomadic peoples living in the grassland. Glass was primarily used as ornament beads. The major discoveries are:

In 1959, at Zhalannor, Manchuria, Hulunbaier, a large number of funerary objects were excavated from the tombs of the ancient Xianbei people, determined by the archeologists. More than 30 beads made of glass, wood, amber, turquoise, shell and agate were found on the necks of the dead.[11]

Seven more pieces of *liao* (glass) beads were found during the cleanup of 13 tombs of the Zhalannor cemetery in 1960, and four more glass beads were found during the cleanup of the cemetery in 1986.[12]

In 1961, at M2 of the group of tombs near Halarer, Hulunbarmen, 18 green glass beads that are tube-shaped, oblate or rhombic were found.[13]

In 1978, at the Xianbei tombs near the Yiming River in Hulunbermen, two round glass beads were found at M5.[14]

In 1979, during the cleanup of the Zhaowan cemetery near Baotou, one translucent ear pendant was found; it was purple–blue and in girt.[15]

Two glass tubes, three bead-segmented tubes and five *pao* ("bubble") ornaments were found at a Xianbei tomb at Liujiazhi, Kezhuoqi, Tongliao city.[16]

(5) In addition to the glass beads and glass ornaments, many more glass artifacts and some fine and beautiful glass utensils were discovered at the tombs and historical sites of the Liao, Jing and Yuan Dynasties. They are:

Fragments of glass artifacts were found at the Changjinggou No. 5 tomb of the Liao Dynasty at Barlinzhuoqi, Chifeng city.[17]

At the tomb of the Liao emperor's son-in-law, located at Dayinzhi near Chifeng city, seven or eight yellow-rusted glass fragments[18] were found; these were originally a cylinder vase with a long neck, small mouth and raised bottom.

At the tomb of the princess of Chen State located in Qinglungshan town, Naimanqi, Tongliao city, several valuable glasswares were discovered, which include one glass plate with a nipple pattern, one long-necked glass vase with a nipple pattern, one carved long-necked glass vase, two high-necked glass vases and two glass mugs with handles. All the seven glass artifacts were daily use utensils and were placed in the back chamber of the tomb.

A dark green glass vase was excavated from the Qingzhou White Tower, at Barlinzhuoqi, Chifeng city. A set of glass belt ornaments for a drum player was excavated in 1978 from the ruins of Shangjing ("upper capital") of the Liao Dynasty, called Liaoshangjing ("ancestral mausoleum"), and then collected by the Barlinzhuoqi Museum. In the collections of the same museum there is another set of floral glass belts with a weight of 264 g, excavated from a Liao tomb at Haoertouhuaganzhigou, Balinzhuoqi. A similar floral glass belt of the Liao Dynasty is in the Ongnutezhuoqi Museum of Chifeng city.

A glass Buddha pestle[21] (a musical instrument), in the shape of cylinder and with designs of Buddha's warrior attendant, was excavated from a tomb of the Jin Dynasty in the Honggeer region, Siziwangqi.

A chain of emerald green glass beads holed in the center and carved on the surface was discovered at a section of the Jin Great Wall in Xingmingxiang town, Shuniteqi, Xilinguolemen.[22]

One glass bead was found at M5 in the southern tomb area of Nanzhuanzhishan, located in the upper capital of the Yuan Dynasty, which is in today's Xilinguolemen.[23]

3. Ancient Glass in the Grassland and Cultural Exchange

From the glass objects found in the grassland of Inner Mongolia we can conclude that, from the Western Zhou Dynasty to the Eastern Han Dynasty, glass beads together with beads of other materials were primarily used by the early grassland inhabitants as neck ornaments in their lives and were buried with them on their necks after death. Also, during this period, glasses were treated as precious ornaments and often chained together with jade, gold, silver, turquoise or bone ornaments, or used as material for inlaying. These early glass objects found in the grassland indicate that glass objects of this period were few in quantity, low in quality and small in size. The light blue glass chain necklace found at the Han Dynasty tomb of Xigoupan represents the highest quality of glass ornaments of this period and it shows that the Hun people had full sets of glass ornaments.

The question is: Why are there no full sets of glass ornaments from before the Western Han Dynasty? Until today, no ancient glassmaking sites have been discovered in the grassland of Inner Mongolia, which illustrates the fact that there was no glass production in the grassland, unlike central and southern China, where many glass objects and some glassmaking sites were found. Therefore, the Huns living in the grassland in northern China during the Spring and Autumn and Warring Periods had no decent glass objects until the Han Dynasty, when cultural exchange between the Huns and central China became more frequent and various products from central China were transported in large quantities to the grassland through wars, trade and marriages. These products became daily necessities and luxuries for Hun nobles, and glass objects were among the luxuries that came from central China. The luxury glass necklace chained together with gold and agate pieces was probably worn by a noble Hun woman at Xigoupan; it shows the influence of central China on the esthetic standards of the Hun people at that time and it has also become an important witness to the cultural exchange between the Huns and central China. The dragon and phoenix motifs on the other ornaments unearthed at Xiguopan highlight this historical fact.

During the Wei–Jin and South and North Dynasties, the major inhabitants who settled in the grassland in northern China were the Xianbei. They were often referred to as a "dream-chasing" people. The Xianbei migrated from the Daxinganling Forest region of northeastern China and lived generation after generation in the grassland before expanding gradually into different tribal branches that occupied various parts of the grassland and established many tribal governments. Among the tribal branches, the Tuobaxianbei was the most remarkable one. In the historical process of their search for civilization and dreams, the Tuobaxianbei demonstrated courage and wisdom not found among other minorities at that time. They eventually became the first northern minority to establish a government in central China. From the forest to the grassland and then to Luoyang, the capital of central China, they absorbed all but advanced cultures in the process of searching for civilization, which they did not have before. Among the Xianbei historical and cultural relics are a lot of glass beads, whose quantity is many times that of the glass beads found in the Hun era. This illustrates the phenomenon that the glass bead ornaments originating in the Han Dynasty in central China were also popular among the Xianbei people during the Wei and Jing periods. Their custom of using glass beads was very similar to that of the Han people in central China. However, one important phenomenon worth considering is that these glass bead ornaments were placed in many Xianbei tombs, which indicates that the glass beads, like the beads made of agate and turquoise, were widely used as small ornaments and so did not signify luxuries for the nobles. Unlike the gold and silver mostly used by the nobles at that time, the glass beads, chained with beads of other materials, became popular ornaments among the people. Here, we may suppose that the Huns and Xianbei in the grassland in northern China probably, because of nomadic way of life, had easier trading access than the people in central China. Therefore, the glass beads were easier to obtain through trading than through local production.

Large glasswares such as daily household necessities became available in the northern grassland from the Liao Dynasty, and simple glasswares like glass belts were probably made locally, inferring from the craftsmanship of simple designs and floral motifs. However, the glass mug with a handle, the long-necked glass vase (Photo 2.5), the plate with nipple-like patterns (Photo 2.3) and the nipple-like pattern glass vase with a handle (Photo 2.4) excavated from the tomb of the princess of Chen State were imported products. Based on the analysis on their shapes and technology, and on the scientific measurements and expert evaluations, these glass artifacts probably came from as far away as Egypt, Persia, Islam, etc. For instance, the emblazonry of the nipple-like pattern on the high-necked glass vase, H61, is very similar to that of a glass vase in the National Museum of Quit. A contemporary chemical analysis of this glass vase showed that the glass contains 20.66% sodium oxide, which indicates that it was probably a glass vase from Egypt or Syria. The subsequent chemical measurement results on six other glasswares also indicated that they were from the Middle East. A high-stem glass cup, excavated from a Liao tomb at Tuerjishan in 2003, was definitely not a local product — evaluated on the basis of its thin wall and exotic style of design. During the 11th century, the Liao Dynasty became a powerful country in northern China and had exchanges — as early as the beginning of the Liao Dynasty, established by the Taizhou reign — with the Hezhouhuihe people, who sent annual tributes to Liao. In the early years of the Tianzhan reign, Tazi (Caliphate) and Persia also sent annual tributes to Liao, and such contacts became more frequent during the Shenzhong reign. The annual tribute was actually a major form of trading at that time, and the seven glass artifacts from the tomb of the princess of Chen State of the Liao Dynasty are all from central Asia.

Glassware became more popular during the Jin and Yuan Dynasties, both of which had far more contacts with other nations than the Liao Dynasty, and thus there was no reason for them not to use glassware. Particularly, during the period of the Mongolian Empire, established by Genghis Khan, its territory extended far to the Mediterranean Sea, where the fine glassware and craftsmen naturally became looting targets for the Mongols. However, it is enigmatic to us that so far not any complete glass artifacts of this period have been discovered among the excavated cultural relics in Inner Mongolia. To answer this question, we need to look into the Mongolian burial custom, which was quite different from that of Qitan of the Liao Dynasty. Qitan, particularly Qitan nobles, pursued a custom of luxury burials and therefore more objects have been discovered in their tombs. However, Mongolian nobles pursued a secret burial custom. They cut a piece of wood longitudinally into two parts, hollowed them according to the shape of the dead person, then put the dead body into one half and covered it with the other half to make a complete piece, and finally sealed it with gold or metal and laid it into the ground without a raised form on the surface. Thus, the tombs became invisible when grass grew all over in the following years. This may be the reason why Genghis Khan's tomb has yet to be found. The secret burial custom of Mongolian nobles is one of the reasons why no complete glass artifact has been discovered in their tombs. On the other hand, why is there still no large quantity of glass artifacts found in the newly discovered tombs of the Mongolian kingdom? The answer lies in the historical fact that the Mongols were engaged in warfare year after year, and unlike gold and silver wares, the fragile glass and porcelain wares were hard to preserve in large quantities. Until the formal establishment of the Yuan Dynasty, when its society was peaceful and contented, it was possible to bury daily household necessities with the dead bodies, which are today discovered in their tombs. Therefore, the complete glass artifacts discovered in the Liao Dynasty tombs are really rare treasures.

References

1. G. J. Tian, Taohuabala Xiongnu Tomb, *Archeol. Res* (in Chinese) **1**, 131–143 (1976).
2. G. J. Tian, Xigoupan Xiongnu Tomb, *Cultural Relics* (in Chinese) **7**, 1–10 (1980).
3. Inner Mongolia Archeology Institute, Wulanchabumen Cultural and Archeology Work Station, *The Han Dynasty Tomb Located at Liangcheng County of Wulanchabumen in Inner Mongolia*. China Archeology Collections (Section of Northeastern China, Shengyang) (Beijing Press, 1997), in Chinese, p. 207.
4. Weijian, *The Han Dynasty Tombs in the Mid-South of Inner Mongolia* (China Encyclopedia Press, 1988) in Chinese, pp. 155–157.
5. Weijian, *The Han Dynasty Tombs in the Mid-South of Inner Mongolia* (China Encyclopedia Press, 1988) in Chinese, pp. 248–262.
6. Weijian, *The Han Dynasty Tombs in the Mid-South of Inner Mongolia* (China Encyclopedia Press, 1988) in Chinese, pp. 43–44.
7. Weijian, *The Han Dynasty Tombs in the Mid-South of Inner Mongolia* (China Encyclopedia Press, 1988) in Chinese, pp. 107–108.
8. Z. J. Fu and T. D. Cheng, Brief report on the Han Dynasty tomb discovered at Beiyingzhi of Liangcheng county, *Inner Mongolia Archeol* (in Chinese) **1** (1991).
9. Yikezhao Cultural Relics Work Station, Report on Wulantaolegai Han Dynasty tomb excavation in Hongjing Banner, *Inner Mongolia Archeol* (in Chinese) **1** (1991).
10. Baotou Archeology Institute, *Report on the Zhaowan Han Dynasty Tomb Located in the Baotou Suburb*, Chinese Archeology Collections (Section of Northeastern China, Shengyang) (Beijing Press, 1997) in Chinese, p. 236.
11. L. Zheng, Zalannor ancient group tombs (*first part*), *Archeology* (in Chinese) **9**, 18–19 (1961).
12. L. Zheng, Zalannor ancient group tombs (*second part*), *Archeology* (in Chinese) **12**, 673–680 (1961).
13. X. R. Pan, Ancient tombs found at Wangongshuomu of Chengbaerhuqi, *Archeology* (in Chinese) **11**, 590 (1962).

14. Z. Z. Li, Report on excavation of the Wangong ancient tomb of Chengba-erhuqi, *Archeology* (in Chinese) **6**, 273–283 (1965).
15. D. H. Chen, The Xianbei tombs in Yiming Region, *Inner Mongolia Archeol* (in Chinese) **2**, 18–22 (1982).
16. Baotou Archeology Institute, *Report on the Zhaowan Han Dynasty Tomb Located in the Baotou Suburb* (Inner Mongolia Archeology, 1981), 1st edn., in Chinese.
17. B. Z. Zhang, Liujiazhi Xianbei group tombs at Kezhuozhongqi of Inner Mongolia, *Archeology* (in Chinese) **5**, 430–438 (1989).
18. Inner Mongolian Archeology Institute, Brief report on the No. 5 Liao Dynasty tomb at Chuangjinggou of Barlinyouqi, *Archeology* (in Chinese) **3** (2002).
19. Preparation Group of the Former Rehe Province Museum, Excavation report on the Liao Dynasty Tomb at Dayinzhi of Chifeng county, *Archeol. Rep.* (in Chinese) **3** (1956).
20. Inner Mongolian Archeology Institute, *The Tomb of the Princess of the Cheng State in the Liao Dynasty at the Zelimumen Museum* (Archeology Press, Beijing, 1993) in Chinese, p. 55.
21. Dexing, Zhanghanjun and Hanrenxin, Buddhist cultural relics found at the Qingzhou White Tower in Balinyouqi of Inner Mongolia, *Cultural Relics* (in Chinese) **12**, 6–25 (1994).
22. G. J. Tian, *The Jing Dynasty Tombs at Honger of Siziwangqi* (Inner Mongolia Archeology, 1981), 1st edn., in Chinese.
23. Inner Mongolian Archeology Institute, Xilingguole Archeology Institute and Suniteyouqi Archeology Institute, *The Quangshen Section of the Great Wall of the Jing Dynasty at Xingmingxiang in Suniteyouqi*, Inner Mongolia Archeology Collections, Third Part (2004) in Chinese.
24. Inner Mongolian Archeology Institute, Report on excavation of the Southern tomb at Nanzhuanzhishan in Yuan Shangdu, *Inner Mongolian Archeol* (in Chinese) **2** (1999).

Chapter 19 第十九章

Glasses of the Northern Wei Dynasty Found at Datong

An Jiayao
The Institute of Archaeology, Chinese Academy of Social Sciences, Beijing 100710, China

1. Introduction

The Northern Wei Dynasty (386–534 AD) was set up by the Tuoba people. The Tuoba, a nomadic tribe who may have been of Turkic or Mongol origin, controlled north China under the dynastic name of Northern Wei. The Tuoba rulers not only learnt from the Chinese, but also from the West.

The Northern Wei Dynasty was an important period in the glass-making history of China. There is a most interesting reference in the chapter on the Darouzhi (Great Yen Chin) people in the ancient Chinese work *Bei Shi* (*History of the Northern Dynasties*). On the basis of this record, we know that in the mid–5th century Bactrians manufactured glass in the vicinity of Datong, Shanxi province. So we are paying close attention to the archaelogical findings at Datong.

2. Materials

Two decades ago, we only knew that seven Northern Wei glass vessels had been uncovered at Dingxian, Hebei. The glass vessels

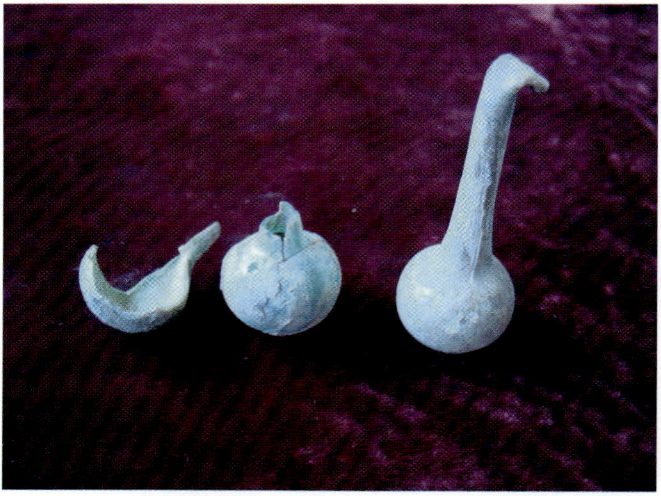

Photo 19.1. Gourd-shaped glass bottles unearthed from the pagoda of the Northern Wei Dynasty at Dingxian, Hebei.

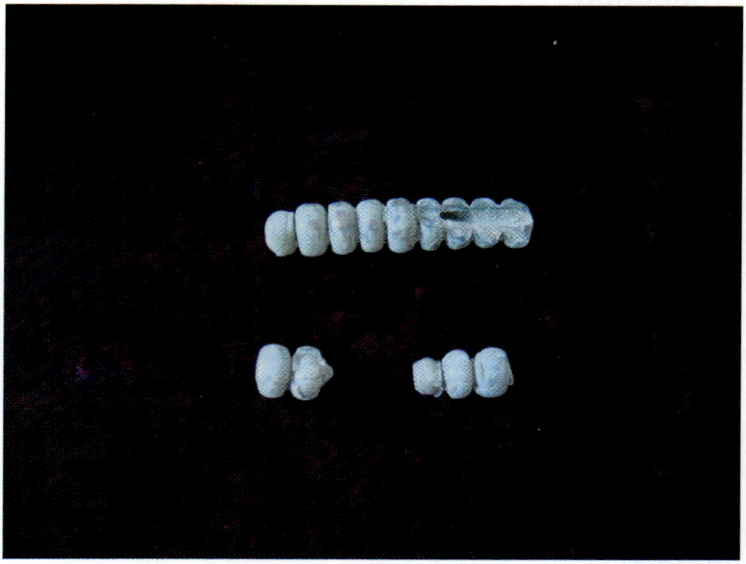

Photo 19.2. Lid fragment and glass beads unearthed from the pagoda of the Northern Wei Dynasty at Dingxian, Hebei.

were found in a stone chest in the foundations of a Northern Wei pagoda, which dated back to 481 AD. They include one alms bowl, two globular bottles, three gourd-shaped bottles (Photo 19.1) and a lot of glass beads (Photo 19.2).

To judge from their quality and method of manufacture, it is conceivable that all seven pieces came from the same source. The most technically advanced of them is the alms bowl, which is made of transparent sky-blue glass with a large number of bubbles, and the corroded surface is white. The mouth curves in, and the rim and the base are round. The walls, about 2 mm, are rather thicker than those of the other vessels; the thickness of the base is 5 mm. The two bottles are azure and transparent, and their walls are extremely thin — (about 1 mm); the glass has many bubbles and the surface is corroded and white. Each has a small mouth with a rounded lip, a short neck and a swollen body with a rounded base. One has a small ring foot. There are three small, gourd-shaped bottles. They are transparent and pale blue, with a globular body and a long neck, and the mouth is modeled into a short hook. One broken vessel is made of azure transparent glass with many small bubbles; it has a flat base and incurved sides, and could have been a wide-mouthed jar.

The objects found at the pagoda include gold and silver coins and vessels, all rare and precious, in addition to the glass vessels. Xia Nai has written an article on the Sasanian silver coins, and is of the opinion that some of these donations for the construction of the pagoda were probably drawn from the Imperial Treasury. This means that the glasses are from Datong city.

In 2001, archeologists excavated a Northern Wei tomb, located 2.2 km south of Datong city. Three glass vessels — a bowl (Photo 19.3), a bottle and a vessel fragment (Photo 19.4) — were found in the tomb. The bowl is blown, lake-blue and transparent. It has a vertical rim with a rounded lip; the body is encircled by a ridge and there is a foot ring; there are pontil marks on the base (H. 5.7–5.9 cm, D. 12.8–12.9 cm). The bottle is blown, lake-blue and transparent, the same as the bowl. It has a small mouth, with the rim rolled in.

Photo 19.3. Glass bowl found in the Northern Wei Dynasty tomb at Datong, Shanxi.

Photo 19.4. Bottle and vessel fragment found in the Northern Wei Dynasty tomb at Datong, Shanxi.

The neck is short, the body is globular and the base is flat (H. 3.1 cm, mouth D. 2.4 cm, body D. 4.5 cm). The vessel fragment, part of a glass ball or a gourd-shaped bottle, is blown, light lake-blue and transparent. The body D is 2.2 cm.

These glass vessels from Datong are similar to those found at Dingxian. They were all blown without a mold. The rims of the alms bowl and the bottle were rounded by heating. The bottle rim seems to have been rolled in to form an annular rim, and the foot rings were formed by an applied thread. These techniques were all traditional Roman and Sasanian characteristics, but they are not seen with regard to earlier glass vessels of China; they were adopted in the Northern Wei and retained thereafter.

3. Methods and Results

I had the chance to examine the glasses from the Dingxian Pagoda years ago. I found that besides the seven glass vessels, there are a lot of glass beads and some vessel fragments from the pagoda. I obtained four samples for chemical analyses (Tables 19.1 and 19.3). They were analyzed by Dr Qian Wei of the Institute of Historical Metallurgy and Materials, University of Science and Technology Beijing, using SEM-EDS (CAMBRIDGE LINK–AN10000), 20 kV.

The results showed that the four samples belong to one group, K–Na–Ca glass, which is high in MgO. Due to the heavy weathering of G16, it is high in CaO.

In 2002, another glass jar was found at Datong. Its shape is similar to that of pottery jars found at Datong. I obtained two samples for chemical analyses (Tables 19.2 and 19.4).

Table 19.1. Glass samples from the Dingxian Pagoda.

No.	Description
G13	Glass beads; green, translucent (fragment)
G14	Glass beads; sky blue, transparent (fragment)
G15	Glass vessel; light green, transparent (fragment)
G16	Blown glass vessel fragment (heavy weathering)

Table 19.2. Glass samples from tomb No. 16, Yingbin Avenue, Datong.

No.	Description
G23	Bottom of glass; sky blue, opaque
G24	Fragment of glass; sky blue, opaque

Table 19.3. Chemical composition of the glass from the Dingxian Pagoda (wt%).

	G13	G14	G15	G16
SiO_2	57.75	60.84	60.10	36.98
CaO	0.98	7.63	9.96	34.18
MgO	9.16	10.34	9.13	6.17
Na_2O	5.98	6.18	5.17	0.57
Al_2O_3	2.35	n.d.	n.d.	4.28
K_2O	9.27	9.33	8.53	2.44
CuO	8.29	4.72	4.10	9.56
Fe_2O_3	4.21	0.66	1.75	1.88
PbO	1.45	n.d.	1.11	1.82
ZnO	n.d.	0.30	0.15	n.d.
MnO	0.56	n.d.	n.d.	0.68
CoO	n.d.	n.d.	n.d.	n.d.
BaO	n.d.	n.d.	n.d.	0.22
Sb_2O_5	n.d.	n.d.	n.d.	n.d.
P_2O_5	n.d.	n.d.	n.d.	1.22

n.d.: not detected.

Table 19.4. Chemical composition of the glass from tomb No. 16, Yingbin Avenue, Datong (wt%).

	G23	G24
SiO_2	55.95	56.69
CaO	10.98	10.39
MgO	1.55	1.51
Na_2O	11.11	11.31
Al_2O_3	6.98	6.04
K_2O	4.73	4.99
CuO	3.19	3.34
Fe_2O_3	1.36	1.35
PbO	2.84	3.07
ZnO	n.d.	n.d.
TiO_2	0.56	0.57
MnO	n.d.	n.d.
CoO	n.d.	n.d.
BaO	n.d.	n.d.
Sb_2O_5	n.d.	n.d.
P_2O_5	0.69	0.63

I examined the three glasses from Datong. They were analyzed by Yao Qingfang of the Chinese National Museum in Beijing, using X-ray fluorescence (DX-95). The results revealed that the three glasses from Datong are made of K–Na–Mg–Ca glass. There is Pb in the glass, but it is lower than 2%. The compositions of the two groups are similar.

4. Discussion

There is a very interesting reference in the chapter on the Darouzhi people in *Bei Shi* (*History of the Northern Dynasties*). At the time of Emperor Taiwu of the Northern Wei (424–452 AD), some people of that country were trading in the capital and said that they were able to cast stone into glass of different colors. They collected ores in the hills and went to the capital to smelt them. On completion the radiance and richness of the glass rivaled that of glass imported from the West. The emperor then commissioned a traveling lodge large enough for 100 or more persons. It was brilliant and radiantly colored and when onlookers saw it all were amazed, believing it to be the work of supernatural forces. From that time on glass became cheaper in the country and people no longer regarded it as precious.

During the reign of Taiwu, the capital was named Pingcheng, which is Datong, Shanxi province, today. Darouzhi was Bactria in Central Asia.

On the basis of this we know that in the mid–5th century Bactrians manufactured glass in the vicinity of Datong. It is therefore quite logical to suppose that the three glasses from Datong and the seven glasses from Dingxian were made by Bactrians at Datong. It is noteworthy that these vessels, being made by blowing, differ greatly from the Han dish and eared cup. Glass-blowing first appeared on the shores of the Mediterranean as early as the 1st century BC, and blown glass vessels were imported into China by the 3rd century AD, but on the whole the transmission of technology in ancient times was often much slower than the spread of merchandise by trade. As the transmission of technology was closely related to the migration of artisans, it is possible that glass-blowing came to China with non-Chinese artisans. If indeed there is a connection between the record in *Bei Shi* and the vessels from the Northern Wei tomb and pagoda, we can conclude that in the 5th century Central Asian craftsmen brought the technique of glass-blowing to China. This was a crucial turning point in the history of Chinese glassmaking. Glass vessels made after the Northern Wei were chiefly made by blowing.

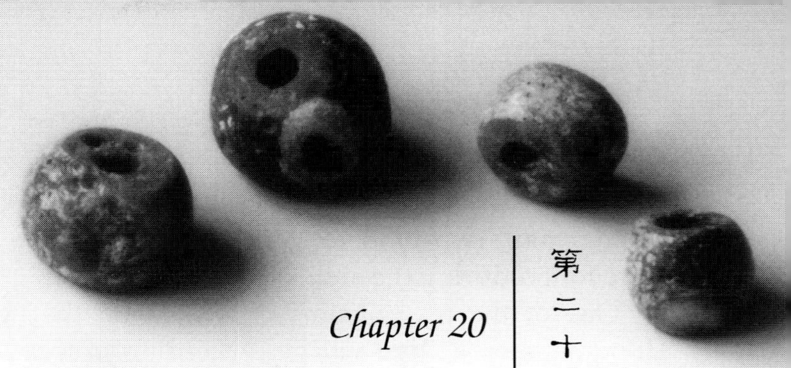

Chapter 20 第二十章

Glass Vessels of the Tang Dynasty and the Five Dynasties Found in Guangzhou

An Jiayao

The Institute of Archeology, Chinese Academy of Social Sciences, Beijing 100710, China

During the late Tang Dynasty and the Five Dynasties, the local governing Liu clan entrenched Lingnan (now the region covering Guangdong and Guangxi). When the Tang Dynasty ended, Liu Yan declared himself an emperor and set up a state named Dayue in 917 AD. Guangzhou was its capital and was called Xinwangfu. Next year, Liu Yan changed the name of the state to Han. Historians call the state Nanhan (Southern Han). The Southern Han continued its rule for 55 years and experienced four generations of the emperor's throne altogether. The first emperor was Liu Yan, on the throne for 26 years. The second was Liu Fen, who was on his throne, for only 2 years and was killed by his brother Liu Cheng. The third was Liu Cheng, on the throne for 16 years; and the fourth was Liu Chang, for 14 years. In the Song Dynasty, Emperor Zhao Kuangying ordered Pan Mei to attack Guangzhou, and Liu Chang capitulated to the Song Dynasty in 972 AD; and the local dynasty, the Southern Han, ended.

In 2003, the Guangzhou Institute of Archeology conducted a comprehensive archeological survey, and partial excavation in Xiaoguweidao district, 15 km southeast of Guangzhou city, in coordination with the reconstruction of a new college park in Guangzhou. Two emperors' mausoleums of the Southern Han were found during the archeological work.[1]

One of the mausoleums is located on the northern side of Houqinggang, Beiting village. According to local chronicles, this mausoleum was accidentally found in 1636, due to a lightning strike. A lot of funerary objects had been stolen from the tomb. The present archeological excavation has confirmed the destruction of this Southern Han mausoleum. The tomb was built with bricks and divided into two chambers; the front chamber had 18 cabinets and the back chamber 12 cabinets. The length of the tomb is 13 m. The wall used for covering the door is 2.64 m in thickness. When cleaning the path of the tomb, 269 glaze ceramic jars lined in order were discovered below the door.

The other mausoleum is believed to be the Kang Mausoleum — the tomb of Liu Yan, who was the first emperor of the Southern Han. It is located on the southern side of Xiangshangang, northeast of Beiting village, with very beautiful natural scenes. The archeological work at the Kang Mausoleum was very interesting. The workers first found a special form of covering soil pile, with a round top and a square base. There were three layers of bricks covering the outer part of the top soil, and white stones were used to lay the square base. The round top is 10 m in diameter and 2.5 m in height, and the length of each side of the base is 12.5 m. Under the square base, there is a 17.5-m-long drain system built with bricks. Two construction bases, with a length of 6 m for each side, were found 50 m north of the covering soil pile. Referring to the historical records about the "worship heaven" of the Southern Han's king, the archeologists primarily inferred that this ruin should be a hillock built for worship heaven activities of the Southern Han. With progress of the excavation, a path of the tomb was discovered, and they thought that it must be the covering soil of the emperor's tomb. A stone tablet was placed in the front chamber of the tomb, the inscription on which clearly recorded that Gaozu died in April of the 15th year of Dayu (942 AD) and was buried in the Kang Mausoleum in September of the first year of Guangtian

(942 AD). Thus it was confirmed that this site is the Kang Mausoleum. It was the tomb of Liu Yan, the first emperor of the Southern Han.

Due to many robberies, there were few intact artifacts, and only fragments of ceramic jars and bowls, stone figure fragments, jade plates, silver rings, copper coins, etc. However, the hundreds of glass fragments unearthed were noticeable and stimulating (Photo 20.1). Currently the restoration work is going on. Determination based on the unearthed fragments indicated that these glasses should belong to glassware of bowls, cups, bottles and so on. Most of the glasses are transparent and greenish, with little difference between deep and light green. The tone of the color is correlated with both the color itself and the thickness of the glass. In addition to the green glass fragments, some blue glass fragments were found (Photo 20.2). Technicians of the Institute of Archeology, Chinese Academy of Social Sciences, have systematized and dressed these fragments, while one glass bottle has been successfully restored (Photo 20.3). This glass bottle has an outspread rim of the mouth, a short neck, a round body with a swollen shoulder and a little convex base with tube-blowing marks. It was made by blowing in a mold and decorated with ridges on its outside. This type of decorative glassware was first discovered in China, and is

Photo 20.1. Glass fragments unearthed from the Kang Mausoleum of the Southern Han.

Photo 20.2. Blue glass fragments unearthed from the Kang Mausoleum of the Southern Han.

Photo 20.3. Restored glass bottle with ridges unearthed from the Kang Mausoleum of the Southern Han.

comparable to the glassware produced by the Arabic world. The glassware with ridges on the outside was unearthed in Siraf, Iran. Siraf was an important coastal city of Iran along the Persian Gulf, and is now called Taheri. The Grand Mosque ruin of the 9th–12th centuries found by archeologists is a representative work of Islamic architecture. One intact glass bottle, 5.4 cm in diameter and with ridges on its outside, was discovered at the Grand Mosque ruin.[2] The collections in museums also include some glass wares with ridges, dating back to the 7th and 8th centuries, and even earlier, from the Sasanian period to the post-Sasanian period before Islamic times. Some early Islamic glass wares are in the Kuwait National Museum; one is a glass bowl dating back to the 7th and 8th centuries, which was made by the blowing technique and decorated with ridges on its wall.[3] A blue glass cup with similar decoration is in the Corning Museum of Glass, USA; it is 7.4 cm in height and its mouth diameter is 8.0 cm. It is worth mentioning that there is a row of nipple patterns on the bottom of the Corning Museum glass cup.[4] Interestingly, one of the glass fragments unearthed from the Kang Mausoleum also has a row of nipple patterns on its bottom. Samples of the glass fragments found at the Kang Mausoleum have been provided to the Institute of Historical Metallurgy and Materials, University of Science and Technology Beijing, for chemical formulation analysis. The measurement has not been completed yet, but it is believable that according to the style of emblazonry and the technology of the fragments the glasses unearthed from the Kang Mausoleum came from the Islamic world of West Asia.

The discovery of West Asian glasses in the Kang Mausoleum was not an accidental event. Although China could produce high quality glassware during the Sui and Tang Dynasties, some beautiful glasses made in West Asia were still the treasures pursued by the nobles. There is a record in the Chinese historical literature, Zizhi Tongjian (*Comprehensive Mirror for Governors*) which tells a story about Emperor Daizong of the Tang Dynasty and a large glass dish which reflected well the great worth of imported glassware.[5] The story goes thus. General Liu Sigong put down a rebellion in Guangzhou, and offered a glass dish to Emperor Daizong. The dish's diameter was 9 *cun* (about 27 cm), and Emperor Daizong thought that it was a treasure. When he found that General Liu had offered another glass dish, with a diameter of 10 *cun* (about 30 cm), to Prime Minister Yuan Zai, Emperor Daizong felt very unhappy. One year later, he talked about this matter and was still indignant. General Liu Sigong was a governor of Lingnan. He put down the rebellion in Guangzhou in 775 AD and confiscated thousands of possessions of merchants. Guangzhou was an important port city during the Tang Dynasty and the Five Dynasties (from the 7th to the 10th century). Among the confiscated possessions, there must have been some glass vessels which had from the Arabic world along the Sea Silk Road, and some of them must have been larger glass vessels. So scholars had been expecting to find the glassware of the Tang Dynasty and the Five Dynasties uncovered in Guangzhou. In 2000, some glasses were found in the Tang layer of excavation of the Nanyue king's palace site in Guangzhou, including glass beads and fragments (Photos 20.4 and 20.5). A glass bowl was restored (Photo 20.6). This bowl has a yellowish and greenish tinge, and has vertical walls and a kick-base with pontil marks. It is 3.8 cm in height, the diameter of its mouth is 7.85 cm, and that of its base is 7.65 cm. This type of glass bowl is a common form of Islamic glass.

Photo 20.4. Glass fragments unearthed at the Nanyue king's palace site of the Tang Dynasty in Guangzhou.

Photo 20.5. Glass beads unearthed at the Nanyue king's palace site of the Tang Dynasty in Guangzhou.

Photo 20.6. Glass bowl unearthed at the Nanyue king's palace site of the Tang Dynasty in Guangzhou.

Due to the novel shape and gorgeous decoration of the imported glass, and also because it was transported from remote areas, crossing mountains and rivers, nobody knew the detailed technology of the Western glass. So, the people of the Sui and Tang Dynasties thought the Western glass not only precious but also mysterious. The essential problem of whether the Western glass is made of natural or artificial material became a long-time enigmatic topic. Yan Shigu, a famous scholar of the Tang Dynasty, gave a comment when he reviewed the book *Hanshu — Xiyuzhuan* (*The History of the Han Dynasty — A Memoir of the Western Regions*).[6] He said that the glasses in red, white, black, yellow, blue, green, crimson, purple and so on produced by the Daqin Kingdom (Rome) were all made of natural material, and that their color and luster were better than that of jade. He said further that the glasses then popularly used were made by melting the stone and adding different medicines, and then casting into wares; they were very brittle, and were actually not real glass. The misunderstanding of Yan Shigu represented a general view of the people of the Tang Dynasty about the Western glass; namely, the Western glass was made directly with natural precious stone, so it was very valuable. Some Islamic glasses that came from West Asia were found in the Tang layer of the archeological excavation; the most representative are 19 sets of glassware unearthed from the underground palace of the Famen Temple (874 AD) at Fufeng, Shaanxi.[7] More Islamic glasses were unearthed during the Song layer excavation; the most famous one is a scent cameo bottle unearthed from the pagoda base of the Jinzhi Temple (977 AD) at Dingxian, Hebei.[8]

The West Asian glass vessels found at the Kang Mausoleum in Guangzhou in 2003 provide new evidence for the glass trade along the Sea Silk Road.

References

1. The National Culture Relics Bureau (ed.), *The Important Archeological Findings in China* (Culture Relics Press, Beijing, 2004), in Chinese pp. 147–153.
2. D. Whitehouse, Excavation at Siraf, *Third Interim Report* (Iran, 1970) Vol. 8, pp. 1–18.
3. S. Carbonim and D. Whitehouse, *Glass of the Sultans* (The Metropolitan Museum of Art and the Corning Museum of Glass, 2002), p. 86.
4. D. Whitehouse, *Sasanian and Post-Sasanian Glass in the Corning Museum of Glass* (The Corning Museum of Glass, New York, 2005), pp. 23–24.

5. Zizhi Tongjian (*Comprehensive Mirror for Governors*) (Zhonghua Shuju, 1956), in Chinese Vol. 225.
6. Hanshu—Xiyuzhuan (*The History of the Han Dynasty — A Memoir of the Western Regions*) (Zhonghua Shuju), in Chinese.
7. J. Y. An, Approach to the Islamic glasses unearthed in China in recent years, *Archeol.* (in Chinese) **12**, 1116–1126 (1984).
8. J. Y. An, The early glass wares in China, *J. Archeol.* (in Chinese) **4**, 430 (1984).

Chapter 21 第二十一章

PIXE Study on the Ancient Glasses of the Han Dynasty Unearthed in Hepu County, Guangxi

Li Qinghui

Shanghai Institute of Optics and Fine Mechanics,
Chinese Academy of Sciences, Shanghai 201800, China

Wang Weizhao

Hepu County Museum, Guangxi Zhuang Autonomous Region,
Hepu 536100, China

Xiong Zhaoming

Archaeological Team of Guangxi Zhuang Autonomous Region,
Nanning 530022, China

Gan Fuxi

Shanghai Institute of Optics and Fine Mechanics,
Chinese Academy of Sciences, Shanghai 201800, China
Fudan University, Shanghai 200433, China

Cheng Huansheng

Institute of Modern Physics,
Fudan University, Shanghai 200433, China

1. Introduction

In the past 20 years, through the close cooperation between archeologists and scientists, some new results about ancient Chinese glass have been obtained.[1] K_2O–SiO_2 glasses of the Han Dynasty (206 BC – 220 AD) were found continually in China, especially in Guangxi, Guangdong and other places in southern and southwestern China. Among these places, emphasis should be placed on Hepu county of Guangxi; both the types and the quantities of the K_2O–SiO_2 glasses in this place are very abundant.[2,3] But the technical origin and the making place of the K_2O–SiO_2 glasses in China are still undetermined.[4–6]

In the Han Dynasty, Hepu county was a very prosperous port and one of the important political, economic and cultural centers in southern China. According to the literature of this period, no later than the reign of Hanwu (140–88 BC), a sea route for trading had been established. This route began in Hepu county and linked India, Sri Lanka and Southeast Asian countries, and was named the Sea Route of the Silk Road by historians. By this route, gems, glasses, spice plants and other goods were imported into China from India and other places.

In this article, some research results about the early glasses unearthed in Hepu county are reported. It is useful to consider the above-mentioned questions and related cultural and technical exchange research between China and other countries during the Han Dynasty.

2. Samples and Experiment

2.1. *Samples*

All the glass samples were provided by the Hepu County Museum and the Archeological Team of Guangxi Zhuang Autonomous Region. They date from the Western Han Dynasty (206 BC–25 AD) to the Eastern Han Dynasty (25–220 AD), and include glass beads with different colors, tubes, ear pendants and some fragments of decorations and vessels. Brief sample descriptions are displayed in Tables 21.1 and 21.2.

Most of the samples in Table 21.1 are intact and their composition should be determined by the nondestructive

Table 21.1. Description of the ancient glass samples obtained in 2004.

Test numbers	Date	Unearthing place	Description
HP-4a	Eastern Han Dynasty	Hehuan Ji-she-ling M18	Ear pendant; diameter of small and big end face 6 mm and 12 mm, height 18 mm
HP-4b			Purple bead; inner diameter 4 mm, outer diameter 9 mm
HP-5		Heguan M10	Heart-shaped glass wafer; 1.4 mm × 1.1 mm
HP-6a		Hehuan Beichajiang Erma Factory M23	Purple bead
HP-6b			Azure bead
HP-6c			Blue bead
HP-7		Hehuan Mechanical Factory M1	Azure bead with shape of Chinese word "*sheng*"; 11 mm × 13 mm × 6 mm
HP-10a	Xinmang period (9–23 AD)	Hehuan Muzhu Mountain M1	Jade-green tubular bead
HP-10b			Purple round bead
HP-10c			Red round bead
HP-10d			Black bead
HP-10e			Purple bead
HP-10f			Green little bead
HP-14c	Eastern Han Dynasty	Hepu Jiu-zhi-ling M5	Green little bead
HP-14e			Blue tubular bead
HP-14f			Jade-green tubular bead
HP-15c	Early Eastern Han Dynasty	Feng-men-ling M26	Navy-blue bead
HP-15d			Blue bead
HP-15e			Grass-green bead
HP-15g			Brown bead

Table 21.2. Description of the ancient glass samples obtained in 2006.

Test number	Original Number	Description
XZHM06-01	Feng-men-ling 03HFM26: 67	Fragment of hexagonal prism bead; jade-green, transparent; diameter of perforation 2 mm, length 2.5 cm, length hexagon 0.5 cm
XZHM06-02	Feng-men-ling 03HFM26: 83	Fragment of blue bead; semitransparent, diameter 0.5 cm
XZHM06-03	Feng-men-ling 03HFM26: 53	Fragment of grass green bead; semitransparent
XZHM06-04	Feng-men-ling 03HFM26: 62	Fragment of brown bead; opaque
XZHM06-05	Feng-men-ling M28: 13	Fragment of blue bead; semitransparent
XZHM06-06	Jiu-zhi-ling M5	Fragment of blue bead; semitransparent
XZHM06-07	Feng-men-ling M23B: 29	Fragment of blue bead; semitransparent
XZHM06-08	Feng-men-ling M23A: 30	Fragment of brown bead; semitransparent

external-beam PIXE (proton-induced X-ray emission) technique, while some fragments could be determined by ICP-AES (inductively coupled plasma atomic emission spectrometry). The samples in Table 21.2 were analyzed by the modified PIXE technique. The structure state of the samples was first analyzed by an X-ray diffractometer of type D/Max 2550V. The morphology of the specimens was determined by an SEM of type EPMA 8705 QH2.

2.2. *Experimental*

External-beam PIXE has proven to be an effective technique for the analysis of archeological artifacts. It allows a quick multielemental determination of elemental concentrations nondestructively, while information on light elements ($Z \leq 12$) in the sample was lost because of X-ray absorption by air. To make up for this weakness of external-beam PIXE, we used ICP-AES for the analysis of light elements, such as Na and Mg. Here, we attempted to find, among numerous chemical compositions of glass, the features that would be characteristic of or specific to the early glass unearthed in Hepu. This is the first step in the search for their origins.

The external-beam PIXE experiments were performed at the NEC 9SDH-2 Pelletron tandem accelerator of Fudan University. The proton beam was extracted through a 7.5-μm-thick Kapton window, and traveled 10 mm in air before reaching the glass sample. The beam spot diameter on the sample was 1 mm and the beam current was 0.05 nA. The original energy of the proton beam was 3.0 MeV; therefore the actual energy of the protons reaching the sample was 2.8 MeV, as a result of energy loss in the Kapton film and air. An ORTEC Si (Li) detector (165 eV FWHM at 5.9 keV), placed at 90° relative to the beam direction, was used. From the measured PIXE spectrum the chemical composition ($Z \geq 13$) of the sample could be obtained by using the deconvolution program GUPIX-96.

The ICP-AES experiment was conducted at the Shanghai Institute of Optics and Fine Mechanics, Chinese Academy of Sciences. Approximately 50–100 mg of each sample were weighed, put into the teflon crucible and decomposed by HF and $HClO_4$. Then HF was expelled until the $HClO_4$ fume was exhausted. The sample was extracted by HCl and moved into the measuring flask to determine the volume, and was ready to be measured. High temperature plasma, produced by the interaction between an induced magnetic field and argon, provides an effective excitation source and a source of charged ions. In a quantitative calibration, we used the synthetic solutions containing Na, Mg and other elements.

During the experiment, by the modified external-beam PIXE technique, the flowing He gas was infused between the sample and the detector. So the lower mass elements such as Na and Mg could be successfully detected. The details of the method can be found in the article by Huansheng Cheng *et al.* in this book,[8] and in other articles.[9,10]

Table 21.3. Chemical compositions of the reference samples determined by the ICP-AES and modified PIXE techniques (wt%).

Sample	Composition	Na_2O	MgO	Al_2O_3	SiO_2	K_2O	CaO	BaO
Window glass	Reference composition	13.0	3.5	1.5	73.00	0.5	8.5	
	ICP-AES	13.26	3.61	1.06	73.20	0.40	7.81	
	PIXE	13.10	3.57	1.79	73.32	0.34	7.64	
BY1	Reference composition	5.00			70.00		15.00	10.00
	ICP-AES	7.29			69.87		14.25	8.59
	PIXE	6.78			66.02		13.72	7.82
BY2	Reference composition	15.00			70.00		15.00	
	ICP-AES	14.75			70.57		14.77	0.21
	PIXE	15.10			70.59		14.10	0.22

3. Comparison Between PIXE and ICP-AES

Before analysis of the early glass, we compared PIXE with ICP-AES. The reference samples were chosen and trial-produced according to the analysis need. The measured data of chemical composition are listed in Table 21.3. We found that the measurement by modified PIXE gave the basically same results as that by ICP-AES if reasonable relative errors were taken into account.

4. Results and Discussions

The chemical compositions of these early glasses are listed in Tables 21.4–21.6. The analytical composition in these tables has been normalized to 100%.

Based on the analytical results, the determined 46 samples can be divided into the following types:

(1) K_2O–SiO_2 glass

About 23 samples belong to this kind of glass, which occupied about 50% of the analyzed samples. The quantity of this kind of

Table 21.4. Chemical compositions of the samples obtained in 2004, determined by the PIXE technique (wt%).

Sample	Al_2O_3	SiO_2	P_2O_5	Cl	K_2O	CaO	TiO_2	MnO	Fe_2O_3	CoO	CuO	BaO	PbO
HP-1-a	4.18	92.06	0.00	0.18	1.44	1.35	0.04		0.56				
HP-1-b	3.84	90.75		0.23	1.23	2.52	0.11	0.38	0.60				
HP-1-c	3.42	91.56		0.12	1.04	2.83	0.10	0.08	0.64				
HP-8-a	3.90	88.14		0.16	2.99	1.15	0.18	1.58	1.54	0.07	0.02		
HP-8-b	4.22	89.63		0.19	1.78	1.16	0.18	1.27	1.18	0.07	0.02		
HP-6-a	7.89	81.39		0.06	8.10	0.19	0.21	0.87	0.82	0.05	0.02		
HP-6-b	7.60	80.90		0.10	7.98	0.69	0.23	0.05	0.82	0.02	1.40		
HP-6-c	7.37	78.34		0.04	10.86	0.28	0.21	0.07	0.76	0.02	1.66		
HP-4-a	4.91	89.72	0.09	0.18	2.20	1.76	0.09	0.10	0.77	0.02			
HP-4-b	4.54	81.09		0.12	8.42	1.12	0.20	2.28	1.86	0.13	0.05		
HP-5	4.62	51.74	0.72	2.27	0.14	0.86		0.05	0.34	0.03	0.03	16.06	23.16
HP-3-b	3.55	90.91	0.12	0.18	2.76	1.46	0.17	0.03	0.75				
HP-10-a	4.15	89.30	0.10	0.21	1.34	3.21	0.18	0.11	0.80		0.01		
HP-10-b	5.97	86.11		0.14	1.35	3.86	0.14	0.88	0.99	0.05	0.09		

(Continued)

Table 21.4. *(Continued)*

Sample	Al_2O_3	SiO_2	P_2O_5	Cl	K_2O	CaO	TiO_2	MnO	Fe_2O_3	CoO	CuO	BaO	PbO
HP-10-c	6.34	68.72	1.18	0.13	14.57	2.15	0.20	0.30	2.01	0.02	3.27		0.61
HP-10-d	6.61	74.33	0.03	0.08	12.80	0.41	0.21	4.11	0.72		0.17		
HP-10-e	7.84	79.54		0.08	8.75	0.37	0.28	1.34	1.25	0.03	0.03		
HP-10-f	8.71	77.46		0.17	9.92	0.46	0.28	0.85	1.59	0.01	0.12		
HP-7	6.77	87.56		0.26	3.33	0.78	0.15		0.67		0.02		
HP-14-c	7.03	80.41		0.36	7.95	0.84	0.23	0.06	0.99	0.02	1.75		
HP-14-e	8.75	77.19		0.39	9.86	0.25	0.30	1.08	1.58	0.05	0.29		
HP-14-f	8.35	82.13		0.49	5.02	0.75	0.23	0.05	0.86	0.02	1.44		
HP-15-a	3.24	91.93		0.11	2.11	1.50	0.18	0.09	0.66	0.02			
HP-15-b	2.96	91.54		0.24	2.70	1.32	0.14	0.06	0.67		0.02		
HP-15-c	4.70	81.83		0.05	10.01	0.25	0.20	1.19	1.22	0.09	0.11		
HP-15-f	6.61	75.78	5.46		2.46	1.12	0.09	0.03	0.85	0.02	2.71		
HP-15-g	3.07	78.43		0.63	3.96	6.02	0.11	0.04	5.04	0.04	0.89		
HP-15-d	5.52	82.34		0.09	7.95	0.60	0.20	1.14	1.27	0.06	0.03		0.40
HP-15-e	5.15	77.00	0.44	0.32	1.66	1.12	0.09	0.05	0.87	0.02	2.34		10.33

Table 21.5. Chemical compositions of glass beads unearthed from the Fengmen Mountain, determined by ICP-AES (wt%)

Sample	Na_2O	MgO	Al_2O_3	SiO_2	P_2O_5	K_2O	CaO	TiO_2	MnO	Fe_2O_3	CuO	ZnO	BaO	PbO	Description
HP-15Ia	0.68	0.38	1.64	79.55	0.50	15.03	1.32	0.15	0.05	0.60	0.01	0.06	0.03		Jade-green, transparent glass bead
HP-15Ib	0.62	0.35	1.64	80.69	0.57	13.92	1.26	0.15	0.05	0.62	0.01	0.06	0.03	0.03	Jade-green bead
HP-15Ic	0.34	0.41	2.91	74.89	0.30	16.32	0.88	0.20	1.33	1.25	0.05	0.07	0.31	0.74	Blue bead
HP-1Id	0.38	0.25	3.23	79.51	0.09	13.44	0.32	0.18	1.21	1.11	0.06	0.09	0.12	0.01	Blue bead
HP-15-e	10.14	0.43	2.23	65.27	0.25	6.39	1.04	0.10	0.04	0.76	2.11	0.61	0.03	10.60	Grass-green bead
HP-15If	8.79	0.41	2.08	76.74	0.25	7.33	1.05	0.08	0.03	0.69	0.06	0.58	0.02	1.89	Grass bead
HP-06-01	0.20	0.30	2.87	78.71		14.04	0.88	0.17	1.01	1.23	0.12		0.13	0.34	Navy-green, opaque bead
HP-06-02	8.25	0.32	2.14	76.94		5.36	1.01	0.10		0.73		0.57	SnO 0.73	10.87	Grass-green, opaque bead
HP-06-03	0.47	0.27	1.60	82.25	0.45	12.96	1.19	0.15	0.04	0.59			0.03		Jade-green, transparent glass
XZHM06-01	0.55	0.30	1.61	81.22	0.46	13.71	1.33	0.15	0.04	0.60			0.03		

Table 21.6. Chemical compositions of the samples determined by the modified PIXE technique (wt %).

Sample number	Na_2O	MgO	Al_2O_3	SiO_2	P_2O_5	Cl	K_2O	CaO	TiO_2	MnO	Fe_2O_3	CuO	PbO	Remarks
XZHM 06-01		0.77	3.11	89.96	0.82	0.19	2.62	1.39	0.16	0.06	0.73			
XZHM 06-01	0.36	0.40	4.68	89.31		0.36	1.78	1.75			1.36			EDX
XZHM 06-02	1.25	0.99	5.33	78.03	0.51	0.13	9.84	0.80	0.18	1.05	1.64	0.03		
XZHM 06-03	1.12	0.69	3.85	68.53			2.81	1.28	0.09	0.02	0.94	3.03	17.60	
XZHM 06-05		0.37	6.98	76.04	0.43	0.02	11.51	0.42	0.28	2.02	1.63	0.02		
XZHM 06-06		0.49	4.61	80.78	0.34	0.06	9.42	0.44	0.21	1.55	1.17	0.08		
XZHM 06-07		0.72	4.79	85.47	0.65	0.14	3.65	1.49	0.13	1.12	1.42	0.02		
XZHM 06-08		3.10	9.54	60.04	1.04	0.22	15.95	4.21	0.57	0.14	2.14	2.81		
XZHM 06-04	Organic material, resembling amber													

glass is the largest for the Han Dynasty in Guangxi and Guangdong. Their main flux is K_2O, mostly between 10% and 15%. The content of Na_2O is less than 2%, and that of MgO less than 1%. The content of CaO is mostly below 1.5%, with the highest being 4.21%. But the content of Al_2O_3 is fluctuant, with the highest being 9.54%. For the samples with K_2O lower than 10% determined by the unmodified PIXE technique, there is perhaps some Na_2O.

(2) PbO–BaO–SiO_2 glass

Only one sample, HP-5, belongs to this kind of glass, with PbO of 23.16% and BaO of 16.06%. This fine sample is colorless and semi-transparent. It is a type that is hard to find among the same kind of glass in China.

(3) PbO–SiO_2 glass

There is also only one sample, XZHM06-03, with PbO of 17.60% and SiO_2 of 68.53%.

(4) Na_2O–K_2O–PbO–SiO_2 glass

Three samples belong to this kind of glass: HP-15-e, HP-15If and HP-06-02. They are characterized by higher contents of both Na_2O and K_2O (higher than 6%). The content of PbO in HP-15-e and HP-06-02 is 10.60% and 10.87% respectively, while that of HP-15If is only 1.89%.

(5) (Na_2O)K_2O–CaO–SiO_2 glass

Sample HP-15-g belongs to this kind of glass, with CaO of 6.02%, K_2O of 3.96% and higher Fe_2O_3 of 5.04%.

(6) Samples with a high content of SiO_2

Based on the results determined by the unmodified PIXE technique, about 11 samples belong to this kind of glass, which contained SiO_2

more than 89% and occupied about 27% of the analyzed samples. They are HP-1-a, HP-1-b, HP-1-c, HP-8-a, HP-4-a, HP-3-b, HP-10-a, HP-10-b, HP-7, HP-15-a and HP-15-b. This kind of sample has good quality and was found in almost all the tombs. It is near-transparent and colorless, or light green, blue and so on. Therefore, it is impossible to make such high quality glasses with high content of SiO_2 (>89%) at the time. Because the Na_2O content could not be measured by non-modified PIXE Method. These glasses maybe belong to soda lime silicate system.

According to the XRD analysis for sample XZHM06-01 (not provided here), the structural state of its main body is noncrystalline. The EDX analytical results on sample XZHM-06-01 were also consistent with that by PIXE. An SEM image of this sample (Fig. 21.1) shows that there are many air bubbles, cracks and holes. But the ICP-AES analytical results on XZHM-06-01 showed that it contains K_2O of 13.71% (see Table 21.5). The ICP-AES analytical results on HP-15Ia and HP-06-03 from the same tomb showed that they also have a higher K_2O content, 15.03% and 12.96% respectively. The authors think that this difference is due to the surface analyzing character of PIXE and EDX, while the result obtained by ICP-AES is the composition of the whole sample. The difference between the chemical composition of the surface and that of the body is due to the weathering of the sample, which led to loss of the flux in the sample surface.[10,11]

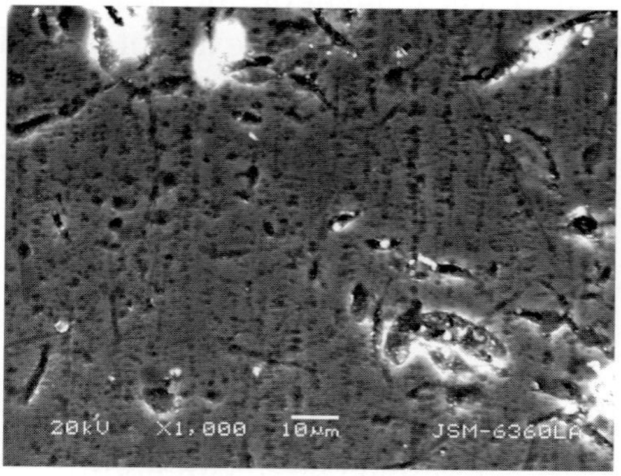

Fig. 21.1. SEM image of sample XZHM-06-01.

(7) Others

During the experiments, according to the rigidity, chemical composition and other properties, some agate samples were distinguished. One organic sample, XZHM 06-04, probably amber, was also found.

For the $PbO–BaO–SiO_2$ and $PbO–SiO_2$ glasses found in Hepu county, it is thought primarily that they were native products of China. According to the known literature, some scholars considered that the $K_2O–SiO_2$ glasses found in Guangxi were made locally in a great measure based on the lead isotope analytical results, artistic character, quantity and ancient literature.[3,4,12] However, other scholars thought that the $K_2O–SiO_2$ glasses of the Han Dynasty were probably imported from Aricamedu in India or countries in Southeast Asia through the Sea Route of the Silk Road.[13] They thought that the $K_2O–SiO_2$ glasses dating back to 600–300 BC were found in India; this period was earlier than the Han Dynasty. Another important reason is that the manufacturing workshop of this period has not been discovered in Guangxi or other places in China up to now.

In fact, $K_2O–SiO_2$ glass of the Warring States period was excavated in Hunan province, and coexisted with $PbO–BaO–SiO_2$ glass.[14] $K_2O–SiO_2$ glass of the Warring States period was also found in Jiangchuan county, Yunnan province,[15] and Wensu county.[16] During the Warring States period (475–221 BC), $K_2O–SiO_2$ glass and $PbO–BaO–SiO_2$ glass coexisted in China. The unearthed sites of these two kinds of glass covered northwest China, the valleys of the Yellow River and the Yangtze River, and other places, such as Sichuan and Guizhou. We think that the technical development of $K_2O–SiO_2$ glass and of $PbO–BaO–SiO_2$ glass have some internal relationship. Their technique origins are probably in the porcelain and metallurgy techniques of China. The $Na_2O–K_2O–PbO–SiO_2$ glasses reported in this article and others[17] may give some clues to this relationship. What about the influence of the external glass-making technology from India, Europe and other places? The final results need more detailed research on the worldwide ancient glasses. Some primary results are reported in another article by the authors.[18]

Acknowledgments

This work is supported by the research of the National Natural Science Foundation of China with Grant No. 50672106 and the Intellectual Innovation Project of the Chinese Academy of Sciences with Grant No. KJCX, SYW No. 12. The authors are grateful to Ma Bo, Zhu Dan and Lin Jiawei of Fudan University and Xu Yongchun of the Shanghai Institute of Optics and Fine Mechanics for their help in the related experiments.

References

1. F. X. Gan, *Development of Early Chinese Glass Technology*, 1st edn. (Shanghai Scientific and Technical Publishers, 2005), in Chinese pp. 1–57.
2. M. G. Shi, O. L. He and F. Z. Zhou *J. Chin. Ceram. Soc.* (in Chinese) **14**(3), 307–313 (1986).
3. Q. S. Huang, *Archeology* (in Chinese) **3**, 264–276 (1988).
4. K. H. Zhao, *Stud. Hist. Natural Sci.* (in Chinese) **10**(2), 145–156 (1991).
5. R. H. Brill and J. H. Martin, (ed.), (Corning Museum of Glass, New York, 1991).
6. J. Y. An, *Acta Archaeologica Sinica* (in Chinese) **4**, 413–448 (1984).
7. Archaeological Team of Guangxi Zhuang Autonomous Region and Hepu County Museum, *The Han Period Burial Site at Fengmenling, Hepu: An Excavation Report, 2003–2005* (Science Press, Beijing, 2006), in Chinese, pp. 131–136.
8. H. S. Cheng, B. Zhang and D. Zhu *et al.* In: F. X. Gan (ed.), *Study On Ancient Glasses Along The Silk Road*, (Fudan University Press, Shanghai 2007) in Chinese, pp. 91–95.
9. Q. H. Li, J. Z. Huang and F. Li *et al.*, *Sci. Conser. Archeol.* (in Chinese) **18**(2), 8–13 (2006).
10. B. Zhang, H .S. Cheng, B. Ma and Q. H. Li *et al.*, *Nucl. Instrum. Methods Phys. Res. B* **240**, 559–564 (2005).

11. J. Z. Li and X. Q. Chen, *J. Chin. Ceram. Soc.* (in Chinese) **14**(3), 293–296 (1986).
12. J. X. Wang, P. Li and Z. Zhang, *Nucl. Tech.* (in Chinese) **17**(8), 499–502 (1994).
13. I. S. Lee, Characteristics of early glasses in ancient Korea, with respect to Asia's maritime bead trade [A], in *Proceeding of the XXth International Congress on Glass* [C] (2004; Kyoto, Japan), O-15-006.
14. F. K. Zhang, Z. H. Cheng and Z. G. Zhang, *J. Chin. Ceram. Soc.* (in Chinese), **11**(1), 67–75 (1983).
15. Yunnan Provincial Museum, *Acta Archaeologica Sinica* (in Chinese) **2**, 97–156 (1975).
16. Q. H. Li, J. Z. Huang, F. Li and F. X. Gan, *Sci. Conser. Archeol.* (in Chinese) **18**(2), 8–13 (2006).
17. C. Y. Zhao, In: Archaeological Team of Guangxi Zhuang Autonomous Region and Hepu County Museum, *The Han Period Burial Site at Fengmenling, Hepu: An Excavation Report, 2003–2005* (Science Press, Beijing, 2006), in Chinese, pp. 182–184.
18. Q. H. Li, F. X. Gan and D. H. Gu, *Stud. Hist. Natural Sci.* (in Chinese) **26**(2), 41–54 (2007).

Chapter 22

第二十二章

Multivariate Statistical Analysis of Some Ancient Glasses Unearthed in Southern and Southwestern China

Fu Xiufeng

Shanghai Institute of Optics and Fine Mechanics,
Chinese Academy of Sciences, Shanghai 201800, China

Gan Fuxi

Shanghai Institute of Optics and Fine Mechanics,
Chinese Academy of Sciences, Shanghai 201800, China
Fudan University, Shanghai 200433,. China

1. Introduction

Owing to its superior natural resources and incomparable geographic location, southern and southwestern China (Fig. 22.1) has been a major, prosperous center of culture and of the economy since the Han Dynasty. In its east, the Chu Culture spread from Hunan and Hubei provinces to the west and southwest of China along the Yangtze River; in its north, the Northwest Silk Road entered Sichuan via Xinjiang and Qinghai; and in its south, the Southwest Silk Road and the Sea Silk Road traversed both Europe and Asia. The culture of the Silk Roads not only promoted intercourse between China and foreign countries, but also reinforced the exchange between southern and southwestern China, and also with central China.

Fig. 22.1. Map of southern and southwestern China.

The south and southwest of China were the cradle and melting pot of several kinds of cultures, in which the Ba Shu culture, Southern Yue culture, Central China culture and Chu culture gathered together and played crucial historical roles. Based on the archaeological data, it has been found that glass adornments had already been used during the Spring–Autumn and Warring States periods in southern and southwestern China. From then on, glass goods were used in the subsequent dynasties.[1] Particularly for the Han Dynasty, great quantities of glass articles have been unearthed; they are of good quality and in different categories, and from all over southern and southwestern China.

The question is whether these glasses were made in China or imported. The exact origin of these glass articles is still uncertain. In the previous studies, researchers have done much work in describing their styles and shapes by comparing them with Western glasses, but studies regarding component analysis and identification are not so numerous. On the other hand, early glass workshop sites have not been found in China yet. However, based on the literatures of archaeology, history and ancient philology, systematic scientific detection could still provide clues to the origin and development of glassmaking techniques. For this study, the method of multivariate statistical analysis was used to analyze the chemical compositions of ancient glasses, and the origin and production areas were discussed through investigation of the distribution of the different kinds of ancient glasses.

2. Samples

There are 85 ancient glass samples, which were collected from the Sichuan University museum, Chongqing museum, Guizhou museum and Hepu (Guangxi) museum. Among the samples, 31 came from Sichuan province and 19 were unearthed in Guizhou.[2] A total of 18 samples from Chongqing were cited in Ref. 3. Samples from Hepu, Guangxi, have been published for the first time. Most of these glass artifacts are single-color beads, eye beads or ear pendants. The chronological span is between the Warring States period and the Han Dynasty. Detailed descriptions of the glasses are shown in Appendix 1, and the chemical compositions of these samples are listed in Appendix 2.

3. Measurements

3.1. *PIXE experimental procedure*

Most of samples are valuable cultural relics; therefore, surface treatments such as polishing, ultrasonic cleaning and fusion are forbidden and nondestructive analytical methods are required. The samples which have the least efflorescent surface were chosen to be detected, and their surfaces were carefully cleaned with anhydrous ethanol.

External-beam PIXE experiments were performed using a 3 MeV pelletron tandem accelerator at the Institute of Physics, Fudan University. The proton beam was extracted through a 7.5-μm-thick Kapton window, and traveled 10 mm in air before reaching the glass sample. The actual energy of the protons reaching the samples was 2.8 MeV, due to energy loss in the Kapton film and air.

The collimated proton beam diameter on the sample was 1 mm and the beam current was kept at 0.05 nA. An ORTEC Si (Li) detector (with 165 eV FWHM resolution at 5.9 keV) recorded the characteristic X-ray spectra. The spectra were gathered for about 15 minutes in order to obtain good statistics. The concentrations of different elements ($Z \geq 13$) were estimated for each sample using the GUPIX-96 software package. But the elements' atomic numbers below 12 could not be measured by routine external-beam PIXE as a result of the X-ray absorption in the air. The principles, characteristics and setup of the PIXE technique can be found in Refs. 4 and 5.

3.2. *Standard samples*

To ensure the reliability of the experimental method for the concentrations of elements, the chemical compositions of standard samples measured by different techniques were compared in this study. The standard value and the values of ICP-AES are shown in Table 22.1, for checking the precision of PIXE. The numbers in parentheses are the relative errors of the measured values against the standard values. From the table, we found that the measurement by PIXE had basically the same results as that by ICP-AES. For PIXE and ICP-AES, the relative errors of most elements were <10%. When the experiments were being carried out, BaF-2 was used to calibrate the lead-barium-silicate glasses, ZF-2 was used to verify the high-lead-silicate glasses, and WG (window glass) was used to check the sodium-calcium-silicate glasses.

3.3. *Evaluation on Na_2O and MgO*

The values of Na_2O and MgO are very important for classifying the ancient glasses in both China and other countries. Although the routine external PIXE cannot provide the value of Na_2O, the concentration of Na_2O could be speculated on rationally in terms of the recipes for ancient Chinese glasses. More than 200 glass samples had already been measured by PIXE, ICP-AES and EDXRF in our recent studies, and we found that there are some glasses — which

Table 22.1. Chemical compositions of standard samples measured by PIXE and ICP.

Sample	Na_2O	MgO	SiO_2	K_2O	CaO	ZnO	BaO	PbO
BaF-2 (*)	1.36		52.26	9.28		9.56	14.47	12.67
BaF-2(I)	1.54		51.98(5%)	8.54(8%)		9.80(2.5%)	13.66(6%)	14.07(11%)
BaF-2(P)			51.2(2%)	8.6(7%)		9.6(5%)	13.5(7%)	11.1(9%)
ZF-2(*)			39.10	4.94				55.41
ZF-2(I)			35.75(8%)	4.27(13%)				59.43(7%)
ZF-2(P)			42.8(9%)	4.7(5%)				50.1(10%)
WG(*)	13.0	3.5	73.00		8.5			
WG(I)	13.26(2%)	3.61(3%)	73.20(3%)		7.81(8%)			
WG(P)	13.1(1%)	3.6(2%)	73.3(4%)		7.7(10%)			

* Indicates the standard value; I — ICP-AES; P — PIXE. WG was measured by improved external-beam PIXE.

almost do not have PbO and BaO, with low K_2O and high CaO — which existed in southern and southwestern China, though in small quantities. This kind of glass was also found in Xinjiang (China) in earlier times, and was considered to be Na_2O–CaO–SiO_2 system glass. By this principle, sample 22 from Sichuan (Appendix 2) belongs to this group, which was not included in the statistical analysis.

The content ranges of Na_2O and MgO in K_2O–SiO_2 system glass and PbO–BaO–SiO_2 glass indicate the other evidence that their contents could not affect the analytical results for the principal components. Several potassium-silicate glass samples from Guangxi measured by ICP-AES are shown in Table 22.2. Obviously, the contents of Na_2O and MgO are below 1%. The EPMA results in Table 22.3 are cited from Ref. 6. The content of Na_2O in lead-barium-silicate glass is mostly less than 4%, and the values of MgO are below 1%, which are far below the sum component (nearly 90%) of PbO, BaO and SiO_2.

4. Multivariate Statistical Analysis

When large quantities of data need to be classified, multivariate statistical analysis, which is based on powerful mathematical and statistical models, provides the method in many fields,[7–10] such as economics, medicine, geology, education and the environment. For the research on ancient ceramics, multivariate statistical analysis has already become the mature and systematic data-processing method.[11,12] B. Zhang *et al.*[13,14] have tried researching on the ancient Xinjiang glasses and some of the southern glasses in China by this method. For this study, cluster analysis and factor analysis were adopted to analyze the ancient glasses from southern and southwestern China. The mathematical theories of these two methods are not taken into consideration here, since they have already been described elsewhere.[15,16]

Cluster analysis is a scientific method for classifying the ancient glasses. By taking the oxide contents of each sample as variables and samples as cases, classification results can be obtained.

Table 22.2. ICP-AES results on several K_2O–SiO_2 glass beads from Guangxi province.

Sample	Color	Na_2O	MgO	CaO	K_2O	Al_2O_3	MnO	Fe_2O_3	TiO_2	BaO	PbO	P_2O_5	ZnO	CuO
HP-15-A	Transparent	0.68	0.38	1.32	15.03	1.64	0.05	0.60	0.15	0.03	0.00	0.50	0.06	0.01
HP-15-B	Green	0.62	0.35	1.26	13.92	1.64	0.05	0.62	0.15	0.03	0.03	0.57	0.06	0.01
HP-15-C	Blue	0.34	0.41	0.88	16.32	2.91	1.33	1.25	0.20	0.31	0.74	0.30	0.07	0.05
HP-15-D	Blue	0.38	0.25	0.32	13.44	3.23	1.21	1.11	0.18	0.12	0.01	0.09	0.09	0.06

Table 22.3. Chemical compositions of several PbO–BaO–SiO_2 glasses measured by EPMA.

No.	Sample	Date	MgO	Na_2O	SiO_2	Al_2O_3	Fe_2O_3	PbO	BaO	CaO	K_2O	CuO
80	Colorless *bi*	300BC	0.15	1.87	36.8	0.28	0.14	42.6	17.4	0.46	0.16	0.02
83	Dragon-shaped, colorless screen	200 BC–100 BC	0.035	2.72	40.5	0.18	0.24	35.2	19.7	0.96	0.22	0.01
84	Light green cicada	200 BC–200 AD	0.04	3.56	42.4	0.23	0.32	33.3	19.2	0.40	0.13	0.05
93	Black bead	400 BC–100 BC	0.61	3.75	37.3	1.19	7.35	37.5	9.40	0.37	0.37	0.42
94	Black bead	400 BC–100 BC	0.53	2.02	41.7	1.90	5.04	34.5	10.1	0.63	0.63	0.35

Moreover, we can get a more comprehensive contrast among ancient glasses from different areas or different systems by changing the variables. For instance, more cluster information can be gained by taking the main elements and trace elements as variables separately.

The essential purpose of factor analysis is to describe, if possible, the covariance relationships among many variables in terms of a few underlying, but unobservable, random quantities called factors. Many variables of ancient glass samples can be reduced to a few factors and presented in a two- or three-dimensional factor analysis diagram for further discussion. Factor analysis simultaneously avoids the weakness of cluster analysis in analyzing the proportion and correlation of variables and the rationality of selecting the variables.

5. Results and Discussion

5.1. *Cluster analysis*

According to Appendix 2, the chemical compositions of all the samples were analyzed by means of hierarchical cluster analysis. The method used was centroid clustering, and the distances between samples were calculated using squared Euclidean distances. Each sample was clustered step by step with the SPSS13.0 (Statistical Package for Social Sciences) program. Finally, in the cluster dendrogram (Fig. 22.2), samples were divided into three groups, namely G1, G2 and G3, corresponding to the different kinds of glass systems, as follows:

G1: K_2O–SiO_2,
G2: PbO(~25 wt%)–BaO–SiO_2,
G3: CaO–PbO(~40 wt%)–BaO–SiO_2.

The different system glasses were manufactured by using different kinds of recipes. G1 samples account for 37.5% of whole samples in which the sum contents of SiO_2 and K_2O are 87.6 wt% on average.

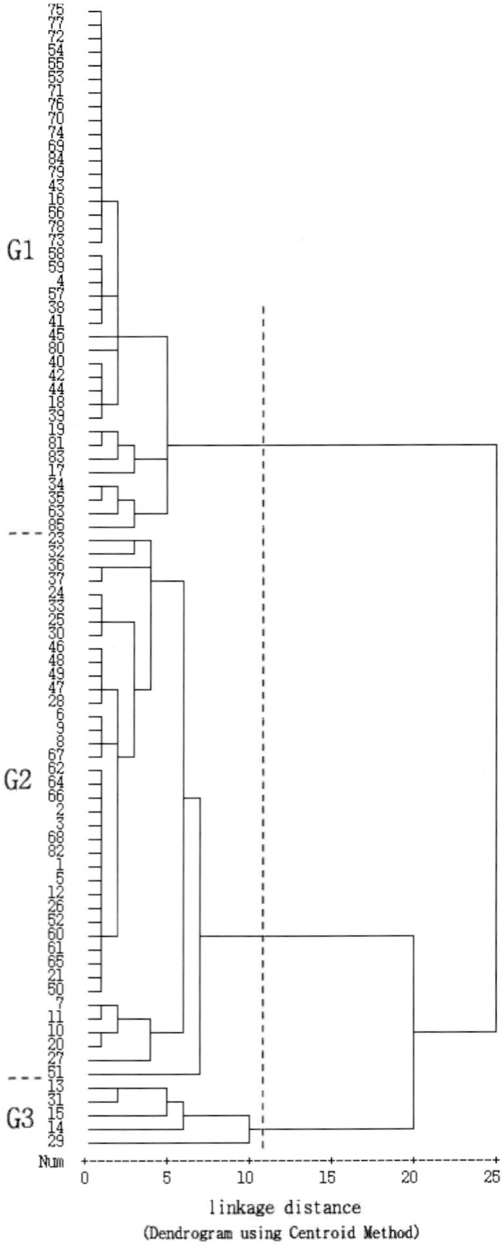

Fig. 22.2. Cluster analysis dendrogram of ancient glasses from southern and southwestern China.

Table 22.4. Distribution ratios of different systems of glasses in each area of southern and southwestern China (%).

Unearthed site	G1	G2	G3
Sichuan (SC)	16.67	66.67	16.67
Chongqing (CQ)	55.56	44.44	0
Guizhou (GZ)	42.11	57.89	0
Guangxi (GX)	94.12	5.88	0

G2 and G3 samples ought to belong to lead-barium-silicate system glass, but the content of PbO increases about 15% distinctively, and it exceeds the content of SiO_2 in G3.

To obtain a more explicit contrast, the distribution ratios of different systems of glasses were calculated respectively (Table 22.4). We can find that the samples from Sichuan areas can be considered to belong to G1, G2 and G3 systems. It is obvious that most kinds of glasses spread in Sichuan areas and the proportion of lead-barium-silicate glasses (G2 + G3) exceeds that of potassium-silicate glasses (G1). The samples from Chongqing and Guizhou mainly belong to G1 and G2 systems, and are nearly equal in quantity. And samples from Hepu, Guangxi, are mainly classified as G1 systems, and account for above 90% of all samples.

As discussed above, it can be determined that the ancient glasses unearthed in southern and southwestern China are mainly lead-barium-silicate system and potassium-silicate system glasses, besides a small quantity of high-lead and high-calcium ones. These two kinds of glasses are often excavated in the same era and the same place. We have learnt that potassium-silicate system glasses were mainly produced in Guangxi and Guangdong, and the Yangtze valley was the cradle of lead-barium-silicate glasses.[1] From the above discussion, Sichuan, Chongqing and Guizhou areas underwent a relatively balanced excavation of these two kinds of glasses. In this way, the distribution of glass systems in southern and southwestern China can be described thus: the single lead-barium-silicate system glasses are in its north and northeast, the single potassium-silicate system glasses are in its south, and in its middle these two systems of glasses coexist.

Sichuan province has been called "the land of abundance" since the Warring States period, and Chengdu, its capital, was the starting place of the Southwest Silk Road. Besides, Hepu, Guangxi, was a crucial transfer site of the Silk Road via the sea. Also, the branch routes of the Silk Road extended to Guizhou, Guangdong, Yunnan, etc. Thus, trade exchanges were frequent in these areas, including the spread of ancient glassmaking techniques. The glass samples unearthed in these areas have the obvious Chinese characteristics, and were often found in civilian graves. From the ancient literature, we found that the foreign glasses at that time were as valuable as gold, so if these glasses came from foreign countries, they should be in noble tombs, but the fact is that very few glass articles have been excavated from noble tombs. Accordingly, the reasonable explanation for this doubt should be independent glassmaking in China.

5.2. *Factor analysis*

Factor analysis can be used to explore the origin of ancient glasses. The rotated component matrix of glass samples (Table 22.5) demonstrates the correlative extent between variables and factors via the absolute value of the relation coefficient.

Four factor variables are named F1, F2, F3 and F4 (see Table 22.5). F1 represents the contribution of SiO_2, PbO, BaO and K_2O, so we call it the principal component factor. F2 explains the contribution of Al_2O_3 and Fe_2O_3. Al^{3+} and Fe^{2+} ions could enter the [SiO_4] tetrahedron and substitute the Si^{4+} ions, so we call F2 the substitute factor. F3 mainly embodies the proportion of CaO and could be called the calcium regulator factor. F4 shows the contribution of CuO, which is a colorant oxide; therefore, we call F4 the copper colorant factor. A three-dimensional spatial diagram of factor analysis without considering F4 is presented in Fig. 22.3. The three glass groups obtained by cluster analysis range in sequence from the top down, along with the reduction of the F1 value. In the era from the late Warring States to the Eastern Han Dynasty, the evaluation of ancient

Table 22.5. Rotated component matrix of factor analysis of ancient glasses from southern and southwestern China.

	Component			
	F1	**F2**	**F3**	**F4**
SiO_2	0.798	−0.166	−0.443	−0.079
PbO	−0.849	−0.074	0.240	0.072
BaO	−0.878	−0.135	−0.083	−0.092
Al_2O_3	0.247	0.758	−0.276	0.240
Fe_2O_3	0.136	0.748	0.117	−0.100
CaO	0.113	0.143	0.885	−0.112
TiO_2	0.799	0.337	−0.088	−0.162
K_2O	0.835	0.099	0.024	−0.085
Cr_2O_3	−0.485	0.102	−0.088	−0.139
CuO	−0.015	−0.037	0.028	0.935
MnO	0.564	−0.048	−0.208	−0.385
P_2O_5	−0.238	−0.034	0.751	0.215
SO_3	−0.184	0.710	0.180	−0.076

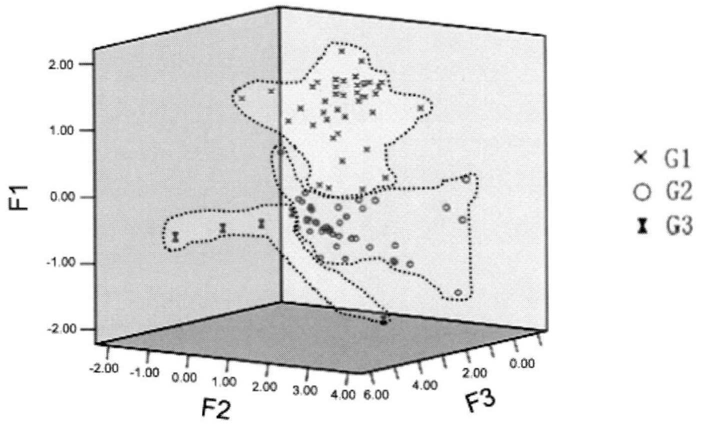

Fig. 22.3. Three-dimensional factor analysis diagram of ancient glasses from southern and southwestern China.

glasses in southern and southwestern China went through a continuous development process from G1 to G2, and then G3.

In the three-dimensional factor analysis diagram, we can hardly find the factor distributing rules of each place. Because the main elements of the same-system glasses are similar and the different-system glasses are often distributed in the same place, it is difficult to certify the production areas and origin information of the ancient glasses, but the trace elements often provide us with useful clues. Obviously, the colorant elements are trace elements of ancient glasses. We got two factors (F1′ and F2′) from factor analysis of five colorant oxides (Fe_2O_3, TiO_2, CuO, Cr_2O_3 and MnO). By taking F1′ and F2′ as the *x*-axis and the *y*-axis respectively, we obtained the two-dimensional colorant factor score diagram (Fig. 22.4).

From Fig. 22.4, it is evident that the colorant factors of Sichuan and Chongqing glasses (o) have a more extensive distribution than those of other places, and most of the samples distributed to the left of the dotted line. The colorant factor coverage of Guangxi glasses (•) is comparatively independent of that of Sichuan province, on

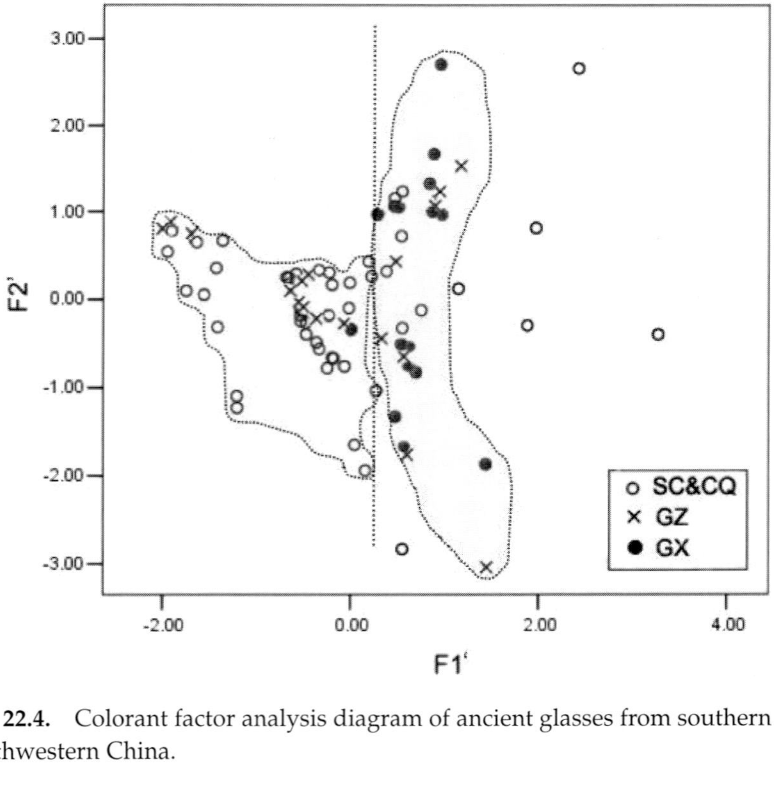

Fig. 22.4. Colorant factor analysis diagram of ancient glasses from southern and southwestern China.

the right of the dotted line. It is educible that the ancient glasses of these two areas were manufactured from local resources independently. The colorant distributing position of Guizhou samples (×) has an overlap region with the Sichuan and Guangxi areas. G2 glasses of Guizhou overlap with those of Sichuan and Chongqing, and G1 glasses of Guizhou overlap with those of Guangxi. It is indicated that in all probability both of the two areas influenced the manufacturing technique of ancient Guizhou glasses. Guizhou province lies in the midland of southwestern China, which is an important crossway for the Southwest Silk Road and the Silk Road via the sea. So it is likely that the ancient glasses of Guizhou were traded in these two places through the branch routes of the Silk Roads. On the other hand, the historical record on Guizhou is not so prolific before the Han Dynasty, and therefore the spread of ancient glasses could be considered as one of the important proofs of intercommunion between Guizhou and other places.

6. Conclusions

From the results of PIXE and multivariate statistical analysis of ancient glasses (475 BC–220 AD) unearthed in southern and southwestern China, it has been found that these glasses belong basically to two main categories — lead-barium-silicate glasses and potassium-silicate glasses. They are consistent with two kinds of Chinese characteristic glasses originating in the Yangtze valley and Guangxi province. Cluster analysis divided lead-barium-silicate glasses into two groups — PbO(~25 wt%)–BaO–SiO_2 system and the CaO–PbO(~40 wt%)–BaO–SiO_2 system. The distribution ratios of different groups in each area of southern and southwestern China were also obtained by cluster analysis. From factor analysis, the results show that ancient glasses of Guangxi and Sichuan areas could be manufactured locally, and that Guizhou glasses were influenced by both of these two sites. The experimental results and data-processing method will lead to some breakthroughs in studying the cultural and technological intercourse along the Southwest Silk Road and the Silk Road via the sea.

Appendix 1

Table 1. Condition of glass samples unearthed in Sichuan province.

Sample	Unearthing site	Date	Description
1	South Nanchong Station M40:1	E-H	Ear pendant, dark blue
2	South Nanchong Station M27:2	E-H	Ear pendant, dark blue
3	South Nanchong Station M27:4	E-H	Ear pendant, light blue
4	South Nanchong Station M27:3	E-H	Ear pendant, dark blue
5	Qingchuan Haojiaping M13:15	WS	Black base of eye bead
6	Qingchuan Haojiaping M13:15	WS	Green part of dragonfly eye of bead
7	Qingchuan Haojiaping M13:15	WS	Yellow part of dragonfly eye of bead
8	Qingchuan Haojiaping M13:15	WS	Eyespot of eye bead
9	Qingchuan Haojiaping M13:16	WS	Green part of dragonfly eye of bead
10	Qingchuan Haojiaping M13:16	WS	Yellow part of dragonfly eye of bead
11	Qingchuan Haojiaping M13:16	WS	Yellow part of dragonfly eye of bead
12	Qingchuan Haojiaping M13:16	WS	Black base of eye bead
13	—	Late WS–H	Green glass bead, tubelike
14	—	Late WS–H	Green glass bead, tubelike
15	—	Late WS–H	Green glass bead, flat
16		H	Glass eardrop, dark blue
17	Li county	H	Glass bead, yellow
18	Li county	H	Crystal-like glass bead
19	Li county	H	Glass bead, dark blue

(*Continued*)

Table 1. (*Continued*)

Sample	Unearthing site	Date	Description
20	—	WS–W-H	Blue base of bead with green-and-white stratified eyes
21	—	WS–W-H	Yellow tessera of bead with green-and-white stratified eyes
22	—	H	Yellow glass bead with erodent veins
23	—	WS–W-H	Blue base of bead with green-and-white stratified eyes
24	—	WS–W-H	Green glass on bead with green-and-white stratified eyes
25	—	WS–W-H	White part of eye bead, as above
26		WS–W-H	Blue glass bead with blue-and-white stratified eyes
27	—	WS–W-H	Nonweathered surface of sample, as above
28	—	WS–W-H	Green glass on eye of bead, as above
29	—	WS–W-H	White part on eyes of black bead with green-and-white eyes
30	—	WS–W-H	Black base of sample, as above
31	—	WS	String of glass beads, green

Table 2. Condition of glass samples unearthed in Chongqing.

Sample	Unearthing site	Date	Description
32	South Bao Cheng Road	WS	Pale green glass *bi*
33	Ba County, Dong Sun Ba	WS	Black glass base of bead
34	Ba County, Dong Sun Ba	WS	Green eyes of bead, as above
35	Ba County, Dong Sun Ba	WS	White eyes of bead, as above
36	Ba County, Dong Sun Ba M49	WS	Beads with blue-and-white stratified eyes
37	Ba County, Dong Sun Ba M49	WS	Light blue eye
38	Zhang Ming Ya grave, Bao Cheng Road	WS	Blue glass eardrop, opaque
39	Zhang Ming Changshan village M12	WS	Yellow glass bead
40	Bao Cheng Road, Zhaohua city, Bao Lun Yuan	WS	Blue glass bead
41	Bao Cheng Road, Zhaohua city, Bao Lun Yuan	WS	White glass bead
42	Bao Cheng Road, Zhaohua city, Bao Lun Yuan	WS	Coffee glass bead
43	Cheng Yu Road	WS	Sky-blue glass bead
44	Cheng Yu Road	WS	Red glass bead
45	Bao Cheng Road, Zhaohua city, Bao Lun Yuan M7	WS	Blue glass bead
46	Bao Cheng Road, Zhaohua city, Bao Lun Yuan M7	WS	Black glass bead (big)
47	Bao Cheng Road, Zhaohua city, Bao Lun Yuan M7	WS	Grass-green glass bead
48	Bao Cheng Road, Zhaohua city, Bao Lun Yuan M7	WS	Black glass bead (small)
49	Bao Cheng Road, Zhaohua city, Bao Lun Yuan M19	WS	Yellow glass bead, semitransparent

Table 3. Condition of glass samples unearthed in Guizhou province.

Sample	Unearthing site	Date	Description
50	Hezhang, Kele M91	Late WS	Glass bead
51	Hezhang, Kele M87	Late WS	Glass bead
52	Qingzhen, Ya-long-ba M14	H	Glass bead, green
53	Qingzhen, Ya-long-ba M1	H	Glass ear pendant, blue
54	Weining, Zhong-shui M42	W-H	Glass bead (medium), blue
55	Weining, Zhong-shui M42	W-H	Glass bead (big), blue
56	Hezhang, Kele M38	W-H	Glass bead, dark blue
57	Hezhang, Kele M38	W-H	Glass bead, light blue
58	Hezhang, Kele M38	W-H	Glass bead, green
59	Anshun, Ninggu M12	E-H	Glass bead
60	Qingzhen, Zhonghou village M100	H	Glass bead, white
61	Qingzhen, Zhonghou village M100	H	Glass bead, blue
62	Qingzhen, Zhonghou village M10	H	Glass bead, green
63	Qingzhen, Ludi village M56	H	Sheeplike glass pendant, green
64	Weining, Zhong-shui-li M24	8–23 AD	Ear pendant
65	Pingba, Jin-jia-ba M12	E-H	Ear pendant
66	Xingren, Jiaole M2	E-H	Ear pendant
67	Qianxi Gantang M18	E-H	Lionlike glass pendant, blue
68	Qianxi M13	E-H	Ear pendant

Table 4. Condition of glass samples unearthed in Guangxi province.

Sample	Unearthing site	Date	Description
69	Hehuan, Ji-she-ling M18	E-H	Counter-shaped bead, purple
70	Hehuan, North Chajiang M23	H	Counter-shaped bead, blue–purple translucent
71	Hehuan, North Chajiang M23	H	Counter-shaped bead, light blue translucent
72	Hehuan, North Chajiang M23	H	Counter-shaped bead, blue translucent
73	Hehuan, Mu-zhu-ling M1	8–23 AD	Counter-shaped bead, black
74	Hehuan, Mu-zhu-ling M1	8–23 AD	Counter-shaped bead, purple
75	Hehuan, Mu-zhu-ling M1	8–23 AD	Counter-shaped bead, Cambridge blue
76	Hepu, Jiu-zhi-ling M5	E-H	Counter-shaped bead, green
77	Hepu, Jiu-zhi-ling M5	E-H	Round tube, blue
78	Hepu, Jiu-zhi-ling M5	E-H	Round tube, light green
79	Feng-meng-ling M26	H	Small fragment, dark blue
80	Feng-meng-ling M26	H	Small fragment, grass-green
81	Feng-meng-ling M26	H	Small fragment, brown
82	Heguan M10	H	Heart-shaped wafer
83	Hehuan, Mu-zhu-ling M1	8–23 AD	Counter-shaped bead, maroon
84	Feng-meng-ling M26	H	Small fragment, blue
85	Feng-meng-ling M26	H	Small fragment, grass-green

H — Han Dynasty (202 BC–220 AD); W-H — Western Han Dynasty (202 BC–8 AD); E-H — Eastern Han Dynasty (25 AD–220 AD); WS — Warring States (475 BC–221 BC).

Appendix 2

Table 1. Chemical compositions of ancient glasses unearthed in southern and southwestern China (wt%).

Sample	SiO_2	PbO	BaO	Al_2O_3	Fe_2O_3	CaO	TiO_2	K_2O	Cr_2O_3	CuO	MnO	P_2O_5	SO_3
1	53.1	24.6	17.9	2.2	0.4	0.1	0.0	0.1	0.0	0.0	0.1	0.7	0.0
2	54.1	22.8	16.6	3.4	0.6	1.7	0.0	0.3	0.0	0.0	0.2	0.3	0.0
3	52.9	22.9	16.8	3.2	0.9	2.0	0.0	0.5	0.0	0.1	0.3	0.5	0.0
4	86.0	0.0	0.3	3.1	2.2	0.9	0.1	5.6	0.0	0.0	1.6	0.0	0.2
5	54.7	25.2	11.8	3.1	0.4	1.3	0.0	0.5	0.0	2.5	0.0	0.5	0.0
6	47.8	31.3	11.7	2.8	0.6	1.0	0.0	0.7	0.0	0.9	0.0	0.0	3.2
7	35.2	35.4	9.8	9.1	2.6	1.1	0.0	0.9	0.1	0.1	0.1	0.1	5.5
8	50.1	28.9	12.3	2.6	0.4	1.3	0.0	0.6	0.1	2.8	0.0	0.2	0.8
9	49.7	30.3	11.5	2.3	0.6	1.1	0.0	0.7	0.0	1.0	0.0	0.6	2.3
10	42.3	33.7	11.4	5.1	1.6	1.1	0.0	0.6	0.0	0.4	0.0	0.4	3.4
11	34.5	37.4	10.9	7.2	2.6	1.3	0.0	0.7	0.0	0.2	0.0	0.4	4.8
12	53.5	25.7	12.3	3.0	0.4	1.1	0.0	0.4	0.0	2.9	0.0	0.7	0.0
13	23.1	39.3	9.1	3.4	0.6	8.1	0.0	0.7	0.0	1.3	0.0	14.4	0.1
14	9.1	42.3	10.1	2.0	0.7	12.7	0.0	0.8	0.0	1.3	0.2	19.8	1.0
15	19.8	33.8	17.2	4.6	2.0	5.4	0.0	1.1	0.1	0.5	0.0	7.7	7.9
16	81.7	0.0	0.1	3.0	1.4	1.7	0.2	10.4	0.0	0.0	1.4	0.0	0.2
17	61.9	11.8	0.1	1.5	0.4	10.0	0.1	14.3	0.0	0.0	0.0	0.0	0.0
18	73.9	0.4	0.0	3.4	1.1	7.0	0.2	9.8	0.0	0.0	0.0	0.2	4.1
19	63.7	0.0	1.1	5.3	0.8	10.9	0.0	12.7	0.1	0.0	0.0	0.3	5.1
20	41.0	31.8	10.0	4.8	0.5	1.9	0.0	2.5	0.0	0.6	0.0	2.4	4.6

(Continued)

Table 1. *(Continued)*

Sample	SiO_2	PbO	BaO	Al_2O_3	Fe_2O_3	CaO	TiO_2	K_2O	Cr_2O_3	CuO	MnO	P_2O_5	SO_3
21	52.9	19.8	12.8	4.8	1.6	1.4	0.0	1.5	0.1	0.1	0.0	1.0	4.2
22	59.1	1.5	0.0	3.8	0.6	24.9	0.0	3.5	0.0	0.1	0.0	0.5	5.9
23	45.5	20.5	1.2	8.4	10.5	6.5	0.3	2.5	0.0	0.6	0.1	1.1	3.0
24	49.8	28.4	6.5	5.0	1.0	3.0	0.0	1.8	0.1	1.0	0.0	1.5	1.9
25	47.8	32.1	3.7	5.7	1.0	3.9	0.0	1.5	0.0	1.1	0.0	1.1	2.2
26	53.5	24.3	11.4	3.6	0.6	2.0	0.0	0.7	0.1	2.6	0.0	1.4	0.0
27	44.6	41.2	0.8	4.1	0.6	2.5	0.0	1.0	0.0	4.1	0.0	1.1	0.0
28	56.9	26.6	9.5	2.7	0.5	1.3	0.0	0.8	0.0	0.4	0.0	0.5	0.8
29	30.9	57.1	1.1	4.6	0.7	3.1	0.1	1.2	0.0	0.2	0.0	1.0	0.0
30	48.8	28.9	0.9	5.6	6.8	4.9	0.2	2.4	0.0	0.5	0.0	0.8	0.5
31	26.9	46.8	4.6	3.7	1.2	5.4	0.0	0.8	0.0	1.3	0.1	9.4	0.0
32	48.4	19.2	6.5	9.7	1.7	2.4	0.2	4.5	0.0	0.1	0.0	0.0	7.4
33	48.4	25.5	6.9	6.8	4.1	3.9	0.1	2.1	0.0	0.3	0.0	1.0	0.9
34	64.8	16.8	3.7	5.8	1.4	1.7	0.0	1.9	0.0	0.4	0.0	0.7	2.9
35	68.3	20.3	4.7	3.7	0.4	0.6	0.0	0.9	0.0	0.1	0.0	0.4	0.6
36	56.1	11.5	8.8	7.9	2.3	1.7	0.0	1.4	0.0	1.5	0.0	7.2	1.8
37	56.7	15.8	11.6	4.2	1.0	0.9	0.0	0.5	0.1	1.5	0.0	7.7	0.0
38	81.5	0.0	0.0	3.0	3.0	1.2	0.5	4.8	0.0	0.1	3.4	0.0	2.4
39	69.7	0.0	0.0	7.3	3.8	8.4	0.5	6.2	0.0	0.0	0.2	1.6	2.3
40	76.6	0.0	0.0	4.3	1.2	9.7	0.2	4.5	0.0	1.1	0.1	0.8	1.5
41	83.4	0.0	0.0	2.1	0.8	7.4	0.1	3.6	0.0	0.0	0.0	0.7	1.7
42	78.3	0.0	0.0	3.9	1.5	7.8	0.1	4.6	0.0	0.1	0.0	1.1	2.4

(Continued)

Table 1. *(Continued)*

Sample	SiO_2	PbO	BaO	Al_2O_3	Fe_2O_3	CaO	TiO_2	K_2O	Cr_2O_3	CuO	MnO	P_2O_5	SO_3
43	79.9	0.4	0.0	4.8	1.6	1.1	0.2	9.8	0.0	0.2	0.8	0.7	0.6
44	75.1	0.0	0.0	4.3	2.2	7.9	0.1	5.0	0.0	0.1	0.2	0.9	4.1
45	76.1	0.0	0.0	13.1	1.3	2.2	0.4	4.4	0.0	0.8	0.0	0.5	0.7
46	58.5	27.8	7.5	3.0	0.3	0.6	0.0	0.4	0.1	1.0	0.0	0.9	0.0
47	59.6	27.4	6.6	3.5	0.2	1.0	0.0	0.4	0.1	0.4	0.0	0.9	0.0
48	58.5	28.8	7.0	2.7	0.3	0.6	0.0	0.5	0.0	0.6	0.0	1.1	0.0
49	59.5	28.5	6.7	2.2	0.3	0.6	0.0	0.5	0.0	0.8	0.0	0.8	0.0
50	58.1	19.7	9.1	6.1	1.9	1.1	0.0	0.6	0.0	0.6	0.1	1.2	1.5
51	38.8	12.7	12.4	11.7	3.9	2.0	0.0	6.4	0.0	4.1	0.0	2.3	5.6
52	51.4	27.1	12.8	5.1	1.0	0.6	0.0	0.7	0.0	0.5	0.0	0.5	0.2
53	77.3	0.0	0.2	5.2	3.0	1.4	0.2	9.3	0.0	0.1	2.2	0.0	1.2
54	77.4	0.0	0.2	6.8	1.4	0.5	0.3	10.8	0.0	0.2	1.1	0.3	1.1
55	76.2	0.1	0.2	6.4	1.4	0.6	0.3	12.0	0.0	0.2	1.4	0.4	0.9
56	81.1	0.0	0.0	3.5	0.6	0.8	0.1	10.4	0.0	2.2	1.1	0.0	0.3
57	84.7	0.0	0.0	5.6	0.8	1.1	0.1	5.4	0.0	1.5	0.6	0.0	0.4
58	88.2	0.0	0.0	4.1	0.9	1.4	0.1	2.1	0.0	2.8	0.0	0.0	0.4
59	88.0	0.0	0.1	4.8	1.9	1.7	0.2	2.6	0.0	0.1	0.2	0.0	0.4
60	58.7	20.2	16.8	2.2	0.5	0.9	0.0	0.2	0.0	0.0	0.0	0.5	0.0
61	55.1	17.7	16.0	6.4	1.1	2.3	0.0	0.7	0.0	0.0	0.2	0.4	0.0
62	54.6	25.6	14.3	2.4	0.6	1.0	0.0	0.3	0.0	0.4	0.0	0.9	0.0
63	72.1	11.7	8.3	3.9	0.5	0.7	0.0	0.4	0.0	0.3	0.0	0.9	1.3
64	54.2	24.5	15.1	2.6	0.7	2.0	0.0	0.2	0.0	0.1	0.2	0.6	0.0

(Continued)

Table 1. *(Continued)*

Sample	SiO_2	PbO	BaO	Al_2O_3	Fe_2O_3	CaO	TiO_2	K_2O	Cr_2O_3	CuO	MnO	P_2O_5	SO_3
65	52.7	16.4	15.0	9.5	1.4	3.5	0.0	0.9	0.1	0.1	0.1	0.5	0.0
66	53.1	26.0	16.4	1.7	0.4	1.6	0.0	0.1	0.1	0.0	0.0	0.6	0.0
67	51.4	30.8	14.3	1.3	0.4	0.4	0.0	0.1	0.0	0.5	0.0	0.8	0.0
68	52.6	22.9	17.6	3.9	0.6	1.0	0.0	0.4	0.1	0.0	0.1	0.9	0.0
69	81.1	0.0	0.0	4.5	1.9	1.1	0.2	8.4	0.0	0.1	2.3	0.0	0.2
70	81.4	0.0	0.0	7.9	0.8	0.2	0.2	8.1	0.0	0.0	0.9	0.0	0.4
71	80.9	0.0	0.0	7.6	0.8	0.7	0.2	8.0	0.0	1.4	0.1	0.0	0.2
72	78.3	0.0	0.0	7.4	0.8	0.3	0.2	10.9	0.0	1.7	0.1	0.0	0.4
73	74.3	0.0	0.0	6.6	0.7	0.4	0.2	12.8	0.0	0.2	4.1	0.0	0.5
74	79.5	0.0	0.0	7.8	1.3	0.4	0.3	8.8	0.0	0.0	1.3	0.0	0.5
75	77.5	0.0	0.0	8.7	1.6	0.5	0.3	9.9	0.0	0.1	0.9	0.0	0.4
76	80.4	0.0	0.0	7.0	1.0	0.8	0.2	8.0	0.0	1.8	0.1	0.0	0.4
77	77.2	0.0	0.0	8.8	1.6	0.3	0.3	9.9	0.0	0.3	1.1	0.0	0.3
78	82.1	0.0	0.0	8.4	0.9	0.8	0.2	5.0	0.0	1.4	0.1	0.0	0.7
79	81.8	0.0	0.0	4.7	1.2	0.3	0.2	10.0	0.0	0.1	1.2	0.0	0.4
80	75.8	0.0	0.0	6.6	0.9	1.1	0.1	2.5	0.0	2.7	0.0	5.5	4.9
81	65.1	0.0	0.0	3.5	0.7	9.6	0.1	14.5	0.0	0.9	0.0	1.0	0.8
82	51.7	23.2	16.1	4.6	0.3	0.9	0.0	0.1	0.0	0.0	0.1	0.7	0.0
83	68.7	0.6	0.0	6.3	2.0	2.2	0.2	14.6	0.0	3.3	0.3	1.2	0.5
84	82.3	0.4	0.0	5.5	1.3	0.6	0.2	8.0	0.0	0.0	1.1	0.0	0.4
85	77.0	10.3	0.0	5.2	0.9	1.1	0.1	1.7	0.0	2.3	0.1	0.4	0.0

All oxides' mass fraction has been converted to the value when their oxide ions were on the maximum valence. For example, "Fe_2O_3" means not only the existence of Fe^{3+} but also the presence of Fe^{2+}.

Acknowledgments

We would like to acknowledge the support of the National Natural Science Foundation of China (grant No. 50672106) and the Knowledge Innovation Program Project of the Chinese Academy of Sciences (No. KJCX-N04) in funding this ongoing research.

References

1. F. X. Gan, *Development of Ancient Chinese Glass* (Shanghai Science and Technology Publishers, 2005), in Chinese.
2. Q. H. Li, B. Zhang and F. X. Gan, Chemical composition analysis of some ancient glasses unearthed in southern China by the PIXE technique. In F. X. Gan (ed.), *Study on Ancient Glasses in Southern China — Proceedings of the 2002 Nanning Symposium on Ancient Glasses in Southern China* (Shanghai Science and Technology Publishers, 2003), in Chinese, pp. 76–84.
3. F. Li, Q. H. Li and F. X. Gan, Analysis of some ancient glasses unearthed in the Sichuan area by PIXE *Nucl. Tech.* (in Chinese) **29** (2006), in press.
4. E. T. Williams, PIXE analysis with external beams: systems and applications, *Nucl. Instrum. Methods B* **3**, 211–219 (1984).
5. X. K. Wu, X. Z. Zeng and F. J. Yang, Archeological applications of PIXE and IXX for paperlike objects at Fudan University, *Nucl. Instrum. Methods B* **75**, 458–462 (1993).
6. R. H. Brill, S. S. C. Tong and D. Dohrenwend, Chemical compositions analysis on some early Chinese glasses. In: F. X. Gan (ed.), *Scientific Research in Early Chinese Glass — Proceedings of the Archeometry of Glass Session of the International Symposium on Glass* (China Architecture & Building Press, Beijing, 1988), in Chinese, pp. 15–35.
7. X. J. Zhao, J. L. Zhang and D. L. Chen, Application of multivariate statistics analysis to classification and identification of paleontologic fossils, *J. Xi'an Mining Inst.* (in Chinese) **16**, 183–185 (1996).
8. J. Shi and Y. Xiong, Multivariate statistical and cluster analysis method applied in exploitation of natural resources, *J. Shandong Univ. Technol. (Sci. Tech.)* (in Chinese) **17**, 81–83 (2003).

9. J. S. Qi and H. B. Xu, Factor analysis and cluster analysis of trace elements in some Chinese medicinal herbs, *Chin. J. Anal. Chem.* (in Chinese) **26**, 1309–1314 (1998).
10. I. Stanimirova, B. Walczak and D. L. Massart, Multiple factor analysis in environmental chemistry, *Analytica Chimica Acta* **545**, 1–12 (2005).
11. I. E. Kieft, D. N. Jamieson and B. Rout, PIXE cluster analysis of ancient ceramics from North Syria, *Nucl. Instrum. Methods Phys. Res. B* **190**, 492–496 (2002).
12. J. Wu and J. Z. Li, Multi-variate statistical analysis of the chemical compositions for bodies and glasses of Jingdezhen blue and white procelain, *J. Ceram.* (in Chinese) **18**, 130–135 (1997).
13. B. Zhang, H. S. Cheng and B. Ma, PIXE and ICP-AES analysis of early glass unearthed from Xinjiang (China), *Nucl. Instrum Methods Phys. Res. B* **240**, 559–564 (2005).
14. B. Zhang, Y. H. Li, Q. H. Li *et al.*, Non-destructive analysis of early glass unearthed in South China by external-beam PIXE, *J. Radioanal. Nucl. Chem.* **261**, 387–392 (2004).
15. Y. T. Zhang and K. T. Fang, *Introduction of the Multivariate Statistical Analysis Method* (Science Press, Beijing, 1982), in Chinese.
16. H. J. Luo, *Ancient Chinese Ceramics and Multivariate Statistical Analysis* (China Light Industry Press, Beijing, 1997), in Chinese.

Chapter 23 第二十三章

Study of the Ancient Glasses Found in Chongqing

Ma Bo

School of Information Science and Engineering,
Fudan University, Shanghai 200433, China

Gan Fuxi

Shanghai Institute of Optics and Fine Mechanics,
Chinese Academy of Sciences, Shanghai 201800, China;
Fudan University, Shanghai 200433, China

Feng Xiaoni and Gao Menghe

Department of the Museum, Fudan University,
Shanghai 200433, China

Shen Shifang

Chongqing Museum, Chongqing 40015, China

1. Introduction

Located in the southwest of China, Chongqing and its peripheral districts are characterized by mountainous landforms with river interludes. This area is a traffic hub connecting the Jianghan Plain and the Sichuan Basin, because it is the exit of the Sichuan Basin. Successive politicians had put considerable emphases on this special area, owing to the old saying "It will be plunged into chaos before the world wars and will be prosperous after the world peace." Eight ancient nationalities inhabited the area, such as the Yi, Pu and Ju. These early ancestors in Chongqing created the splendid ancient civilization of Ba, including the ancient glassware.

A large number of glass artifacts have been unearthed in Chongqing and its peripheral districts, such as Qingchuan, Zhaohua, Yujiaba of Yunyang, Maliuwan of Wanzhou, and Chongqing city.[1–4] The glass artifacts were made in the era from the Warring States to the Six Dynasties. For this article, the proton-induced X-ray emission (PIXE) technique was used to characterize the compositions of the samples of ancient glass; the samples were provided by the Chongqing Museum, the Archeological Team of Fudan University in the Three Gorges, and the Wanzhou Cultural Relics Administrative Office.

2. Analysis of Ancient Glassware Composition

2.1. *Source of the samples*

The ancient glass samples from Nos. 1 to 27, provided by the Chongqing Museum, were unearthed from the southern part of the Baocheng Railway, near the Chengyu Highway, and Banan, Kaixian, etc. within Chongqing city. The samples from Nos. 28 to 38, provided by the Archeological Team of Fudan University in the Three Gorges and the Wanzhou Cultural Relics Administrative Office, were unearthed at Maliuwan in Wanzhou and Luo Renfa's Tomb in Chongqing (Photos 23-1 and 23-2). The period runs from the Warring States to the Six Dynasties. A detailed description of the samples is given in Table 23.1.

2.2. *Method of the experiment*

The external-beam PIXE experiments were performed at the NEC 9SDH-2 Pelletron tandem accelerator of Fudan University. The proton beam was extracted through a 7.5-μm-thick Kapton window,

Photo 23.1. Aqua translucence glass Bi and three Eye beads, made in the Warring States, unearthed in Baxian and Kaixian in Chongqing.

Photo 23.2. A string of glass beads of the Six Dynasty unearthed from Chegyu highway.

and traveled 10 mm in air before reaching the glass sample. The beam spot diameter on the sample was 1 mm and the beam current was 0.01 nA. The original energy of the proton beam was 3.0 MeV; therefore the actual energy of the protons reaching the samples was 2.8 MeV, as a result of energy loss in the Kapton film and air. An ORTEC Si (Li) detector (165eV FWHM at 5.9 keV), placed at 90° relative to the beam direction, was used. From the measured PIXE

Table 23.1. The description of the samples unearthed in Chongqing and its peripheral districts.

No.	Exp. order	Number of sample	Date	Description of samples	Unearthing site
1	24022002	51SBN	Warring States	Aqua translucent glass *bi*. Diameter 11.9 cm, aperture 3.8 cm, depth 0.3 cm. Flat–round, a central hole, decorated with valley profile, pattern and shape of same as that made in same period of Warring States	Southern part of Baocheng Railway
2	24022003	55SBNN296-a	Six Dynasties	Dark blue translucent glass ear pendant. Height 2.3 cm, diameter 1.2–1.6 cm. Column-like, top small and bottom big with a central hole, compact and delicate, still bright colors	Southern part of Baocheng Railway
3	24022004	555SBNN296-b	Six Dynasties	Big part of blue ear pendant (same as No. 2)	Southern part of Baocheng Railway
4	24022005	55SBNN296-c	Six Dynasties	Middle of blue ear pendant (same as No. 2)	Southern part of Baocheng Railway
5	24022006	55SCM12N223-a	Six Dynasties	White transparent bead (part of string)	No. 7 tomb at Baolun garden, Zhaohua along Baocheng Railway

(Continued)

Table 23.1. (*Continued*)

No.	Exp. order	Number of sample	Date	Description of samples	Unearthing site
6	24022007	55SCM12N223-b	Six Dynasties	Yellow opaque bead (part of string)	No. 7 tomb at Baolun garden Zhaohua along Baocheng Railway
7	24022008	55SCM12N223-c	Six Dynasties	Red opaque bead (part of string)	No. 7 tomb at Baolun garden, Zhaohua along Baocheng Railway
8	24022009	55SCM12N223-d	Six Dynasties	Brown opaque bead (part of string)	No. 7 tomb at Baolun gardon Zhaohua along Baocheng Railway
9	24022010	54CBD1-a	Warring States	Black part of green-and-white eye bead in black-and-white glass body. Diameter 2.1 cm, aperture 1.2 cm. Glass beads round, black body, a central hole, with white beads geometric profile for the lining, containing a number of groups on multilayered white and green circle stripes, exquisite beauty	Dongsunba in Baxian, Chongqing

(*Continued*)

Table 23.1. *(Continued)*

No.	Exp. order	Number of sample	Date	Description of samples	Unearthing site
10	24022011	54CBD1-b	Warring States	Transparent. Green part of green-and-white eye bead in black-and-white glass body (same as No. 9)	Dongsunba in Baxian, Chongqing
11	24022012	54CBD1-c	Warring States	Transparent. White part of green-and-white eye bead in black-and-white glass body (same as No. 9)	Dongsunba in Baxian, Chongqing
12	24022013	54CBD2-a	Warring States	Sky-blue part of sky-blue-and-white eye bead in black glass body	Yujiaba in Kaixian, Chongqing
13	24022014	54CBD2-b	Warring States	Deep blue part of sky-blue-and-white eye bead in black glass body	Yujiaba in Kaixian, Chongqing
14	24022015	54CBD2-c	Warring States	Blue part of sky-blue-and-white eye bead in black glass body	Yujiaba in Kaixian, Chongqing
15	24022016	54CBD2-d	Warring States	Black part of the sky-blue-and-white eye bead in black glass body	Yujiaba in Kaixian, Chongqing
16	24022017	55SZBN211-a	Six Dynasties	Blue transparent bead (part of string)	Cheng-Yu Highway
17	24022018	55SZBN211-b	Six Dynasties	White transparent bead (part of string)	Cheng-Yu Highway

(Continued)

Table 23.1. (*Continued*)

No.	Exp. order	Number of sample	Date	Description of samples	Unearthing site
18	24022019	55SZBN211-c	Six Dynasties	Colorful opaque face of bead (part of string)	Cheng-Yu Highway
19	24022020	55CBDM49D49-a	Warring States	Blackish blue opaque part of Cambridge-blue eye in blue glass body	Dongxunba in Baxian, Chongqing
20	24022021	55CBDM49D49-b	Warring States	Cambridge-blue transparent part of Cambridge-blue eye in blue glass body	Dongxunba in Baxian, Chongqing
21	24022022	51SCYY52-a	Six Dynasties	White transparent bead (part of string)	No. 7 tomb at Cheng-Yu and Baocheng Highway
22	24022023	51SCYY52-b	Six Dynasties	Cambridge-Green opaque bead (part of string)	Cheng-Yu Highway
23	24022024	51SCYY52-c	Six Dynasties	Red translucent bead (part of string)	Cheng-Yu Highway
24	24022025	55SZBM7N217-a	Six Dynasties	Blue transparent small glass bead, diameter 1.5 cm (part of string)	Cheng-Yu Highway
25	24022026	55SZBM7N217-b	Six Dynasties	Blue opaque big bead, diameter 2 cm (part of string)	No. 12 tomb at Changshan, Zhangming

(*Continued*)

Table 23.1. (*Continued*)

No.	Exp. order	Number of sample	Date	Description of samples	Unearthing site
26	24022027	55SZBM7N217-c	Six Dynasties	Green opaque bead (part of string)	Tomb at Changshan, Zhangming
27	24022029	55SZBM7N217-d	Six Dynasties	Black opaque glass bead (part of string)	No. 12 tomb at Changshan, Zhangming
28	24022036	02CWMLM4:02	Eastern Han Dynasty	Cambridge-green transparent droplet bead, central perforation	R. F. Luo's tomb in Wanzhou, Chongqing
29	24022037	02CWMLM1:01	Eastern Han Dynasty	Pearl-white translucent cylindrical bead, corroded, central perforation	R. F. Luo's tomb in Wanzhou, Chongqing
30	24022038	02CWMLM4:01	Northern and Southern Dynasties	Blue transparent bead	R. F. Luo's tomb in Wanzhou, Chongqing
31	24022039	01CWMWM2:01	Eastern Han Dynasty	Dark blue bead, central hole. Diameter 0.7 cm, aperture 0.2 cm, height 0.5 cm	Maliuwan in Wanzhou, Chongqing
32	24022040	01CWMWM2:03	Eastern Han Dynasty	Blue transparent ear pendant, residues. Height 1.1 cm, largest diameter 0.8 cm	Maliuwan in Wanzhou, Chongqing

(*Continued*)

Table 23.1. *(Continued)*

No.	Exp. order	Number of sample	Date	Description of samples	Unearthing site
33	24022042	01CWMWM2:02	Eastern Han Dynasty	Green transparent drum bead. Diameter 0.4 cm, aperture 0.1 cm, height 0.3 cm	Maliuwan in Wanzhou, Chongqing
34	24022043	01CWMWM2:04	Eastern Han Dynasty	Pearl-white translucent cylindrical tube. Diameter 0.55 cm, aperture 0.1 cm, length 1.2 cm	Maliuwan in Wanzhou, Chongqing
35	24022044	01CWMWM2:08	Eastern Han Dynasty	Green transparent bead. Diameter 0.4 cm, aperture 0.1 cm, height 0.2 cm	Maliuwan in Wanzhou, Chongqing
36	24022045	01CWMWM2:09	Eastern Han Dynasty	Blue transparent bead. Diameter 0.4 cm, aperture 0.1 cm, height 0.2 cm	Maliuwan in Wanzhou, Chongqing
37	24022046	01CWMWM2:10	Eastern Han Dynasty	Blue transparent bead. Diameter 0.3 cm, aperture 0.08 cm, height 0.2 cm	Maliuwan in Wanzhou, Chongqing
38	24022047	01CWMWM2:11	Eastern Han Dynasty	Green transparent bead. Diameter 0.25 cm, aperture 0.1 cm, height 0.15 cm	Maliuwan in Wanzhou, Chongqing

spectrum the chemical composition of the sample could be obtained using the deconvolution program GUPIX-96.[5]

2.3. *Experimental Results*

Table 23.2 shows the compositions of the glass samples in percentage of weight measured in air by the PIXE method. It can be seen that the chemical compositions of ancient glass samples can be divided into three categories.

(1) The series of $PbO–BaO–SiO_2$ glasses

The samples include: three eye beads made in the Warring States period and unearthed at Baxian and Kaixian of Chongqing (these samples are Nos. 9, 10, 11, 12, 13, 14, 15, 19 and 20); one glass *bi* disk (sample No. 1); and the glass beads unearthed at Changshan village of Zhangming in Chongqing, the tomb of Luo Renfa at Wanzhou, and Maliuwan (these samples are Nos. 27, 28, 29, 30, 33 and 38). The body of the eye bead and its eyeball are composed of lead-barium-silicate glass. Ten of the samples contain PbO only; they are typical lead-barium-silicate glass and lead-silicate glass made in China, which occupy nearly 50% of this batch of glass artifcats.

(2) The series of $K_2O–SiO_2$ glasses

Three beads were unearthed at Maliuwan of Wanzhou in Chongqing (Nos. 32, 36 and 37). These samples contain about 10% K_2O, and the contents of Na_2O and CaO are less than 4%, respectively. They belong to the K_2O–SiO2 series of glasses.

(3) The series of alkali lime silicate ($R_2O–CaO–SiO_2$) glasses

The samples with 4–5% K_2O and 6–9% CaO contents look like potassium calcium-silicate glass. The glass samples are 16, 18 and 24.

Table 23.2. Chemical compositions of the samples from Chongqing and its peripheral districts (wt%).

No.	Exp. order	Al_2O_3	SiO_2	P_2O_5	SO_3	Cl	K_2O	CaO	TiO_2	Cr_2O_3	MnO	Fe_2O_3	CoO	CuO	BaO	PbO
1	24022002	9.526	47.760	0.000	7.252	1.266	4.449	2.322	0.208	0.000	0.000	1.721	0.000	0.104	6.440	18.953
2	24022003	2.632	91.574	0.000	0.150	0.225	2.490	1.111	0.125	0.000	0.522	1.135	0.038	0.000	0.00	0.00
3	24022004	3.009	81.271	0.000	2.393	0.221	4.761	1.238	0.541	0.000	3.389	2.952	0.169	0.055	0.00	0.00
4	24022005	4.022	88.639	0.000	0.980	0.265	2.564	1.345	0.164	0.000	0.570	1.402	0.012	0.037	0.00	0.00
5	24022006	1.485	97.644	0.000	0.613	0.000	0.118	0.069	0.000	0.000	0.000	0.070	0.000	0.000	0.00	0.00
6	24022007	7.223	68.879	1.580	2.312	1.143	6.148	8.299	0.493	0.000	0.153	3.747	0.000	0.025	0.00	0.00
7	24022008	0.192	98.844	0.000	0.678	0.068	0.123	0.047	0.000	0.000	0.000	0.049	0.000	0.000	0.00	0.00
8	24022009	18.539	43.624	3.170	10.681	3.919	4.012	13.657	0.299	0.037	0.033	1.868	0.033	0.128	0.00	0.00
9	24022010	6.709	47.637	0.944	0.857	1.589	2.021	3.802	0.049	0.014	0.025	3.997	0.175	0.295	6.758	25.128
10	24022011	5.702	64.056	0.682	2.865	1.190	1.920	1.635	0.035	0.000	0.000	1.337	0.000	0.399	3.606	16.573
11	24022012	3.699	67.820	0.391	0.605	0.709	0.865	0.591	0.000	0.000	0.000	0.398	0.011	0.056	4.683	20.170
12	24022013	2.955	58.035	0.901	0.000	0.821	0.347	0.623	0.000	0.051	0.000	0.300	0.011	0.995	7.404	27.556
13	24022014	3.456	58.709	0.879	0.000	1.526	0.344	0.961	0.000	0.052	0.000	0.217	0.000	0.424	6.485	26.948
14	24022015	2.629	58.010	1.063	0.000	0.886	0.508	0.561	0.000	0.000	0.000	0.271	0.000	0.633	6.910	28.529
15	24022016	2.195	58.550	0.829	0.000	1.638	0.505	0.573	0.000	0.042	0.000	0.259	0.012	0.739	6.624	28.035
16	24022017	4.231	75.694	0.825	1.507	1.139	4.445	9.564	0.146	0.014	0.063	1.225	0.025	1.121	0.00	0.00
17	24022018	2.069	82.425	0.726	1.680	1.227	3.576	7.311	0.088	0.000	0.034	0.830	0.017	0.017	0.00	0.00
18	24022019	3.888	77.332	1.069	2.383	1.273	4.549	7.688	0.142	0.000	0.027	1.516	0.027	0.106	0.00	0.00
19	24022020	7.832	33.747	28.922	1.761	0.936	1.352	1.651	0.000	0.000	0.000	2.264	0.000	1.477	8.694	11.364
20	24022021	4.092	55.741	7.599	0.000	1.726	0.448	0.852	0.000	0.099	0.000	1.016	0.043	1.483	11.410	15.492

(*Continued*)

Table 23.2. *(Continued)*

No.	Exp. order	Al_2O_3	SiO_2	P_2O_5	SO_3	Cl	K_2O	CaO	TiO_2	Cr_2O_3	MnO	Fe_2O_3	CoO	CuO	BaO	PbO
21	24022022	21.684	77.080	0.000	0.302	0.000	0.073	0.085	0.000	0.000	0.000	0.777	0.000	0.000	0.00	0.00
22	24022023	21.883	76.975	0.000	0.386	0.039	0.000	0.000	0.000	0.000	0.000	0.718	0.000	0.000	0.00	0.00
23	24022024	2.094	87.796	0.000	0.722	0.434	1.857	6.957	0.016	0.028	0.000	0.096	0.000	0.000	0.00	0.00
24	24022025	4.291	74.381	0.852	4.072	0.991	4.949	7.800	0.138	0.030	0.160	2.155	0.053	0.129	0.00	0.00
25	24022026	12.950	75.954	0.481	0.721	0.893	4.365	2.168	0.438	0.000	0.034	1.239	0.000	0.756	0.00	0.00
26	24022027	12.165	77.259	0.473	1.406	1.032	2.532	2.941	0.595	0.000	0.073	1.476	0.000	0.047	0.00	0.00
27	24022029	3.472	54.356	1.576	0.000	2.170	0.600	0.697	0.000	0.035	0.000	0.339	0.000	0.460	12.324	23.972
28	24022036	3.046	57.336	1.539	0.000	1.310	0.202	0.846	0.000	0.000	0.000	0.312	0.000	0.000	13.068	22.340
29	24022037	3.645	58.724	1.074	0.000	1.709	0.465	1.311	0.000	0.060	0.053	0.493	0.018	0.034	14.647	17.768
30	24022038	3.708	59.744	1.092	0.000	0.000	0.473	1.334	0.000	0.061	0.054	0.501	0.018	0.035	14.901	18.077.
31	24022039	6.053	76.985	0.00	0.000	0.797	4.228	6.627	0.00	0.099	2.849	0.117	0.00	0.307	0.205	0.215
32	24022040	3.665	79.033	0.000	0.332	0.066	11.585	1.747	0.177	0.000	1.698	1.595	0.084	0.017	0.00	0.00
33	24022042	7.812	73.227	0.041	0.000	0.144	0.195	3.974	0.000	0.000	0.023	1.568	0.023	0.045	4.737	8.211
34	24022043	10.941	78.151	0.000	0.512	0.484	2.896	3.364	0.685	0.000	0.208	1.970	0.019	0.771	0.00	0.00
35	24022044	13.330	72.053	0.244	0.531	0.461	2.904	4.465	0.710	0.000	0.092	2.991	0.090	2.130	0.00	0.00
36	24022045	4.600	81.335	0.000	0.283	0.088	9.953	1.324	0.126	0.018	0.896	1.280	0.064	0.032	0.00	0.00
37	24022046	4.076	80.852	0.066	0.287	0.086	10.047	1.328	0.192	0.000	1.188	1.768	0.073	0.036	0.00	0.00
38	24022047	11.889	63.828	3.588	0.00	0.00	7.335	2.353	0.29	0.00	0.262	1.949	0.037	1.597	0.00	6.872

But the sodium and magnesium elements could not be detected in our experiment of 2002. Therefore, we analyzed the three glass samples (Nos. 31, 35, 36) unearthed at Maliuwan of Wanzhou in Chongqing by PIXE under the protection of He gas in 2006, the measured results show that two of them (Nos. 31, 35) are alkali-calcium-silicate glass ($Na_2O(K_2O)$–CaO–SiO_2), and the other is potassium-silicate glass. The detailed analysis of their chemical compositions is shown in Table 23.3.

The glass with a K_2O content of less than 4% belongs to the $Na_2O(K_2O)$–CaO–SiO_2 glass system. There are other five samples belonging to this system, and they are Nos. 2, 3, 4, 17, 23, 31 and 35 (Nos. 2, 3 and 4 are the same sample).

There are five samples of the alkali lime-silicate glass with a high content of Al_2O_3 (>7%); in code numbers are 6, 8, 25, 26 and 34.

Some of the samples are mineral materials. For instance, samples 5 and 7 having high content of SiO_2 (more than 97.64%), are similar to agate. Two samples with code numbers 21 and 22 having a content of Al_2O_3 larger than 21%, should belong to the aluminum silicate mineral.

3. Results and Discussion

There are three kinds of glasses in Chongqing, based on analysis of the chemical compositions of glass samples, namely lead-barium-silicate glass, potassium-silicate glass and alkali lime-silicate glass. All the glasses measured in our experiment have similar characteristics to the ancient glasses discovered in Qinghai province.[7]

The lead-barium-silicate glass unearthed in Chongqing is a special category of the eye beads that were invented and fabricated by the Chinese in the period of the Warring States. Similar eye beads have been found in other places, such as the glass beads unearthed at the graves of the late Warring States at Mijiatan, Chenxi, Huanhua of Hunan province,[8] the eye beads found at the graves of the Western Han Dynasty in the south bank of Chongqin city,[1] and so on. The difference however, is that the eye beads supplied

Table 23.3. Chemical compositions of the samples from Maliuwan of Wanzhou in Chongqing (wt%).

No.	Number of sample	Na_2O	MgO	Al_2O_3	SiO_2	P_2O_5	SO_2	Cl	K_2O	CaO	TiO_2	MnO	Fe_2O_3	CuO	BaO	PbO
31	01CWMWM2:01	8.15	1.08	4.92	70.29	0.94	0.64	0.77	3.53	5.90	0.25	0.06	2.85	0.27	0.08	0.26
35	01CWMWM2:08	10.35	1.02	11.73	65.38	0.30	0.37	0.54	2.46	3.43	0.54	0.06	2.10	1.54	0.20	0.00
36	01CWMWM2:09	0.00	0.60	4.59	81.11	0.40	0.09	0.16	9.31	1.21	0.09	0.93	1.43	0.02	0.08	0.00

by the Chongqing Museum have a relatively simple structure, with only two layers in their eye part. It is assumed that Chongqing and its peripheral districts were significantly influenced by the culture of Chu state and the Central Plains after the period from the Eastern Zhou Dynasty to the Han Dynasty, so there is the possibility that these beads were imported from Jiang-Han and the Central Plains.

The chemical composition of alkali lime-silicate glasses were similar to those of glass in ancient Babylon and Egypt,[6] probably imported along the following paths:

(1) Going From West Asia to Central Asia, and then Suiye around the current Tokmok, Kyrgyzstan to the Balkash Lake district and the Yili River valley, along the southern part of the Altai Mountain, via the Eerqisi River valley across the northern part of the Tianshan Mountain, and then diverging into two ways to the east: one way was from Chishui city down to Wuwei, then to Huangzhong (now Xining); the other way was from the south path of the Desert Silk Road, across Qinghai, to the Hexi Corridor, and finally southward to Huangzhong.

The chemical composition of the ancient glasses unearthed in Chongqing are similar to those of the glasses in the Qinghai district, which may be related to the immigration of the Sanmiao nationality. In the early ancient times, Sanmiao, the ancestor of Qiang, migrated from Northeastern China to Qinghai. During the Xia and Shang Dynasties, the inhabitants of Qinghai were called "West Qiang," and Qinghai was also known as "Qiang Land." Qinghai province, close to Gansu, Xinjiang, and Tibet, was the crossroads of the Desert Silk Road and the Southwestern Silk Road, so it was the called "Central Land of Huangzhong," where there was a mixture of Qiang, Xiao Rouzhi (Minor Yen Chin) and Han inhabitants. The Qiang people then moved southwestward to Sichuan, after the Western Zhou Dynasty. We think that the similar combination of chemical composition of the ancient glasses of Chongqing and the Qinghai district may be attributed to the migration of the Qiang ancestors.[9]

(2) Transferring through the route from Shendu (Indian subcontinent) to Shu (Sichuan). Early on, before Zhang Qian's diplomatic travels to the Western Regions, the Southwestern Silk Road bridging Shu and Shendu, passed through Chongqing, which may have been the passage for transporting alkali lime-silicate glasses between ancient India and Chongqing. The transport path of the glasses was probably from the southwest coast of India, along the Ganges River valley, to Mo Paer of northeast India, then going across the Qingdunjiang River to Menggong of north Myanmar, across the Irrawaddy River, and finally via Yunnan to Sichuan. This Silk Road was established much earlier than the others.[10]

4. Conclusions and Suggestions

The samples unearthed in Chongqing and its peripheral districts can be divided into three categories, according to their chemical composition: the $PbO–BaO–SiO_2$ glass system, the $K_2O–SiO_2$ glass system and the $R_2O–CaO–SiO_2$ glass system. Besides, there are natural minerals, including agate and aluminum silicate minerals.

The series of $PbO–BaO–SiO_2$ glasses, dating back to the Warring States, are lead-barium-silicate glass fabricated in China. They indicate that Chongqing was influenced by the culture of Chu and the Central Plains early in the period of the Warring States. Later, in the Han Dynasty, the Eastern Han Dynasty and the Six Dynasties, the series of $R_2O–CaO–SiO_2$ glasses appeared in Chongqing, and were mainly imported from the northwest or southwest of China. Several categories of samples have been unearthed from a single grave, indicating that the Chongqing ancestors already treasured the ancient glasses, which served as important items to be buried with the dead.

Finally, the provenances of the $K_2O–SiO_2$ glasses and those with a high content of Al_2O_3 silicate glasses are to be studied further.

Acknowledgments

We would like to thank the Chongqing Museum and the Wanzhou Cultural Relics Administrative Office, Chongqing and the archeology team of Fudan University for providing the samples. We are also grateful to Prof. H. S. Cheng, Dr. B. Zhang, D. Zhu and J. W. Lin of the Institute of Modern Physics, Fudan University, for supporting the experiments.

References

1. T. W. Gong and Y. H. Zhuang, Two tombs of the Western Han Dynasty unearthed in Nan'an of Chongqing district, *Cultural Relics* (in Chinese) **7**, 28–29 (1982).
2. The Eastern Zhou Dynasty cemetery at Yujiaba of Yunyang in Chongqing: the excavation report in 1997, *J. Archeology* (in Chinese) **11**, 86–87 (2002).
3. Commission for the Preservation of Archeological Monuments of Sichuan, Cultural Bureau of Fuling District, Four tombs of the Warring States at Xiaotianxi of Fuling in Chongqing, *Archaeology* (in Chinese) **1**, 14–17 (1985).
4. H. J. Feng, Y. Y. Yang and J. Y. Wang, Sichuan ancient coffin burial with boat, *Acta Archaeologica Sinica* (in Chinese) **2**, 77–95 (1958).
5. Q. H. Li, B. Zhang, F. X. Gan, H. S. Cheng, B. Ma and D. H. Gu, A number of the chemical compositions of the glass unearthed in southern China with PIXE analysis, in: F. X. Gan ed., *Study on Ancient Glass in Southern China* (Shanghai Scientific and Technical Publishers, 2003), in Chinese, pp. 76–84.
6. R. H. Brill, *Chemical Analyses of Early Glasses* (The Corning Museum of Glass, New York, 1999), Vol. 2, pp. 382–386.
7. X. Y. Ren, Short discussion on the glasses of the Han Dynasty unearthed in Qinghai, in: F. X. Gan ed., *Study on Ancient Glass Along the Silk Road.* (Fudan University Press, 2007), in Chinese, pp. 170–175.
8. Huaihua Cultural Relics Administrative Office and Chenxi Cultural Relics Administrative Office, Brief report on the excavation of the

Eastern Zhou Dynasty tomb unearthed at Mijiatan of Nanchenxi, *Archeol. Culture Relics* (in Chinese) **2**, 3–13 (1998).
9. X. B. Liu, Movement of ancient Qiang people during the pre-Han Dynasty. *Ethno-national Studies* (in Chinese) **2**, 39–43 (2002).
10. Y. X. Jiang, *Ancient Silk Road in the Southwest of China,* Vol. 2 (Sichuan University Press, 1995) in Chinese, Vol. 2, pp. 42–57.

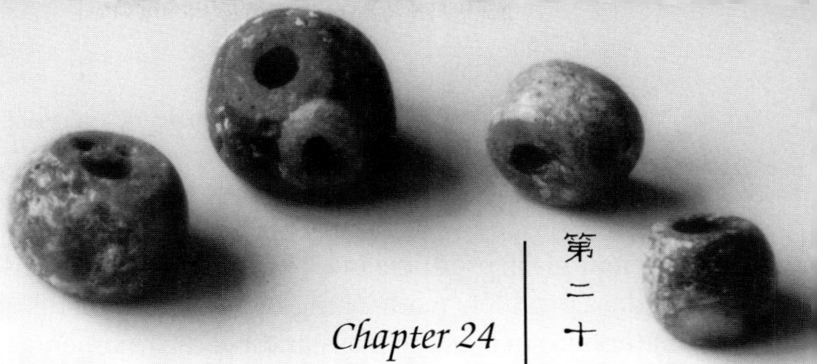

Chapter 24 第二十四章

Study of the Earliest Eye Beads in China Unearthed from the Xu Jialing Tomb in Xichuan of Henan Province

Gan Fuxi

Shanghai Institute of Optics and Fine Mechanics,
Chinese Academy of Sciences, Shanghai 201800, China
Fudan University, Shanghai 200433, China

Cheng Huansheng

Institute of Modern Physics, Fudan University, Shanghai 200433, China

Hu Yongqing

Henan Research Institute of Cultural Relics and Archeology,
Zhengzhou 450000, China

Ma Bo

School of Information Science and Engineering,
Fudan University, Shanghai 200433, China

Gu Donghong

Shanghai Institute of Optics and Fine Mechanics,
Chinese Academy of Sciences, Shanghai 201800, China

1. Introduction

Ancient glass products in China have three origins. The first is that of glass products made using the domestic technique and local raw materials, such as the glass and beads ornaments embedded in the ancient swords excavated from the Chu tombs of the Spring and Autumn Period to the Warring States in inner China (this mainly refers to the Yellow River valley and the Yangtze River valley).[1] The second is that of glass products made using the foreign technique but local raw materials, such as the glass beads of the Western Zhou Dynasty to the Spring and Autumn Period unearthed in Xiyu (the Western Regions; this mainly refers to the Xinjiang area).[2] The third origin is that of glass products made in and imported from foreign countries. The Western glass from Mesopotamia, ancient Egypt, Greece, Rome and the Islamic glass all have a similar chemical composition, and belong to the soda lime silicate glass. These glasses are easily distinguished.[3] Many glass artifacts had been imported from western countries after Zhang Qian visited the Western Regions diplomatically during the Han Dynasty; such glasses have been unearthed in many places in China. However, very few Western glasses of the pre-Qin period, especially in the early Warring States, were discovered in Central China. The chemical compositions of the fragments of eye beads unearthed from the Zeng Houyi tomb in Sui Xian county of Hubei province,[4] Hou Gudui in Gushi county of Henan province,[5] and Xu Jialing in Xichuan county of Henan province[6] have been reported separately. Reference 1 (in its Table 4.1) gives the result of the analysis of each sample's chemical composition. Recently, the Henan Research Institute of Cultural Relics and Archaeology provided 11 intact eye beads unearthed at Xu Jialing in Xichuan county of Henan province to us for measurement. This paper reports the structure and chemical composition of those samples measured by nondestructive analysis techniques. The results are very amazing and significant.

2. Description of the Ancient Eye Beads and the Excavation Situation

The measured 11 samples of ancient glass inlaid beads or so called eye beads were unearthed from the M10 ancient tomb of the early

Warring States Period (500 BC) at Xu Jialing in Xichuan county of Henan province by the Henan Research Institute of Cultural Relics and Archaeology in October 1991. These samples have similar shapes, with a sky-blue body, a dark-blue pupil and an ochre circlar pattern on the inlaid white eyeball. They were intact beads without efflorescence, all made in the early Warring States Period. The detailed description is given in Table 24.1 and six perfect eye beads are shown in Fig. 24.1 (HNZZ-01, 03, 04, 05, 08, 11).

Xichuan is located in the southwest of Henan province, and the southeastern extension area of the Qinling Mountains. Danjiang River, one of the main branches of Hanshui, goes through the county from northwest to southeast. The Chu people founded its capital in Danyang during the Western Zhou Dynasty, controlled the Jiangsu and Anhui areas and part of the Yellow River valley, and challenged to Central China. It became one of the five powerful states in the Spring and Autumn Period and one of the seven biggest states in the period of the Warring States. The Danjiang River valley was the primary area where the Chu people lived. The Xu Jialing cemetery is located in the Danjiang River alluvial plain and the west of Shunyangchuan. It now adminstratively belongs to the Yanjiang village of Cangfang town, in Xichuan of Henan province. The cemetery is popularly named the "Phoenix Head." It is situated on the northeast of Longshan Mountain and the west bank of Danjiang River. The topography gradually increases from east to west and the M10 tomb is located in the northwest. The tomb is probably that of a member of the ancient literati or a senior official, from examining the burial objects, such as bronze ware and jade carvings.[7]

3. Method of the Experiments

The external beam PIXE experiments were performed at the NEC 9SDH-2 Pelletron tandem accelerator of Fudan University. The proton beam was extracted through a 7.5-μm-thick Kapton window, and traveled 10 mm in air before reaching the glass sample. The beam spot diameter on the sample was 1 mm and the beam current was 0.01 nA. The original energy of the proton beam was 3.0 MeV; therefore the actual energy of the protons reaching the samples was

Table 24.1. Description of the ancient glass eye beads from the Xu Jialing tomb in Henan province.

Exp. No.	Sample No.	Description of the sample (size: cm)	State
HNZZ-01	M10:3	Body: nearly column; cross-section: circle, two ends: nearly plane; a central hole; height: 1.9; max. diameter: 2.1; aperture: 0.52.	Perfect
HNZZ-02	M10:41	Body: nearly column; cross-section: circle, two ends: nearly plane; a central hole; height: 1.8; max. diameter: 2.0; aperture: 0.52.	Slightly damaged
HNZZ-03	M10:4-1	Body: bead; cross-section: circle, two ends: nearly plane; a central hole; height: 1.5; max. diameter: 1.88; aperture: 0.48.	Imperfect
HNZZ-04	M10:4-2	Body: bead; cross-section: circle, two ends: nearly plane; a central hole; height: 1.49; max. diameter: 1.87; aperture: 0.38.	Perfect
HNZZ-05	M10:4-3	Body: bead; cross-section: circle, two ends: nearly plane; a central hole; height: 1.48; max. diameter: 1.81; aperture: 0.34.	Perfect
HNZZ-06	M10:4-4	Body: bead; cross-section: circle, two ends: nearly plane; a central hole; height: 1.42; max diameter: 1.81; aperture: 0.28.	Imperfect
HNZZ-07	M10:4-5	Body: bead; cross-section: circle, two ends: nearly plane; a central hole; height: 1.41; max. diameter: 1.75; aperture: 0.28.	Slightly damaged
HNZZ-08	M10:4-6	Body: bead; cross-section: circle, two ends: nearly plane; a central hole; height: 1.39; max. diameter: 1.79; aperture: 0.25.	Imperfect
HNZZ-09	M10:95-1	Body: bead; cross-section: circle, two ends: nearly plane; a central hole; height: 1.46; max. diameter: 1.71; aperture: 0.25.	Imperfect

(*Continued*)

Table 24.1. *(Continued)*

Exp. No.	Sample No.	Description of the sample (size: cm)	State
HNZZ-10	M10:95-2	Body: bead; cross-section: circle, two ends: nearly plane; a central hole; height: 1.37; max. diameter: 1.61; aperture: 0.22.	Slightly damaged
HNZZ-11	M10:95-3	Body: bead; cross-section: circle, two ends: nearly plane; a central hole; height: 1.15; max. diameter: 1.4; aperture: 0.16.	Perfect

Fig. 24.1. Ancient glass eye beads from the Xu Jialing tomb in Henan province.

2.8 MeV as a result of energy loss in the Kapton film and air. An ORTEC Si (Li) detector (165 eV FWHM at 5.9 keV), placed at 90° relative to the beam direction, was used. From the measured PIXE spectrum the atomic composition of the elements, of which their atomic number larger than 12 ($z \geq 12$) in the sample could be obtained using the deconvolution program GUPIX-96. For measuring the Na content of the glass, the sample should be cycled by He gas in order to avoid atmosphere absorption. The detailed experimental process can be found in Ref. 8. The X-ray diffraction (XRD) spectra, obtained by a Bruker D8 Advance X-ray diffractometer using Cu Ka radiation ($\lambda = 0.115406$ nm), were employed to identify the phase constitutions in the samples. The accelerating voltage and the applied current were 40 kV and 40 mA, respectively.

4. Experimental Result

Because these eye bead samples are very similar, we selected only three of them (HNZZ-01, 03, 05) for XRD measurement. Three parts (the body, the pupil and the inlaid white part) were measured for each of them, and the results are shown in Figs. 24.2 to 24.4.

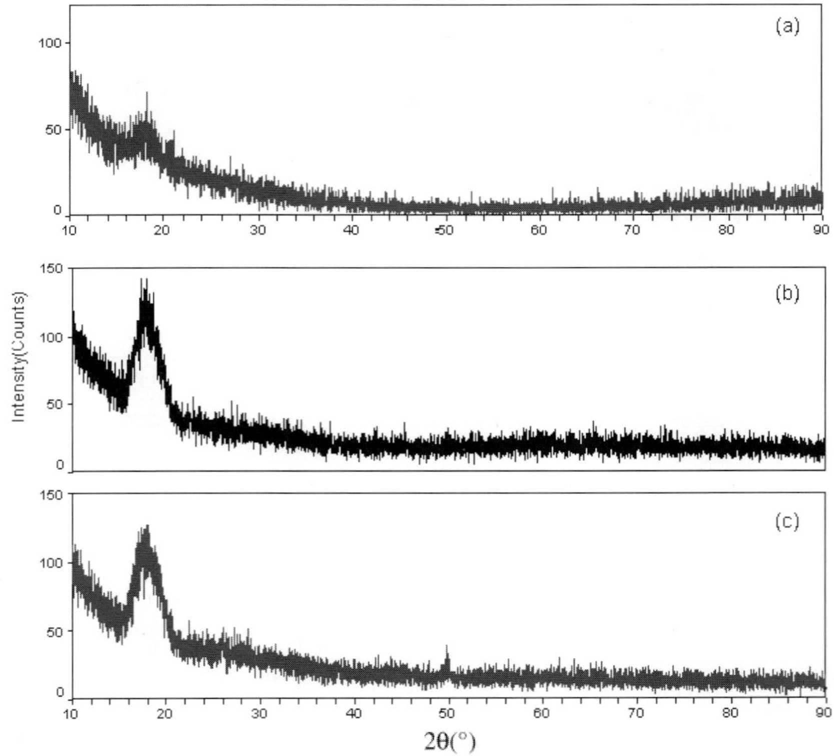

Fig. 24.2. X-ray diffraction pattern for the pale-blue body of eye beads: (a) HNZZ-01, (b) HNZZ-03, (c) HNZZ-05.

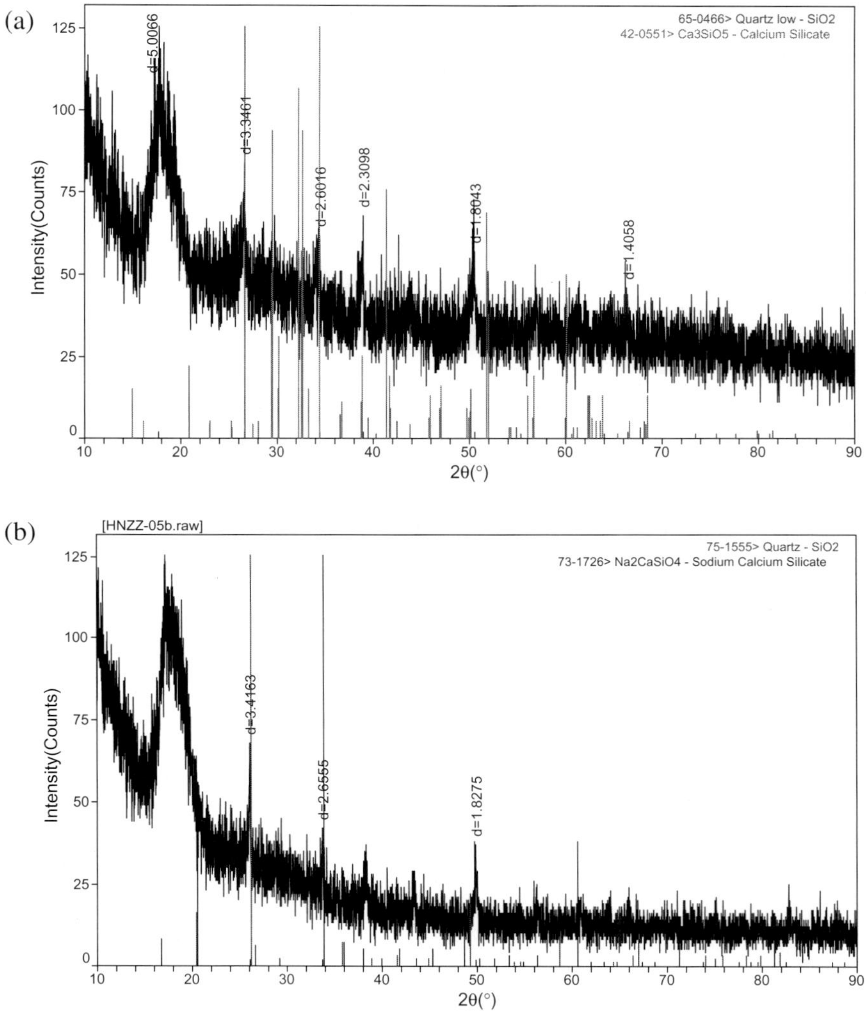

Fig. 24.3. X-ray diffraction pattern of the dark-blue pupil of eye beads: (a) X-ray diffraction pattern of HNZZ-03 dark-blue pupil of eye beads, in which blue is for α-quartz (SiO_2) and red for calcium silicate, 42-0551, $Ca_3(SiO_5)$; (b) X-ray diffraction pattern of HNZZ-05b dark-blue pupil of eye beads, in which blue is for quartz (SiO_2) and red for sodium–calcium silicate, 73-1726, Na_2Ca (SiO_4).

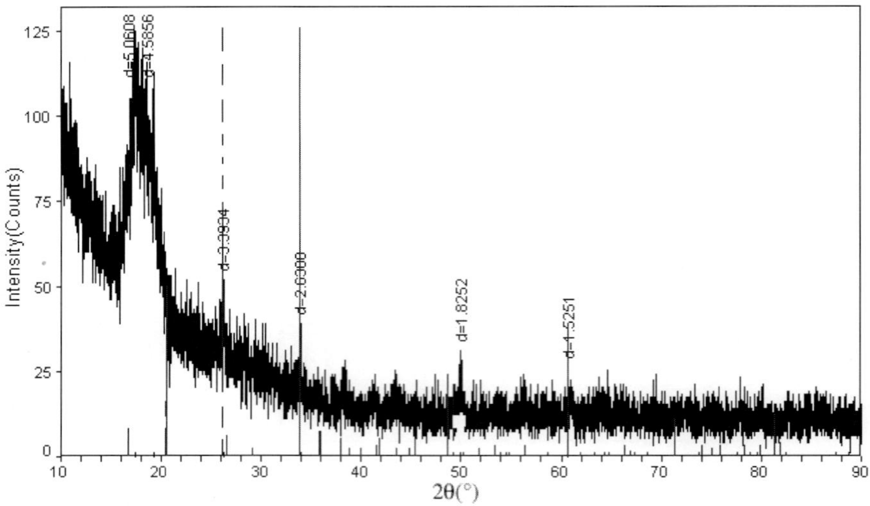

Fig. 24.4. X-ray diffraction pattern of the inlaid white part of the HNZZ-05c eye bead, in which blue is for quartz (SiO_2) and red for sodium–calcium silicate, $Na_2Ca\ (SiO_4)$.

The broad band around $2\theta = 20°$ is the scattering at a small angle, which indicates the inhomogeneity in the sample. The intensity enhancement of small angle scattering band shows the increases in the size and quantity of dispersed clusters in the sample. The sharp diffraction peaks show that a crystalline cluster exists in the sample. The vertical lines stand for some possibly existing crystals. The chemical compositions of all 11 samples are measured by using PIXE method. Different parts of each sample are measured respectively for comparison. The results are shown in Table 24.2.

5. Discussion

The glasslike artifacts dating from the Western Zhou Dynasty and the early Spring and Autumn Period which were unearthed in Central China (often called *liaoqi* or *liuli* by archeologists) are not glass state objects, according to the present study of the origin and evolution of the chemical compositions of the earliest glasses in this region.[9]

Because the furnace temperature could not reach a sufficiently high degree at that time, the products were actually the sinter of quartz with a little glass phase material. The products show glass luster on their surface, so we call them as "Yousha" (glazed sand), called faience in Western countries. With increasing furnace temperature, the content of the glass phase material in the products is increased and becomes a mixture of quartz sand and glass, which is named "bosha" (frit).[10] The glasses made in the Yangtze River valley in the middle and later parts of the Warring States Period belonged to the typical lead–barium silicate system ($PbO–BaO–SiO_2$). Therefore, we are most concerned with what the chemical composition and structural state of the glasslike artifacts dating from the late Spring and Autumn period to the early Warring States Period are and where they come from. The eye beads unearthed from the Xu Jialing tomb in Xichuan county of Henan province belong just to this period.

The XRD results in Fig. 24.2 show that the body of these eye beads behaving is a characteristic of the glass state. The broad band in the smaller angle pattern is the small angle scattering and shows that the sample is inhomogeneous, while the different intensities of the smaller angle scattering band show that the degree of optical inhomogeneity of the samples is different. From the XRD results in Figs. 24.3 and 24.4, we can see that there exist the micro-crystal clusters in the pupil and the inlaid white part. The main crystal clusters in the samples are α-quartz, sodium–calcium–silicate $Na_2Ca(SiO_4)$ and calcium-silicate $Ca_3(SiO_5)$ referring to the mineral XRD card. These compounds come from the partly melted materials.

The chemical composition of the samples is given in Table 24.2. It can be seen from the chemical composition that the body, the pupil and the inlaid white part all belong to the soda lime silicate glass system ($Na_2O–CaO–SiO_2$). The contents of Na_2O, CaO, SiO_2 and Al_2O_3 are in the range of 5–10%, 7–15%, 65–76% and 3–5%, respectively, in which the content of K_2O is less than 2.5%, and MgO is less than 1.5%. The chemical composition of these samples has a distinct difference from that of the glass beads and glass inlaid in the swords unearthed in Jiangling of Hubei province

Table 24.2. Chemical composition of different parts of the 11 eye beads (wt%).

Exp. No.	Position of sample and its color	Na_2O	MgO	Al_2O_3	SiO_2	P_2O_5	SO_3	Cl	K_2O	CaO	TiO_2	MnO	Fe_2O_3	CoO	NiO	CuO	ZnO	PbO	BaO	I	Hg
HNZZ-01	Body; blue	9.53	0.19	3.4	72.8	0.31	0.45	0.61	2.15	8.54	0.04	0.03	0.58	0.03	0.00	1.21	0.03	0.00	0.0	0.00	0.00
	Pupil; blue	9.91	0.97	3.5	70.5	0.22	0.56	0.68	1.17	8.62	0.11	0.04	2.86	0.25	0.00	0.48	0.04	0.00	0.0	0.00	0.00
	Inlaid white and ochre circle	7.37	1.17	3.8	75.0	0.23	0.67	0.83	1.01	8.73	0.11	0.04	0.76	0.04	0.00	0.13	0.04	0.00	0.0	0.00	0.00
HNZZ-02	Body, blue	10.4	0.56	3.9	72.5	0.11	0.60	0.67	1.62	7.87	0.08	0.00	0.62	0.00	0.00	0.84	0.06	0.00	0.0	0.00	0.00
	Pupil; blue	7.54	0.42	4.0	71.2	0.29	0.85	0.81	1.12	8.78	0.07	0.05	4.00	0.16	0.00	0.53	0.11	0.00	0.0	0.00	0.00
	Eye ball; blue	10.0	0.56	3.4	71.8	0.15	0.50	0.70	0.88	7.89	0.11	0.04	3.15	0.17	0.00	0.50	0.08	0.00	0.0	0.00	0.00
	Inlaid white and ochre circle	6.49	0.31	3.5	69.0	0.42	0.93	1.11	0.89	9.85	0.00	0.00	6.08	0.00	0.24	1.16	0.00	0.91	0.0	0.00	0.00
HNZZ-03	Body; blue	8.89	0.35	3.0	76.7	0.69	0.60	0.64	1.06	6.73	0.05	0.00	0.48	0.00	0.00	0.72	0.00	0.00	0.0	0.00	0.00
	Pupil; blue	6.65	0.77	5.7	71.3	0.33	0.54	0.75	1.16	8.21	0.24	0.05	3.27	0.14	0.00	0.72	0.09	0.00	0.0	0.00	0.00
	Ochre circle	0.73	0.00	3.3	57.5	0.25	0.81	1.19	2.23	26.6	1.27	0.42	3.88	0.55	0.00	0.41	0.00	0.00	0.0	0.00	0.00
	Inlaid white	6.17	0.33	5.1	67.0	0.54	0.84	0.77	1.05	12.7	0.28	0.09	1.29	0.08	0.00	0.21	0.00	0.00	0.0	3.38	0.00
HNZZ-04	Body; blue	8.87	0.49	3.0	75.8	0.29	0.52	0.59	1.26	7.16	0.07	0.00	0.42	0.00	0.00	1.47	0.00	0.00	0.0	0.00	0.00
	Pupil; blue	7.14	0.15	2.6	72.9	0.00	0.46	0.74	1.34	8.20	0.15	0.00	4.25	0.24	0.00	1.51	0.23	0.00	0.0	0.00	0.00
	Inlaid white and ochre circle	6.13	0.00	3.3	70.3	0.78	0.78	0.82	1.19	12.1	0.24	0.15	0.89	0.04	0.00	0.04	0.00	0.00	0.0	2.99	0.00
HNZZ-05	Body; pale blue	10.2	0.93	4.5	69.7	0.59	0.25	0.52	1.43	8.63	0.13	0.03	0.88	0.00	0.03	1.83	0.00	0.28	0.0	0.00	0.00
	Pupil; blue	8.03	0.96	3.0	73.3	0.30	0.33	0.88	0.74	8.09	0.07	0.06	3.21	0.28	0.00	0.72	0.05	0.00	0.0	0.00	0.00
	Inlaid white	4.90	0.33	3.8	68.5	0.45	0.49	0.99	1.43	13.0	0.59	0.10	1.13	0.20	0.10	0.10	0.00	0.00	0.0	3.28	0.00

(Continued)

Table 24.2. *(Continued)*

Exp. No.	Position of sample and its color	Na_2O	MgO	Al_2O_3	SiO_2	P_2O_5	SO_3	Cl	K_2O	CaO	TiO_2	MnO	Fe_2O_3	CoO	NiO	CuO	ZnO	PbO	BaO	I	Hg
HNZZ-06	Body; blue	10.3	0.55	2.9	71.9	0.22	0.47	0.62	1.37	8.58	0.08	0.00	0.68	0.00	0.00	2.26	0.00	0.00	0.0	0.00	0.00
	Pupil; blue	10.7	0.63	2.9	70.9	0.80	0.36	0.44	1.51	8.74	0.10	0.04	2.17	0.11	0.00	0.44	0.11	0.00	0.0	0.00	0.00
	Ochre circle	6.59	0.00	2.8	69.7	0.44	0.55	0.72	1.45	13.5	0.24	0.08	0.85	0.05	0.02	0.06	0.03	0.00	0.0	2.68	0.00
	Inlaid white	7.79	0.00	3.4	69.6	0.73	0.58	0.72	1.31	11.9	0.19	0.07	0.77	0.02	0.00	0.04	0.02	0.00	0.0	2.58	0.00
HNZZ-07	Body; blue	10.6	0.75	3.6	73.1	0.36	0.40	0.57	1.32	7.41	0.07	0.00	0.49	0.00	0.00	1.16	0.03	0.00	0.0	0.00	0.00
	Pupil; blue	9.85	0.38	3.6	72.0	0.21	0.23	0.50	1.69	7.81	0.08	0.00	2.72	0.12	0.00	0.62	0.11	0.00	0.0	0.00	0.00
	Inlaid white	2.23	0.00	3.0	72.7	0.29	1.11	1.02	1.00	12.6	0.32	0.16	1.36	0.16	0.00	0.00	0.00	0.00	0.0	3.82	0.00
HNZZ-08	Body; blue	5.51	0.69	3.0	77.6	0.41	0.59	0.71	1.21	7.63	0.00	0.00	0.93	0.00	0.00	1.56	0.00	0.00	0.0	0.00	0.00
	Pupil; blue	8.78	0.46	3.3	72.2	0.48	0.35	0.62	1.17	8.06	0.12	0.00	3.27	0.18	0.00	0.87	0.17	0.00	0.0	0.00	0.00
	Inlaid white	6.54	0.00	3.3	65.4	0.19	1.03	0.99	0.99	14.5	0.55	0.11	1.41	0.10	0.00	0.00	0.00	0.00	0.0	4.69	0.00
HNZZ-09	Body; blue	4.63	0.84	3.2	78.4	0.13	0.42	0.62	1.49	8.04	0.00	0.00	0.56	0.00	0.00	1.55	0.07	0.00	0.0	0.00	0.00
	Pupil; blue	7.26	0.57	2.9	74.8	0.30	0.65	0.60	0.83	7.66	0.07	0.06	3.14	0.27	0.00	0.86	0.00	0.00	0.0	0.00	0.00
	Inlaid white	3.98	0.00	3.3	69.9	0.42	0.91	0.91	1.20	12.9	0.30	0.12	1.36	0.06	0.00	0.06	0.00	0.00	0.0	4.27	0.00
HNZZ-10	Body; blue	10.5	0.44	2.9	73.9	0.53	0.36	0.52	0.91	7.58	0.05	0.00	0.54	0.04	0.00	1.67	0.00	0.00	0.0	0.00	0.00
	Pupil; blue	5.46	1.01	3.1	69.0	0.00	0.51	0.81	0.98	7.66	0.00	0.00	6.95	0.52	0.00	3.55	0.25	0.00	0.0	0.00	0.00
	Inlaid white	5.03	0.00	2.6	66.1	0.00	0.74	0.89	0.93	15.7	0.59	0.15	1.27	0.23	0.00	0.07	0.00	0.00	0.0	5.28	0.00
HNZZ-11	Body; blue	10.5	0.70	2.9	72.9	0.24	0.39	0.71	1.30	8.84	0.09	0.03	0.45	0.03	0.00	0.82	0.00	0.00	0.0	0.00	0.00
	Pupil; blue	10.5	0.73	2.9	72.1	0.40	0.51	0.61	1.05	8.54	0.10	0.00	1.75	0.26	0.00	0.26	0.18	0.00	0.0	0.00	0.00
	Inlaid white	4.40	0.12	3.2	67.1	0.49	0.53	1.35	1.41	14.0	0.41	0.05	1.17	0.23	0.00	0.09	0.04	0.00	0.0	5.01	0.00

(which are potash lime silicate glass).[1] The ratio of K_2O to Na_2O has a big difference between them, the former $K_2O/Na_2O \leq 0.2$ and the latter $K_2O/Na_2O > 5$.

The chemical composition of these glass eye beads closely resembles to that of the ancient Western glasses from the Mesopotamian and Egyptian area.[11,12] They are all soda lime silicate glasses. The natron was main flux agent. The only difference between them is that the content of Na_2O is little bit lower than that of the Western glasses. This is the main reason that the pupil of the glass eye-beads has a low quality and incompletely melts.

The impurity of the body of the glass eye-beads is very low, which indicates that pure raw material was used in the preparation process. The content of CuO (~1.5%) is more than that of Fe_2O_3 (<1%). The body of the eye beads is pale-blue, which was caused by the ions of Cu^{2+}. No CoO is found in the body glass, but in the dark-blue pupil of the eye beads there is 0.1–0.3% CoO. The Co^{2+} is a strong blue colorant, which must have been intentionally made and added by the glass masters. This is the earliest discovery of cobalt blue in ancient glass and ancient porcelain glaze unearthed in China (i.e. using CoO as the blue colorant). Using CoO as the colorant for glass in ancient China started in the Eastern Han Dynasty.[12] However, in the ancient Mesopotamian and Egyptian areas, that was done from *ca.*1000 BC.[11] Based on the chemical composition and colorant of these glass eye beads, we can guess that this batch of glass eye-beads was most probably imported from the West.

6. Conclusion

The chemical composition and structural state of the 11 eye beads unearthed at Xu Jialing in Xichuan county of Henan province have been studied by the method of PIXE and XRD. The body of these eye beads has the glass phase. The size of crystal clusters in the blue pupil is much bigger, and obvious crystal characteristic diffraction peaks are observed. The peaks mainly indicate the quartz and sodium silicate and calcium silicate microcrystals. The eye beads can be regarded as products between frit and glass.

They have the same chemical composition as glasses from Western countries, all belonging to the soda lime silicate glass system (Na_2O–CaO–SiO_2). CoO colorant was used to make the pupil of the glass eye beads dark-blue. This is the earliest example of using CoO as the colorant in ancient artifacts unearth in China so far. The use of CoO colorant in ancient West Asia and Egypt was more than 1000 years earlier than in ancient China. These experimental results show that the glass eye beads dating from the early Warring States Period which were unearthed at Xu Jialing in Xichuan county of Henan province are the earliest ancient glass imported from the West to Central China.

Acknowledgments

This work was supported by the National Natural Foundation of China (Grant No. 5067-2106) and the Intellectual Innovation of the Chinese Academy of Sciences (Grant No. KJCX3-SYW. No. 12).

References

1. F. X. Gan, H. S. Cheng and Q. H. Li, Origin of ancient Chinese glasses: study on the earliest ancient Chinese glasses, *Science in China* (E), 2006, 49: 701–713.
2. F. X. Gan, Q. H. Li and D. H. Gu *et al.*, The earliest glass bead unearthed from Baicheng and Tacheng in Xinjiang, *J. Chin. Ceram. Soc.* (in Chinese) 2003, 31: 663–668.
3. F. X. Gan, The development of ancient glass technology in the West. In: Gan F X *et al.*, *Development of Chinese Ancient Glass* (Shanghai Science and Technology, 2005) in Chinese, 38–52.
4. The Museum of Hubei Province, *Zeng Houyi Tomb* (Cultural Relic, Beijing, 1989) in Chinese, p. 658.
5. F. K. Zhang, Z. H. Cheng and Z. G. Zhang, Study on ancient Chinese glasses, *J. Chin. Ceram. Soc.* (in Chinese) 1983, 11(1): 70–71.
6. Q. H. Li, F. Li and F. X. Gan, Chemical composition analysis of some ancient glass artifacts of the Warring States period, *Conservation of Cultural Relics and Archeology* (in Chinese) 2006, 18(2): 8–13.

7. Henan Institute of Cultural Relic and Archeology, Nanyang Institute of Cultural Relic and Archeology and Xichuan Museum, *The Chu Tomb in Heshangling and Xujialing of Xichuan* (Zhengzhou University Press, 2004) in Chinese, pp. 1–2.
8. H. S. Cheng, B. Zhang and D. Zhu *et al.*, Application of external beam-proton-induced x-ray emission analysis in archeology and cultral relics research. In: F. X. Gan (ed.), *Study on Ancient Glass Along the Silk Road* (Fudan University Press, 2007) in Chinese, pp. 91–95.
9. F. X. Gan, Origin and evolution of ancient Chinese glasses, *Chin. J. Nature* (in Chinese), 2006, **28**(8): 187–193.
10. X. F. Fu, F. X. Gan, Ancient Chinese faience and frit, *J. Chin. Ceram. Soc.* 2006, **34**(4): 427–431.
11. R. H. Brill, *Chemical Analyses of Early Glasses* (The Corning Museum of Glass, Corning, New York, 1999), pp. 29–55.
12. R. H. Brill, Ancient glass, Scientific America, 1963, **209**: 120–131.
13. F. X. Gan *et al.*, *Development of Chinese Ancient Glass* (Shanghai Science and Technology Publishers, 2005), in Chinese, pp. 292–295.

Gan Fuxi. Born in Hangzhou (Zhejiang), China, January 1933, a professor of Shanghai Institute of Optics and Fine Mechanics and Fudan University. He received his Ph.D from the Academy of Science USSR. His research interests cover the fields of optical and laser materials, optoelectronics, and optical storage technology as well as ancient glasses. He was elected the Member of Chinese Academy of Sciences in 1980 and Fellow of Third World Academy of Sciences in 1993. An honorable council member of China Association for Science and Technology, and honorable president of Chinese Ceramic Society. He received the Life Time Award of the International Commission on Glass.

Robert Brill. Born in Irvington, New Jersey, USA, May 1929, a research scientist at the Corning Museum of Glass, U.S.A. He received his Ph.D in Physical Chemistry from Rutgers University, USA. He served as the director of the Museum from 1972–1975. As the founder of the Committee on Archeometry of Glass, International Commission on Glass (ICG) and chairman of the committee, he received the ICG's William E.S. Turner Award 2004. He has focused his studies on glass found along the Silk Road for recent ten years.

Tian Shouyun. Born in Shanxi, China, November 1941, a senior engineer of Shanghai Institute of Optics and Fine Mechanics. He graduated from Taiyuan Institute of Technology in 1965 and finished his post-graduate study at Shanghai University of Science and Technology in 1967. A concurrent editor of journal of Acta Optica Sinica and a member of the Science and Technology Translators' Association of CAS-FIT. He served as chief of the Institute Office, responsible for international cooperation and exchange, and has translated hundreds of science reports.

Index